Science, Technology and Medicine in Modern History

General Editor: John V. Pickstone, Centre for the History of Science, Technology and Medicine, University of Manchester, England (www.man.ac.uk/CHSTM)

One purpose of historical writing is to illuminate the present. At the start of the third millennium, science, technology and medicine are enormously important, yet their development is little studied.

The reasons for this failure are as obvious as they are regrettable. Education in many countries, not least in Britain, draws deep divisions between the sciences and the humanities. Men and women who have been trained in science have too often been trained away from history, or from any sustained reflection on how societies work. Those educated in historical or social studies have usually learned so little of science that they remain thereafter suspicious, overawed, or both.

Such a diagnosis is by no means novel, nor is it particularly original to suggest that good historical studies of science may be peculiarly important for understanding our present. Indeed this series could be seen as extending research undertaken over the last half-century. But much of that work has treated science, technology and medicine separately; this series aims to draw them together, partly because the three activities have become ever more intertwined. This breadth of focus and the stress on the relationships of knowledge and practice are particularly appropriate in a series which will concentrate on modern history and on industrial societies. Furthermore, while much of the existing historical scholarship is on American topics, this series aims to be international, encouraging studies on European material. The intention is to present science, technology and medicine as aspects of modern culture, analysing their economic, social and political aspects, but not neglecting the expert content which tends to distance them from other aspects of history. The books will investigate the uses and consequences of technical knowledge, and how it was shaped within particular economic, social and political structures.

Such analyses should contribute to discussions of present dilemmas and to assessments of policy. 'Science' no longer appears to us as a triumphant agent of Enlightenment, breaking the shackles of tradition, enabling command over nature. But neither is it to be seen as merely oppressive and dangerous. Judgement requires information and careful analysis, just as intelligent policy-making requires a community of discourse between men and women trained in technical specialities and those who are not.

This series is intended to supply analysis and to stimulate debate. Opinions will vary between authors; we claim only that the books are based on searching historical study of topics which are important, not least because they cut across conventional academic boundaries. They should appeal not just to historians, nor just to scientists, engineers and doctors, but to all who share the view that science, technology and medicine are far too important to be left out of history.

Titles include:

Julie Anderson, Francis Neary and John V. Pickstone
SURGEONS, MANUFACTURERS AND PATIENTS
A Transatlantic History of Total Hip Replacement

Roberta E. Bivins
ACUPUNCTURE, EXPERTISE AND CROSS-CULTURAL MEDICINE

Linda Bryder
WOMEN'S BODIES AND MEDICAL SCIENCE
An Inquiry into Cervical Cancer

Roger Cooter
SURGERY AND SOCIETY IN PEACE AND WAR
Orthopaedics and the Organization of Modern Medicine, 1880–1948

Jean-Paul Gaudillière and Ilana Löwy (*editors*)
THE INVISIBLE INDUSTRIALIST
Manufacture and the Construction of Scientific Knowledge

Christoph Gradmann and Jonathan Simon (*editors*)
EVALUATING AND STANDARDIZING THERAPEUTIC AGENTS, 1890–1950

Alex Mold and Virginia Berridge
HEALTH AND SOCIETY SINCE THE 1960S
Voluntary Action and Illegal Drugs

Ayesha Nathoo
HEARTS EXPOSED
Transplants and the Media in 1960s Britain

Neil Pemberton and Michael Worboys
MAD DOGS AND ENGLISHMEN
Rabies in Britain, 1830–2000

Cay-Rüdiger Prüll, Andreas-Holger Maehle and Robert Francis Halliwell
A SHORT HISTORY OF THE DRUG RECEPTOR CONCEPT

Thomas Schlich
SURGERY, SCIENCE AND INDUSTRY
A Revolution in Fracture Care, 1950s–1990s

Eve Seguin (*editor*)
INFECTIOUS PROCESSES
Knowledge, Discourse and the Politics of Prions

Crosbie Smith and Jon Agar (*editors*)
MAKING SPACE FOR SCIENCE
Territorial Themes in the Shaping of Knowledge

Stephanie J. Snow
OPERATIONS WITHOUT PAIN
The Practice and Science of Anaesthesia in Victorian Britain

Carsten Timmermann and Julie Anderson (*editors*)
DEVICES AND DESIGNS
Medical Technologies in Historical Perspective

Science, Technology and Medicine in Modern History

Series Standing Order ISBN 978–0–333–71492–8 hardcover
Series Standing Order ISBN 978–0–333–80340–0 paperback
(*outside North America only*)

You can receive future titles in this series as they are published by placing a standing order. Please contact your bookseller or, in case of difficulty, write to us at the address below with your name and address, the title of the series and one of the ISBNs quoted above.

Customer Services Department, Macmillan Distribution Ltd, Houndmills, Basingstoke, Hampshire RG21 6XS, England

Women's Bodies and Medical Science

An Inquiry into Cervical Cancer

Linda Bryder

Professor in History, University of Auckland, New Zealand

New Zealand and Australian edition first published 2009 by Auckland University Press
as 'A History of the "Unfortunate Experiment' at National Womens's Hospital'.

This edition published 2010 by
PALGRAVE MACMILLAN

Palgrave Macmillan in the UK is an imprint of Macmillan Publishers Limited, registered
in England, company number 785998, Houndmills, Basingstoke, Hampshire RG21 6XS.

Palgrave Macmillan in the US is a division of St Martin's Press LLC, 175 Fifth Avenue,
New York, NY 10010.

Palgrave Macmillan is the global academic imprint of the above companies and has
companies and representatives throughout the world.

Palgrave® and Macmillan® are registered trademarks in the United States, the United
Kingdom, Europe and other countries.

ISBN: 978–0–230–23603–5 hardback

This book is printed on paper suitable for recycling and made from fully managed and
sustained forest sources. Logging, pulping and manufacturing processes are expected
to conform to the environmental regulations of the country of origin.

A catalogue record for this book is available from the British Library.

A catalog record for this book is available from the Library of Congress.

10 9 8 7 6 5 4 3 2 1
19 18 17 16 15 14 13 12 11 10

Printed and bound in Great Britain by
CPI Antony Rowe, Chippenham and Eastbourne

Contents

Acknowledgements

Many times during the long journey of discovery that led to the production of this book I wondered whether I should be entering a field which is clearly still so sensitive to so many people. However, I do believe that it is an important story which should be told from the perspective of a medical historian, and this view was reinforced by the feedback from medical historians internationally following seminar presentations on the research. Encouragement came from academics at the Wellcome Unit for the History of Medicine, University of Oxford; the London School of Hygiene and Tropical Medicine; the Oxford Brookes Centre for Health, Medicine and Society, Past and Present; the Manchester Centre for the History of Science, Technology and Medicine, University of Manchester; the Centre for the Social History of Health and Healthcare, University of Strathclyde and the Glasgow-Caledonian University; the Department of Public Health, University of Copenhagen; and members of the Australian and New Zealand Society of the History of Medicine. I wish to acknowledge and thank the Royal Society of New Zealand for financial support for this research through the Marsden Fund.

Individually, I wish to thank Dr John Hood, formerly vice-chancellor of the University of Auckland and then vice-chancellor of the University of Oxford who, during a casual conversation about my research into National Women's Hospital when I was in Oxford, suggested that the Cartwright Inquiry would make an excellent subject for a monograph in its own right. I wish to thank Sir Iain Chalmers, also of Oxford, whose interest was stimulated by inquiries he had received from fellow epidemiologists about my studies and who kindly agreed to read the manuscript. I am grateful to Professor Janet McCalman and Dr Sally Wilde, fellow medical historians, for their incisive comments and feedback, along with the anonymous readers of my manuscript for the publishers. I am particularly indebted to Dr Sam Elworthy, Director of Auckland University Press, and Professor John Pickstone, Wellcome Research Professor at the University of Manchester and editor of the Palgrave Macmillan series in Science, Technology and Medicine in Modern History, for their support, encouragement and useful advice throughout. I also wish to thank my copy editor, Ginny Sullivan, for her careful scrutiny of the manuscript. Among my friends, special thanks go to Peter Nobbs for lending a friendly ear while I was working on this project. Finally I could not have brought this research to fruition without the ongoing support of my family, Derek, Dennis and Martin Dow, to whom I dedicate this book.

Linda Bryder

1.

Introduction
An Inquiry into Cervical Cancer

New Zealand's National Women's Hospital, situated in Auckland, was set up in 1946, a time when confidence in modern medical science soared throughout the Western world. Medical advances during the Second World War included the development of antibiotic drugs to combat serious infections, as well as blood transfusion and other improvements in surgical techniques which made major operations safer. It was confidently expected that further developments would follow. This was the golden age of medicine. Hospitals, with their modern equipment and laboratories, were associated in the public mind with heroic medical science, and medical practitioners and researchers enjoyed a higher social status than ever before.[1] National Women's Hospital, destined to become the largest women's hospital in Australasia, was established as a result of massive fund-raising by women's groups who sought to extend the benefits of modern biomedical science to women. Just over 40 years later this same hospital was the site of a huge public scandal and a government inquiry.[2]

In June 1987 Auckland's *Metro* magazine published what has become a watershed in New Zealand's medical history. The *Journal of General Practice* described it as a 'bombshell'[3] and the *New Zealand Woman's Weekly* announced it had 'opened what must be the most controversial and widely publicised can of worms in New Zealand medical history'.[4] Another magazine, *North and South,* commented that the article was 'one of the most influential pieces of investigative journalism ever published in this country'.[5] Twenty-one years later the *New Zealand Herald* stated it had exposed 'the biggest medical scandal of the century'.[6]

'An Unfortunate Experiment at National Women's' was written by Sandra Coney, a journalist and feminist activist, and Phillida Bunkle, a senior lecturer in Women's Studies at Victoria University of Wellington. They made a convincing case against Dr Herbert Green, associate professor of obstetrics and gynaecology at the University of Auckland Medical School, who they believed had caused a number of women to develop cervical cancer from carcinoma in situ (CIS) by withholding conventional treatment in order to study the natural history of the disease.[7] The *Metro* article set the scene by citing a patient who compared National Women's Hospital to Auschwitz in its medical experimentation.[8]

Coney explained in her prize-winning book published the following year:

> A disastrous research programme had been carried out at National Women's Hospital in Auckland and covered up for years. Women with pre-malignant abnormalities in the cells in the neck of the womb had not received conventional treatment for the condition. The statistician had calculated that these women had developed the maiming and potentially fatal invasive cervical cancer at an appalling twenty-five times the rate of women treated conventionally. They had had normal treatment withheld because one doctor, Associate Professor Herbert Green, believed that the abnormal cells were harmless. He argued that the pre-malignant disease, called carcinoma in situ or CIS, did not progress to invasive cervical cancer.[9]

The response to the magazine article was 'instant and spectacular'.[10] Within two weeks of its publication, the Minister of Health had set up an Inquiry headed by Silvia Cartwright (later Dame Silvia), a family and district court judge. The committee sat for six months, and submitted its report to the Minister in July 1988. Cartwright concluded that the medical profession had 'failed in its basic duty to patients'.[11]

Upon the report's publication, a local Labour MP, Richard Northey, referred to the 1947 Nuremberg Code on patient consent, which had arisen out of Nazi experiments on Jews and the mentally disabled, and declared that it was 'absolutely atrocious that such ill-treatment should have occurred' at National Women's.[12] Dr Alan Gray of the Cancer Society of New Zealand spoke of Green's 'total disregard for the long-term welfare of his patients',[13] and a leading article in the *Australian Medical Journal* stated, 'If a similar treatment were proposed which involved animals, it no longer would be sanctioned by any hospital ethics committee in the world.'[14] Another article in the *New Zealand Nursing Journal* by Jocelyn Keith, nurse tutor at the Victoria University of Wellington School of Nursing, entitled, 'Bad Blood: Another Unfortunate Experiment', compared Green's research to the Tuskegee Syphilis Experiment in Alabama, USA, a study conducted between 1932 and 1972 on the effects of untreated syphilis, involving 600 black Americans. Keith wrote that long after this experiment had stopped, Cartwright 'completed her inquiry into the allegations concerning the treatment of cervical cancer at National Women's Hospital in Auckland'. It was, she said, 'a damning indictment . . . . made even more damning when you realise that the unfortunate experiment at National Women's was quietly proceeding while the Tuskegee Study and the Kennedy hearings were all over the world press.'[15] The *Nursing Journal* described Green's research as 'a secret and life-threatening experiment on women'.[16]

A comment by Fertility Action, a feminist group headed by Sandra Coney which had given evidence to the Inquiry, was widely repeated in the press: 'While the medical profession at National Women's and elsewhere maintained closed ranks and an unbroken silence, the women continued to come to the hospital like lambs to the slaughter.'[17] Not surprisingly then, a letter to the editor of a local paper following the report's release declared, 'New Zealanders owe an enormous debt to Sandra Coney, Phillida Bunkle, Sylvia [*sic*] Cartwright and those who helped create a climate to openly investigate medical wrong-doing. Never before have we been permitted to see such naked arrogance and contempt for women . . . . If indifference to the rights of people they profess to serve is not checked by lay people then doctors' cavalier attitudes will continue to flourish.'[18]

I was in the United Kingdom when the 1987 *Metro* article appeared, working as a research fellow in the history of science at the Queen's College in Oxford where I was resident from 1981 to 1988. A friend had given me a year's subscription to *Metro* so I saw the article when it came out, and like everyone else was horrified by what had taken place in my home town. In mid–1988 I took up a lectureship in New Zealand history at the University of Auckland. My first major research project following my return to New Zealand was the history of the Royal New Zealand Plunket Society, a voluntary infant welfare organisation set up in 1907.[19] As an extension of that and with a growing interest in the history of reproductive health, I then decided to research the history of National Women's Hospital, a significant institution in New Zealand's history and in the history of medicine. It was after all the site of important medical developments internationally through the work of Sir William Liley and Sir Graham Liggins. Liley had performed the first intrauterine blood transfusion in the world, a groundbreaking treatment for Rhesus haemolytic disease. Liggins had pioneered the administration of corticosteroids to women about to have premature babies; this prevented the babies' lungs from collapsing upon birth, a treatment that was subsequently adopted internationally. National Women's Hospital also lent itself to a study of the politics of childbirth. While my primary interest lay in the history of reproductive health, I knew that as part of the research into the hospital I would have to deal with the so-called 'unfortunate experiment' and the Cartwright Inquiry which emerged from it. My initial chapter outline envisaged that this would feature midway through the text as literally an unfortunate episode in the hospital's history. Sandra Coney's book, an amplification of the *Metro* article, would be the principal source, providing, as the cover promised, the 'full story behind the Inquiry into Cervical Cancer Treatment'. In her report Cartwright commended the authors of the *Metro* article for their 'extraordinary determination to find the truth'. She said that, 'The factual basis for the article and its emphasis have proved to be correct.'[20]

While working on this project, I spent some months on sabbatical leave in Oxford. There I took advantage of the extensive run of medical journals and other publications at the Radcliffe Science Library to inform myself of the background into the medical condition that was the subject of the Inquiry – carcinoma in situ of the cervix. The literature I accessed there came as a surprise and forced me to revise my own view of the 'unfortunate experiment' and the Cartwright Inquiry. On returning to New Zealand, I followed this up with a careful reading of the Inquiry transcripts and the considerable media coverage. While 72 interviews conducted by Cartwright with patients remain closed files, what happened to these women can be gleaned from lawyer Rodney Harrison's use of their case notes during his cross-examination of Green at the Inquiry, from Judge Cartwright's report and from Sandra Coney's book on the Inquiry. The transcripts of the eleven women who came forward to give evidence publicly, submissions to the Inquiry, and the many letters written to Cartwright, Health Minister Michael Bassett and medical superintendent Gabrielle Collison provide further evidence of patient experience. I quickly decided that the story which emerged merited a book in its own right, as it would threaten to overwhelm a general history of the hospital and its work in reproductive and neonatal health and medicine. In the early stages of researching the general hospital history, I had contracted Dr Jenny Carlyon to conduct interviews with many of those involved with the hospital. The mass of written material concerning the Cartwright Inquiry made it unnecessary to conduct further or more focused interviews. For this study I draw primarily on written records, both published and unpublished, which are extensive as well as extremely varied. Before embarking on the history, I knew very little about the dramatis personae in this story. The narrative which unfolded, written from the perspective of a social historian of medicine, is more complex than had hitherto been apparent and sheds new light on what is undoubtedly an important episode in New Zealand's social and medical history.

In her report Silvia Cartwright made several claims about the approaches of Green and his colleagues at National Women's Hospital to patients with carcinoma in situ of the cervix. She argued that from the 1960s they ignored the 'world view' that carcinoma in situ was a precancerous condition, that 'normal' or 'conventional' treatment was withheld from some patients with carcinoma in situ, that patients were not told they were part of a trial, that Green misinterpreted his data or manipulated his statistics to prove his hypothesis, and that he and his colleagues ignored world opinion when they questioned the value of population-based cervical screening.

Taking these allegations as a starting point, the first part of this study reviews the medical literature relating to carcinoma in situ in the Western world from the 1950s to the 1980s, and the ways in which Green and his colleagues responded and contributed to that literature. It questions some of the assumptions of the Cartwright Report and the popular understanding of medical practice in relation to CIS at National Women's Hospital. These chapters pose the following questions. Did Green's research and treatment fly in the face of scientific facts, as postulated by Coney?[21] Was Green simply trying to 'prove a personal belief', as suggested by Cartwright?[22] Were the doctors at National Women's Hospital alone in the world in questioning the value of population-based cervical screening, as also suggested by Cartwright?[23] Chapters 1 to 6 of this book examine the medical controversies that underlay the Cartwright Inquiry and the charges of abuse of patient trust, placing this account in the context of the history of post-Second World War medicine generally and the rise of bioethics, and address whether Green was the maverick he was made out to be.

Chapter 7 discusses the history of the women's health movement in New Zealand from the 1970s, and explains how this new social movement clashed with National Women's Hospital, which had a predominantly male medical staff, and with its ongoing research programmes as a postgraduate school of obstetrics and gynaecology. Chapter 8 shows how this translated into the Cartwright Inquiry. The women's health movement aimed to change the balance of power between doctors and their female patients generally. Coney herself was adamant that the Inquiry was not just about Green, who had retired in 1982, but about all doctors. As she later wrote, 'we hoped to broaden the inquiry beyond the specific events at NWH [National Women's Hospital] into a general critique of the practice of medicine, the observance of patients' rights and of the treatment of women within the health care system'.[24] They succeeded: despite the divisions and disputes within the medical profession, the whole of the profession was put on trial. Motivated by feminist principles, Coney explained how she and Bunkle regarded health as the 'cutting edge of sexual politics, the place where women were often at their most powerless'.[25] The Inquiry was for them an opportunity to promote the major goal of the women's health movement, which was to reclaim for women control over their own bodies from the predominantly male medical profession.

Chapters 9 and 10 explore the immediate impact of the Inquiry primarily as expressed through the media, and chapter 11 shows how the representations of what occurred at National Women's became increasingly distorted over time. Following the Inquiry, the medical profession hoped to restore public confidence by endorsing the Cartwright Report wholeheartedly; by charging those involved in the 'unfortunate experiment' with professional misconduct; and by ensuring the selection of a woman to take up the chair of obstetrics and gynaecology at the University of Auckland. Those who critiqued the Cartwright Inquiry were dismissed as chauvinistic, defensive, reactionary and insensitive. The Inquiry was used by nurses and midwives as an opportunity to enhance their professional status at the expense of doctors who had lost the confidence of the public. Above all, however, the Inquiry was about consumer power. The Inquiry had apparently shown that doctors could not be trusted to protect the interests of patients, something the women's health movement had believed all along. The government and health authorities alike accepted that it was necessary to bring in a third party to protect patients' interests and welfare. Hospitals had become dangerous places, especially for women.

The changes that occurred in New Zealand in medicine from the 1970s to the 1990s were not isolated events but part of an international trend. Consumer power, along with women's and patients' rights, was altering relations in medicine.[26] The changes in New Zealand were dramatically highlighted through the Cartwright Inquiry. Coney, Bunkle and Cartwright had picked up on a long-standing medical dispute about the appropriate way to treat CIS and run with one side, ironically not the side usually favoured by feminists of questioning the 'medical model' but rather that which upheld the efficacy of interventionist medicine. One American feminist historian later wrote that an imperative of feminist politics of health was to be critical of medical knowledge and practice, to question interventionist technological solutions to medical problems and to seek alternatives.[27] I would argue, however, that Fertility Action overlooked this perspective as they sought a greater goal of bringing to heel a patriarchal medical institution, the National Women's Hospital and its Postgraduate School of Obstetrics and Gynaecology. *Women's Bodies and Medical Science* addresses the complexities of how this happened and why it succeeded. The examination of the Cartwright inquiry provides a lens through which to explore the relationship between women 's bodies,technology and medicine in the late twentieth century.

2.

Carcinoma in Situ
Meanings and Medical Significance

In 1966 Associate Professor Herbert (Herb) Green took a proposal to the National Women's Hospital Medical Committee to treat more conservatively than previously women who had been diagnosed through a positive cervical smear with carcinoma in situ or abnormal cells in their cervix. Twenty-one years later, at the Inquiry into Cervical Cancer at National Women's Hospital, this was identified as the start of the 'unfortunate experiment' in which some patients received inadequate treatment. Discussing Green's 1966 proposal, Judge Silvia Cartwright wrote, 'In 1966 Green and his colleagues on the Hospital Medical Committee were well aware of the world view that CIS was a precancerous condition.'[1] She claimed that he was trying to 'prove a personal belief' that carcinoma in situ was not a precursor to invasive cancer, 'ignoring virtually all the existing literature which assessed the likelihood of progression to invasion.'[2]

What was the 'world view' of CIS of the cervix? This chapter traces interpretations of the meaning of this medical condition in the international medical literature from the 1950s to the 1990s. An understanding of these debates helps to contextualise Green's 1966 proposal and his management of patients with CIS. Green's colleague, Dr William (Bill) McIndoe, referred to Green's 'strange ideas' relating to CIS, and later Dr Barbara Heslop, medical professor at the University of Otago Medical School, depicted him as a lone researcher in the South Pacific, isolated from the wider world.[3] The following discussion will question these perceptions of Green by reviewing international debates about the meanings and medical significance of a diagnosis of CIS, and his place within them. These debates were not just academic but had significant implications for treatment.

The International Debates, c. 1945–1966

In the years immediately after the Second World War, optimism ran high throughout the Western world that modern medical science would ultimately conquer all major diseases, and cancer was no exception. In America, a revitalised American Cancer Society adopted a 'search and destroy' approach to cancer, and with recent diagnostic developments for cervical cancer, this disease became the model.[4]

The American Society for the Control of Cancer had been founded in 1913. Dominated by surgeons, it aimed to combat the current fatalistic attitude to cancer by arguing that if it was caught early enough, cancer was curable.[5] As historian Patrice Pinel has noted, this was the first public health policy based exclusively on treatment, as no prevention-based strategy was in sight.[6] Cancer of the uterus was the most prevalent fatal cancer among American women in the early twentieth century, and in 1926 Dr James Ewing, a leading pathologist and a founder of the American Society for the Control of Cancer, lamented that there were 'no specific symptoms for early diagnosis' of this disease.[7] However, around this time medical researcher Dr George Papanicolaou was studying cells in the vaginal fluid, acquired by taking a small smear from the surface of the cervix. He argued that these cells could reveal an early form of cancer. Papanicolaou was a Greek doctor who had emigrated to the United States in 1913, where he was appointed assistant in the pathology laboratory at New York Hospital. He worked on developing this new diagnostic tool in the 1920s and 1930s, but his findings were greeted with some scepticism until he published a scientific paper in 1941 and a monograph in 1943, leading to the new science of 'cytology', the study of cells in the vaginal fluid.[8]

In 1945 the American Society for the Control of Cancer was renamed the American Cancer Society (ACS). The Pap smear (as the diagnostic tool developed by Papanicolaou became known from the 1950s) suited the goals of the recharged society. It was scientific, emanating from the laboratory, simple to perform and process, cheap, and surgeons believed that the cervix was relatively easy to operate on. Importantly, the Pap smear would fulfill the society's declared goal of the early detection of cancer.[9]

A stumbling block for the new diagnostic tool was that there was no proof that CIS would progress to cancer if untreated. An American pathologist, Dr Lewis Robbins, later commented when interviewed for a history of the Cancer Society:

> The Pap smear isn't diagnosing cancer, it's diagnosing a precursor. Why did they call it cancer then? Because nobody would pay any attention if they called it dysplasia. If you call it carcinoma in situ, then they will examine it, do something with it . . . . I remember a battle royal at Roswell Park in 1946 . . . . One physician said carcinoma in situ is not cancer, but we have to call it cancer. The pathologist said we can't call it cancer if it doesn't metastasize or if it hasn't already metastasized. But we did.[10]

Dr Charles Cameron, the society's director of medical and scientific research, embraced the Pap smear enthusiastically and persuaded the society to host the first National Cytology Congress in 1948. The congress hailed this new diagnostic tool for carcinoma of the cervix as unique because, they declared, the Pap smear could detect cancer before it was visible. This congress was a turning point in the wider acceptance of the Pap smear as a cancer screening procedure.[11] In 1950 the International Congress of Obstetricians and Gynaecologists passed a resolution to alter the international classification of cancer of the cervix by adding a further stage, Stage 0, to include CIS, a condition detected by the Pap smear. Thus, while carcinoma in situ of the cervix had been named in the 1930s,[12] it was not until the postwar period that a public health campaign was launched. Medical historian Charles Rosenberg has argued that a disease does not exist 'until society decides that it does – by perceiving, naming and responding to it'.[13] This cancer-producing disease (CIS) definitely existed in 1950s America.

In 1954 Papanicolaou suggested a grading system from one to five for smear results. While the categorising of slides continued to be variable and subjective, as will be discussed, the new discipline of cytology encouraged pathologists to be decisive. As Dr William McIndoe of New Zealand's National Women's Hospital explained in 1960, 'the treatment advised and prognosis offered depends on the category in which he [the pathologist] places the lesion'.[14]

Another American doctor, Arthur Holleb, recalled that while many of the medical people in the Cancer Society remained sceptical about the value of the smear test in the early 1950s, the laymen 'were all for it'. This was a time when 'money was rolling in [to the ACS], and we had the research commitment to spend at least 25%'.[15] Sociologists Monica Casper and Adele Clarke have argued that, because the ACS was supporting the smear test, pathologists and others, hoping to share in the newfound largesse of the ACS and its close ally the National Cancer Institute, also began to support it. In 1943 the ACS had an annual budget of $102,000; by 1945 the budget was over $4 million and rising.[16] Cameron told the annual meeting of the ACS in 1951, 'We need not wait for more evidence' of the benefits of the Pap smear.[17] He began to travel across the United States, extolling its virtues at public meetings, even though, as medical historian Barron Lerner later noted, 'definitive data that the Pap smear saved lives hardly existed at this time'.[18] Lerner explained how this 'search and destroy' approach suited post-Second World War American society: 'Replete with both war metaphors and calls for individual responsibility for one's health, Cameron's messages were also characteristically American.' Cameron expressed confidence that with this 'precision weapon', doctors would conquer cervical cancer.[19] From the 1950s women were encouraged to have an annual Pap smear test, and if found to have dysplastic lesions in the cervix, or CIS, were given a hysterectomy (the surgical removal of the uterus) or had the lesions removed by cone biopsy (local treatment under anaesthesia involving the removal of a cone of tissue from the mouth of the cervix). The development of the Pap smear test was celebrated as a great American success story, and Papanicolaou as a 'giver of life'.[20] As British medical sociologist Tina Posner put it, here was a case in medicine where an otherwise fatal condition could apparently be stopped in its tracks by the heroic application of medical science and modern medical technology.[21]

The American Gynecological Society held a conference on CIS in 1952. The opening speakers described this 'dramatic' term as a misnomer.[22] One of the speakers was Dr John McKelvey from the University of Minnesota Medical School, later professor and head of the Department of Obstetrics and Gynecology at Minnesota. He pointed out that if CIS was to be regarded as 'preinvasive' then one must be able to say with reasonable certainty that it would destroy the host unless removed. He referred to research by Dr Hans-Ludwig Kottmeier in Sweden, who reported on 42 cases of CIS he had followed for ten years without treatment. Only three developed clinical carcinoma. McKelvey wondered 'how many of these lesions being reported as carcinoma in situ were not carcinomas at all. To call atypical lesions malignant and treat them as cancers is only to fog our own critical intellectual recesses and to harm the patient by putting the fear of cancer permanently in her mind.'[23] Another speaker compared the discovery of large numbers of cases of CIS with the much smaller death rate from cervical cancer, and estimated that about 10 to 20 per cent of the diagnosed cases of CIS would develop into cancer.[24] While some participants at the conference recommended a conservative approach to cases detected, others advocated a radical approach. Dr Richard TeLinde of the Johns Hopkins School of Medicine, Baltimore, said that his hospital had treated more than 150 cases of CIS by hysterectomy. TeLinde declared, 'Carcinoma in situ gives us an opportunity for cure and I think we should not neglect it.'[25] This conference signalled the start of a lively international debate, though the trend in practice was towards an interventionist approach to the disease.

By the early 1960s, doctors throughout the Western world expressed great confidence that cervical cancer could be eradicated. At the American Medical Association's annual meeting in 1960, chairman Curtis Lund declared that an epitaph could be written for cervical carcinoma: 'The lethal methods are ready, only the proper application remains.' The lethal method was the Pap smear test. He claimed that the lesions found through screening women with no symptoms were 100 per cent curable. He added that since half of all patients with invasive carcinoma survived and about 60 per cent were now discovered before they developed invasive cancer (all of whom survived), 'then the cure rate in our community is 80 per cent, an astounding figure!'[26]

The enthusiasm was shared around the world and formed the basis of widespread cervical screening programmes. Screening was introduced to British Columbia, Canada, in 1949, and by 1962 about a third of women over the age of 20 had been screened. Pointing to a significant reduction in the incidence of invasive carcinoma of the cervix, Canadian gynaecologist David Boyes and his co-workers believed that such a screening programme was 'capable of virtually eliminating invasive cervical carcinoma from the population'.[27] In Britain, Hugh McLaren, professor of obstetrics and gynaecology at the University of Birmingham, began his 1963 book on the prevention of cervical cancer 'on a cheerful note for we now have a method of prophylaxis which, if properly used, may largely eliminate carcinoma of the cervix from our communities. We can now say to womankind, "If you submit to regular tests you won't get cancer of the cervix."[28] In 1966 John Stallworthy, a New Zealander who held the chair of obstetrics and gynaecology at the University of Oxford, England, confidently declared, 'women now living need never die from carcinoma of the cervix'. He believed that this was a concept of 'stupendous importance' and presented the profession as well as the public with a challenge which could not be ignored. He predicted that 'posterity will regard the virtual elimination of cancer of the cervix as one of the great triumphs of pre-ventive medicine', and that future generations would look back to the twentieth century to study how it was accomplished.[29]

At the same time, women's groups began to frame CIS and cervical cancer as a feminist issue. A 1964 article in a British women's magazine described cervical cancer as 'the hidden disease in healthy women' and one which was largely pre-ventable.[30] A Scottish medical officer of health, Dr Marie Grant, agreed: 'Cancer of the cervix must now be regarded largely as a preventable disease, and, with the remedy in our own hands, surely few can fail to be impressed with the urgency of the problems presented to us.'[31] It became increasingly common for women attending family planning clinics or gynaecological clinics in Britain to be given a Pap smear. In 1965 representatives of 57 voluntary organisations met to form the National Cervical Cancer Prevention Campaign as a lay organisation to publicise the importance of cervical smears. The president was Joyce Butler, MP and member of the Co-operative Labour League. At a planning meeting in February 1965, one of the speakers referred to the great part married women played in Britain's economy both currently and during the Second World War, and argued that they 'deserved' a screening programme.[32] The campaign produced an information sheet announc-ing certain cure. It pointed out that approximately 3500 women died each year in Great Britain from cancer of the cervix and uterus, which constituted 1 per cent of women's deaths from all causes. By allocating money to a cervical smear pro-gramme, it estimated that 'over 3,000 lives could be saved each year'.[33]

Meanwhile, in 1965 Walter Sanford Ross, who was later to write the offi-cial history of the American Cancer Society, penned a book on cancer entitled *The Climate of Hope*; originally published in America, it was reprinted in New Zealand in 1967. This included a chapter written by Dr Emerson Day, director of the Strang Clinic, New York City, from 1950 to 1963, who was, according to the blurb, 'intensely dedicated to the idea that cancer is a conquerable disease . . . if the disease is sought out and aggressively pursued'. Day himself identified cancer of the cervix as 'the outstanding example of successful examination leading to cure of cancer'. He explained that 'With the Papanicolaou test, which is pain-less and inexpensive, as well as highly sensitive, we can approach 100 percent control of this disease.' He added that deaths from this form of cancer had already dropped by 50 per cent in the last 20 years.[34]

Yet others remained sceptical about the powers of the Pap test to eliminate cervical cancer and wondered what they were in fact treating. In 1962 Dr John Graham, chief gynecologist at Roswell Park Memorial Institute and associate professor of gynecology at the University of Buffalo Medical School, repeated the observations of ten years previously that the prevalence of CIS was well above the range of clinical cancer, 'indicating that a significant proportion of these lesions come and go without producing the disease cancer'.[35] Graham noted that although CIS of the cervix was not cancer, at times it was classified as such, and gave the example of New York State where CIS was listed as cancer in the annual statistics.[36] Dr Leopold Koss, a pathologist at the Memorial Hospital for Cancer and Allied Diseases New York and a student of Papanicolaou, commented in 1963 on 'The problem of the "disappearing" carcinoma in situ', referred to the apparently high regression rate, and asked 'When is cancer not a cancer?' He added that, 'The approach that seemed most reasonable under these circumstances was a long-term follow-up study with minimal or no treatment.' Koss was subsequently recognised as a world authority on CIS.[37]

Another sceptic of the 'certain cure' hypothesis was an Australian doctor, Malcolm Coppleson, who was based at the King George V Memorial Hospital and the Royal Prince Alfred Hospital in Sydney. In 1967 Oxford University Press published Coppleson's book outlining his work on 'preclinical carcinoma of the cervix uteri', written with a colleague, Dr Bevan Reid from the Queen Elizabeth Research Institute, University of Sydney.

On the invasive potential of CIS, Coppleson and Reid wrote that, 'evidence indicates that the majority of women to whom the histological diagnosis of carcinoma *in situ* is assigned would never have developed a frankly invasive carcinoma, even if left untreated'.[38] Citing the 1965 work of two American gynaecologists (W. M. Christopherson and J. E. Parker), Coppleson and Reid agreed with the authors' conclusion that, 'There is no doubt that cases of carcinoma *in situ* were diagnosed that were not in fact biologically cancer.'[39] They added that even when the diagnosis was unequivocal, the clinical outcome could not be predicted, citing Dr John Graham of the Roswell Park Memorial Institute who had listed in 1962 the findings of seven studies. These studies had followed 262 cases without treatment: 35 became invasive but ten of these 'invaded' in the first year, leading the authors to conclude that these women probably had cancer at the time of the initial diagnosis. Graham had suggested that 'only 10 to 20 per cent of patients with carcinoma *in situ* would develop invasive cancer if untreated'.[40] Like others, Coppleson and Reid noted a clear disparity between the observed frequency of 'in situ cancer' and the incidence of 'invasive cancer'.[41] On management, they wrote, 'the label of "cancer" is there and comprehensively, almost predictably, management tends in the radical direction. We believe the ease of diagnostic methods and the vigour and skill of those employing them have presently outstripped the availability of basic knowledge or their assimilation and correct interpretation.'

They explained that a theme developed in their book was that there was 'room for doubt of many of the more serious interpretations placed on cells or tissues removed from the cervix' by cone biopsy. They continued, 'These thoughts have become the basis of our conservative approach to many of these problems.'[42]

In his foreword, Professor John Stallworthy described it as a 'remarkable book' and 'a great achievement by two young scholars'. He added that the work on which it was based had earned the Edgar-Gentilli Prize of the Royal College of Obstetricians and Gynaecologists (RCOG) for 1967 and would 'certainly establish for them an international reputation as authorities on the subject of epithelial dysplasia, carcinoma *in situ* and cervical cancer'. He commended them for the 'scientific scepticism [with which] they tested views old and new and evolved a philosophy of their own'. In particular, Stallworthy appeared to have been persuaded by their arguments that the situation was not as straightforward as he had indicated in his 1966 text, referred to earlier. He regarded their approach as 'opportune and challenging at a time when community-screening programmes are not only identifying increasing numbers of women with "doubtful" cervices, but are submitting many of them to the danger of unnecessary mutilating surgery'.[43] This 'unnecessary mutilating surgery' will be addressed in more detail when the management of patients is discussed in chapter 3.

Those who were urging a more conservative or cautious approach to treating CIS cannot be distinguished by nationality or by medical discipline (such as pathologists versus gynaecologists) but were scattered throughout the medical world. At the 1967 annual meeting of the [American] Central Association of Obstetricians and Gynecologists, Dr Eamonn De Valera from Ireland told its members: 'There is in our part of the world an increasing realization that many cases of carcinoma in situ do not progress to invasive cancer.'[44] As a mark of De Valera's own standing within the profession, the association elected him an honorary member the following year.

Dr Angela Raffle and Sir Muir Gray later wrote in their 2007 book *Screening: Evidence and Practice* that there was 'growing concern [in the 1960s] among rigorously minded academics about the sweeping claims for [cervical] screening that had been made by apparently authoritative bodies, based on no particular evidence'.[45] One concerned academic was Dr George Knox, professor of social medicine at the University of Birmingham, England, where he worked with Professor Thomas McKeown, well known for his questioning approach to modern medical intervention and for bringing 'something of a conceptual revolution in the disciplines of history and medicine' by questioning the contribution of medicine to improving health status over the previous 200 years.[46]

Professor Knox contributed articles on cervical cytology to two influential Nuffield Provincial Hospitals Trust series on current medical research and screening in 1966 and 1968.[47] He began his 1966 article by explaining that the uncertain value of the Pap smear was 'a matter for concern'. He pointed out that many regarded its value as established and attributed to it the capability of virtually eliminating cervical cancer. These people, he wrote, identified any remaining difficulties as problems of organisation and acceptance. 'But', he continued, 'the fact that the procedure [Pap smear] has not yet won the determined support necessary for an effective preventive measure is not due entirely to ignorance, apathy and expense. There is an informed skepticism of the claims made for it.'[48] In addressing this 'informed skepticism', Knox revisited some of the debates of the 1952 American conference, specifically the evidence that CIS would advance to cancer if left untreated. He noted that most of those who argued in the affirmative cited the work of O. Petersen from the Radium Centre in Copenhagen, who had followed a group of women with CIS without treatment for fifteen years and found that 35 per cent developed invasive lesions.[49] However, Knox explained, this group of women had presented with symptoms such as bleeding from the vagina and many of them might already have had cancer at the time of registration; therefore these cases could not be compared with women with no symptoms who were being screened. Despite reservations about the initial diagnosis, Petersen's study was cited many times as evidence of the invasiveness of CIS. At the time of the Cartwright Inquiry, Sandra Coney referred to it as 'the classic study [which] provided compelling evidence [of invasiveness]'.[50]

Knox also referred to British Columbia, which reported a higher rate of cervical cancer among unscreened than screened women. 'However no data are given about age, social group, parity, race or other selective data which may have determined entry into the screened and unscreened groups and it is not even clear whether or not the occurrence of symptoms may have excluded women from the group labelled as screened.'[51] He argued that the available population and pathological data could be used to suggest not one but two diseases – histologically recognised CIS, a benign disease arising between 20 and 30 years of age, and invasive cancer arising through some other cervical lesion which was not usually detected, or from an uncommon special variety of CIS which was not distinguishable from the benign lesion.[52]

In suggesting the possibility of different diseases, with some forms of CIS being benign, Knox was diametrically opposed to Dr Ralph Richart, professor of pathology at Columbia University College of Physicians and Surgeons, New York, and director of the Division of Obstetrics and Gynecology, Pathology and Cytology at the Sloane Hospital for Women, New York City. Richart was responsible for introducing a new set of terms into gynaecological oncology in 1967. Cervical intra-epithelial neoplasia (CIN) categories replaced the terms 'mild dysplasia' (now CIN1) to 'carcinoma in situ' (now CIN3).[53] Richart's objective was to stress that all abnormal cells of the cervix were potentially cancerous and should be viewed as a spectrum or continuum from mild dysplasia to invasive cancer and treated at the earliest opportunity.[54] Richart, whom Coney and Bunkle had contacted when writing the *Metro* article, was invited to be an expert witness to the Cartwright Inquiry.

Unconvinced that Richart had the evidence to show such a continuum, Knox concluded his 1966 article with the statement that neither histology, cytology nor population data provided a comprehensive description of the natural history of CIS. He said the information was simply not there and no amount of ingenuity could extract it. Indeed, 'Interpretations of its natural history based upon such data should be regarded as demonstrating only the assumptions which went into their assembly.' He pointed out that if CIS as recognised by screening techniques was not usually progressive and was not a common source of invasive lesions, then its removal was a cost to be set against the benefit of treating invasive cancers.[55]

In 1968 the Nuffield Trust set up a working party on screening, under the chairmanship of Thomas McKeown. In his contribution to the resulting publication, Knox argued that it was unethical to continue the development of screening programmes without first attempting to answer 'crucial questions' about the natural history of the disease.[56] In their joint conclusion to the study, McKeown and Knox wrote regarding cervical cancer, 'Again the problem concerns the natural history of the disease. There is no firm evidence on the frequency with which (a) carcinoma in situ progresses to invasive cancer and (b) invasive cancers [are] preceded by abnormal smears.'[57]

Thus the 1950s and 1960s was a time of tremendous enthusiasm for the powers of modern technology and medical science to conquer cervical cancer through screening and early aggressive treatment. Some experts, however, believed that the enthusiasm had overtaken reality and that the medical significance of lesions discovered through a Pap smear was far from clear. The assumption that all positive smears would, if untreated, advance to invasive cancer had led, they argued, to over-treatment and unnecessary harm in many cases. One clinician and academic who picked up on that international literature was Associate Professor Herb Green of New Zealand's National Women's Hospital.

'Carcinoma in Situ' at National Women's Hospital

In 1946 the Auckland Hospital Board set up a new public women's hospital in Auckland as a result of lobbying by women's organisations for more maternity beds and for a hospital that would train doctors in childbirth and women's diseases. There was a shortage of maternity beds resulting from the Labour government's 1939 commitment to provide all women with fourteen days' free stay in a hospital during childbirth, and the baby-boom of the immediate postwar period. A new trend in medicine saw the combination of the specialisms of obstetrics (childbirth) and gynaecology (women's diseases).[58] Women's organisations also launched at this time a massive fund-raising campaign to finance a chair of obstetrics and gynaecology in the new hospital. The hospital became known as

'National Women's Hospital' because it trained doctors from around the country and prepared them for membership of the Royal College of Obstetricians and Gynaecologists, for which they had to sit an examination in Britain.

In the period 1949–51, 68 deaths from cervical cancer occurred at National Women's Hospital, the only hospital in the Auckland province that could provide radiation, the current treatment for the disease.[59] The national mortality rate in 1950 from cervical cancer was 8.6 per 100,000 women (77 women) and from other unspecified parts of the uterus 5.9 per 100,000 (53 women). This was not nearly as high as breast cancer, which claimed 249 lives in 1950 or 27.8 per 100,000 women. That year, cervical cancer accounted for only 2.9 per cent of all cancer deaths in New Zealand.[60] Cervical cancer was, however, a horrible disease, as Professor Hugh McLaren explained in Britain in 1963: 'The misery of the patient is shared by the household for after all what can her family do about blood-stained, foul-smelling discharges or urinary incontinence in the terminal stage of carcinoma of the cervix?'[61] Not surprisingly, Auckland embraced the Pap smear enthusiastically. In 1949 the Hospital Medical Committee discussed the appointment of an additional laboratory assistant to examine vaginal smears.[62]

In 1954 Dr Harvey Carey was appointed postgraduate professor of obstetrics and gynaecology and head of the hospital. Carey, who had been senior registrar in obstetrics and gynaecology at the Royal Postgraduate School at Hammersmith Hospital, London, England, prior to his appointment at Auckland, was an enthusiast for medical research and medical technology. Shortly after his arrival, Dr Brewster Miller from Toronto, a keen promoter of the Pap smear in Canada,[63] visited National Women's Hospital. His visit led to discussions about setting up a laboratory at the hospital to analyse Pap smears, which was eventually achieved in 1958. Immediately following Miller's visit, the hospital's medical committee agreed that cervical smears should be obtained from all women over the age of 30 attending the gynaecological outpatients' department and that general practitioners attending the hospital's refresher courses should be taught the importance of taking smears. The committee looked to the Obstetrics and Gynecology Department of the Royal Victoria Hospital, Montreal, Canada, as the model. They decided to appoint an assistant pathologist to undertake the work, and Miller encouraged them to approach the American Cancer Society to train the pathologist at Papanicolaou's laboratory in New York.[64]

In 1955 Dr John Sullivan was appointed pathologist at National Women's Hospital and in 1957 he was sent on a tour of the United States to learn about cytology. He visited three centres including the Laboratory of Exfoliative Cytology, Chicago, run by Dr George Wied, a disciple of Papanicolaou. In his report to the Hospital Medical Committee, Sullivan wrote that Wied was responsible for introducing cytology as a science. Sullivan explained that he had discussed CIS with the pathologists at the three centres he visited, 'all of whom were agreed that carcinoma in situ is an early progressive malignant lesion which if untreated will eventually become invasive'.[65] In 1959, 130 doctors attended the first New Zealand Cancer Conference held at National Women's Hospital under the auspices of the New Zealand Branch of the British Empire Cancer Campaign (BECC). The guest speaker was George Wied.[66]

In the 1950s New Zealanders generally shared the enthusiasm of their American and British counterparts that the introduction of Pap smears, and the early treatment of the lesions so discovered, would abolish cervical cancer in the population. In 1958 Carey and the pathologist at Green Lane Hospital, Stephen Williams, explained that, given poor cure rates for cervical carcinoma, early diagnosis 'holds out more hope of improving the salvage from this type of malignancy'. By that time about 20,000 women had been screened at National Women's and 67 positive smears detected. Carey and Williams concluded that, 'On the basis that only 42 per cent of clinically diagnosed cases of carcinoma of the cervix in New Zealand survive five years and that all cases of pre-clinical carcinoma, if left untreated, would become invasive, 40 lives have been saved by this cytological survey'. They added that most of these women were in the 30–50 age group, carrying important family responsibilities, which made the achievement even more valuable.[67] From 1960 clinicians at the hospital took smears from all women attending the antenatal and gynaecological clinics, not just those over 30.

In 1961 Carey attempted to extend screening beyond the hospital by enlisting the support of the BECC and general practitioners. He wrote to the Auckland Division of the BECC, explaining that 1.5 per cent of New Zealand women developed carcinoma of the cervix during their lifetime and that 50 per cent of these women died of carcinoma of the cervix despite the best treatment available. He told them, 'These deaths are preventable and it lies within the power of the BECC to apply to the Auckland area the techniques and knowledge whereby these deaths may be prevented.' He pointed to the mass miniature x-ray campaign for tuberculosis, which had been launched in New Zealand in 1955, as a model for population screening.[68]

Carey persuaded the BECC to conduct a cervical screening trial in Thames. The Thames Country Women's Institute formally launched the campaign in April 1962. In the first year of the study, doctors tested 2817 patients and found a positive cytology rate of 7.7 per 1000 patients (22 women). [69] Carey also persuaded the Wanganui branch of the British Medical Association to conduct a cervical smear campaign in 1962, in which they took smears from 6014 women and discovered 46 cases of CIS.[70] The research committee of the New Zealand Council of the College of General Practitioners began a national cervical smear survey in 1962, with 337 doctors registering as participants. These doctors screened 17,000 women in the first year of the survey, which compared favourably with similar surveys elsewhere, such as in British Columbia.[71]

In 1965 the National Women's Hospital's cytology laboratory was processing more than 30,000 smears a year, and in total they recorded 130,600 smears examined since 1958. Reporting from the work of the laboratory, Drs Roy Darby (the medical cytologist) and Stephen Williams (the pathologist) believed that they could eliminate invasive cervical carcinoma from the population and looked to developments in British Columbia for inspiration.[72]

Harvey Carey resigned his post as postgraduate professor in 1963 to take up a position in Australia, and Dr Dennis Bonham was appointed to the chair. Born in England, Bonham took his medical degree at the University of Cambridge and worked at University College Hospital London before moving to Auckland. When he arrived in 1963, he was impressed by the volume of cervical screening taking place. Like many others he was enthusiastic about the potential to conquer cervical cancer. He publicly stated in 1964 that cervical cancer could be abolished by the end of the century.[73] However, clinicians in Auckland as elsewhere were already discussing what was actually being detected through screening and what the appropriate treatment was for the abnormalities discovered.

Jefcoate Harbutt, Herb Green and D-team

In 1955 National Women's Hospital established D-team, a specific group of gynae-cologists dedicated to treating genital cancer. This initiative mirrored the example of Newcastle, England, where all cases of cancer of the vulva were referred to Dr Stanley Way and his team. With relatively few cases nationwide, it appeared to be a logical means of gaining experience. Dr Jefcoate Harbutt headed the new team at National Women's on the understanding that if cases of CIS or cervical cancer were admitted from the private practice of a member of the senior staff, he would maintain a close liaison with Harbutt in the treatment of the case. If major surgery were required, both members would take part in the operation, and the role of surgeon and assistant would be decided by mutual arrangement.[74] D-Team was to treat both private and public hospital patients.[75]

Dr Jefcoate Harbutt was a leading gynaecologist in New Zealand and had an international standing. He was a founding member of the International Federation of Gynecology and Obstetrics in 1954 and later an executive member.[76] Elected to membership of the RCOG in 1936, he became a fellow in 1956 and served as president of the New Zealand council of the college. He also served as president of the New Zealand Obstetrical and Gynaecological Society and worked at National Women's Hospital from the time it opened. At the RCOG's first New Zealand Congress in 1955, Harbutt presented a paper entitled, 'New Concepts in the Treatment of Carcinoma of the Cervix'. Reaffirming the hospital's deci-sion to screen gynaecological patients, Harbutt pointed out that vaginal smears 'give us the opportunity of making a diagnosis in pre-clinical asymptomatic cases and their value has been substantiated. . . . selective screening of cancer-prone patients should be part of the service in any gynaecological clinic.' However, he added a cautionary note: 'Unfortunately, in attempting to cure all patients with this condition [CIS] there is a risk of performing untimely and too radical surgery unless therapy is based on a thorough knowledge of pathology.' He thought there should be a 'minimum of interference' in the 20–29 age group, citing one author-ity 'who reserves total hysterectomy for patients 30–35 and over', and another who claimed that, 'The problem is to establish the reality of ever earlier forms of malignancy and to protect the patient from incomplete or unnecessary therapy.' Harbutt concluded that this showed how little unanimity there was about the status of CIS, and emphasised that 'the problem regarding diagnosis and treat-ment [was] as yet unsolved'.[77] He also told Carey in 1955 that the treatment of CIS was 'still controversial'.[78]

In 1956 Carey appointed Dr Herb Green to be his assistant and to work on D-team with Harbutt. Green later explained that his interest in cervical cancer had been fired by one of his former school teachers who died of the disease, and the distress she suffered because of it. This motivated him to 'want to seek a means by which this particular type of cancer could be prevented'.[79] Green had graduated in medicine at Otago University in 1945, after completing bachelor degrees in arts and science, which included two years of pure mathematics and one year of applied mathematics.[80] In 1948 he sat the RCOG Diploma in Obstetrics, and achieved one of the three top scores in the exam.[81] In 1950 he gained his membership of the RCOG. As a prerequisite for his membership, he wrote a paper on the early diagnosis of cervical cancer.[82] Green had previous experience of National Women's as a house surgeon and registrar in 1948–50. As registrar he produced the hospital's first two annual clinical reports, showing an aptitude for statistical analysis.[83]

Green went to the UK for postgraduate studies in 1951 and while there worked for a time (1954–55) at the Queen Elizabeth Hospital in Newcastle upon Tyne with Dr Stanley Way, who was then the leading expert in England on vulva cancer and an acknowledged expert on cervical cancer. Way ran his own cytopathology laboratory, and was at this time a strong advocate of the use of hysterectomy as the appropriate treatment for CIS. Green returned to New Zealand in 1955 and took a job as consultant obstetrician at Wanganui Hospital. It was during his attendance at a short postgraduate course at National Women's Hospital in May 1956 that Carey approached him to join the medical staff. Green later explained, 'Despite a substantial drop in income, I returned to National Women's Hospital because of my interest in the academic side of medicine and my desire to continue research into cervical cancer.'[84] Carey was known for earmarking potentially bright researchers and persuading them to join him at the hospital. In 1957 he convinced Dr William Liley to return to New Zealand from the Australian National University where he was completing his PhD in physiology.[85] He also persuaded Dr Graham (Mont) Liggins to work at the hospital upon his return from studies in Britain in 1959. Liggins later remarked, 'Harvey in typical fashion said he wanted me to join the staff but he didn't really have a job, he'd get one in due course – that was how he got everybody there he wanted . . .'.[86]

Like Stanley Way in Newcastle, Green was a clinician who developed a special interest in histopathology (the laboratory study of diseased tissues). Dr James Gwynne was pathologist at National Women's Hospital in 1957–58, during Sullivan's study tour of America referred to above. Gwynne later explained that he developed a close professional relationship with Green, particularly with cases of CIS and microinvasion of the cervix. Noting that these conditions were histological – detectable only by laboratory rather than physical examination – Gwynne explained that Green, 'who was a most thorough and conscientious doctor, had set out to learn all about the histology of the cervix, and by the time I came to National Women's Hospital he knew as much if not more about the subject than most pathologists'. He related how Green came to the laboratory almost on a daily basis, 'when we would sit down at the microscope together and discuss the day's cases and mostly come to agreement about the diagnoses'. He added that Green had had a similar relationship with Sullivan.[87]

Like his colleagues at National Women's, Green was enthusiastic about the diagnostic potential of smears. In 1957 he complained at a Hospital Medical Committee meeting that some senior members of staff were not carrying out smears of gynaecological patients. He claimed that two such members had been heard to say that they would not do a smear when the cervix looked normal. He scolded, 'This defeats the whole aim of the screening value of this examination. Many cases could be quoted already from this hospital where the cervix looked normal and the smear showed malignant cells. . . . every patient over 30 years of age should be smeared irrespective of the appearance of the cervix.'[88]

While Green was enthusiastic about taking smears, under the influence of Harbutt he began to question the routine treatment of CIS by hysterectomy. He later explained that this scepticism arose from both clinical practice, particularly the discovery of no sign of cancer in a uterus removed as a result of a smear, and as a result of reading international published research.[89] Green presented a paper to the Hospital Medical Committee in 1958, reviewing hysterectomies performed at the hospital over the previous two years. The committee agreed that this was 'a most enlightening document'. Carey explained that the main object of the paper was to emphasise that it was sometimes preferable to think twice before operating.[90]

Three months later, Green and Harbutt presented a resolution to the Hospital Medical Committee that the official hospital policy regarding the treatment of CIS should be a cone biopsy rather than hysterectomy, provided the next smear turned out to be negative and the pathologist was satisfied that the cone biopsy had included all the 'carcinomatous tissue'. They added that the cone biopsy and follow-up should be carried out by a senior member of staff. Two members, Drs Graham Aitken and Peter Restall, voted against the motion.[91] They did not support the move away from hysterectomy, and were not prepared to hand over their patients to Green and Harbutt's care, despite the existence of D-team. In July 1958, two other members of the senior medical staff, Drs Bruce Faris and Bruce Grieve, moved that 'the treatment of carcinoma Stage 0 [CIS] be in the hands of the clinician and where circumstances arise that radical treatment is not considered necessary that Dr Green be given the opportunity to help follow up the case'.[92] The individual clinician, and not Green, was to decide on the course of treatment. Even in cases where the clinician decided radical treatment was not necessary, Green was only given the opportunity 'to help' with the follow-up.

Three years later, Green asked the medical committee to discuss 'Disharmony in relation to clinical matters'. He pointed out that he had not 'received whole-hearted co-operation in having certain cases – in which he was vitally interested – referred to him'. Following a lengthy discussion, the committee accepted a reso-lution from Dr Alastair Macfarlane, seconded by Green, 'that in the future, as has been recognised in the past, all staff members will co-operate with those doing work in special subjects in handing over appropriate cases, but that this proce-dure shall be by courtesy of the senior clinician originally involved, and shall not operate as the right of the person engaged in that particular field of work'.[93] Ultimate responsibility continued to rest with the individual clinician and not with D-team.

In the early 1960s, Green began to transfer the notes of all cases diagnosed with CIS onto IBM punch cards, an early form of computerisation, to facilitate research. In doing so he noticed that cases were not being followed up as regu-larly as they should be. Green recognised the importance of follow-up in clinical practice, whatever the treatment. He complained that:

> There seemed to be a general impression that there was no need for follow-up once a woman had a hysterectomy. It is just as important to follow-up after hysterectomies as after cone biopsies as quite a number of post-hysterectomies with positive smears have been found. With many of the cases diagnosed and treated in this hospital the follow-up is left to the outside practitioner. Sometimes this is not done as well as could be desired and it was suggested that in these cases the patients be asked to return to clinics.[94]

Referral of cases to D-team continued to be a problem. In an attempt to encour-age other clinicians to refer cases of CIS to him, Green proposed to the Hospital Medical Committee in 1966 that, as an additional safeguard against progression to cancer, he would use the services of the new hospital colposcopist, Dr William McIndoe.

Bill McIndoe, Colposcopy and Herb Green's 1966 Proposal

In 1962 Dr William (Bill) McIndoe was appointed consultant in obstetrics and gynaecology to National Women's Hospital. He had previously trained as an electrical engineer and gained a science degree before studying medicine. He graduated in medicine from Otago University in 1950, and subsequently decided to specialise in obstetrics and gynaecology. He gained his membership of the RCOG in 1961 and his fellowship in 1977. In preparation for his membership, he showed an interest in cytology, producing a paper entitled 'The Diagnosis of Pre-clinical Carcinoma of the Cervix with Particular Reference to the Place of Cytology' in 1960.[95]

Dr Ernst Navratil of Graz, Austria, visited National Women's Hospital in 1958 during his tenure as the McIlrath Guest Professor in Malcolm Coppleson's hospital in Sydney.[96] Navratil described the use of colposcopy in the diagnosis of CIS. The colposcope was an instrument which allowed more than ten-times magnification of the cervix as an aid to diagnosis and treatment. It was invented in Germany in the 1920s but rejected in America as a diagnostic tool until the 1970s, largely because it was seen as a challenge to the Pap smear test.[97] Carey bought a Zeiss colposcope for National Women's Hospital in 1960.[98] Until McIndoe's arrival, however, no-one at the hospital was trained to use it. Shortly after taking up his post, McIndoe outlined the advantages of the colposcope to the medical committee. He suggested that its use in tandem with cytology 'would lead to a further improvement in diagnostic acumen and increase the value of the cytology work'. The committee agreed, and suggested that he be sent to Sydney where Malcolm Coppleson had been operating a colposcopy clinic since 1955.[99] McIndoe applied in 1964 to the Auckland Medical Research Fund (AMRF) to carry out colposcopy research.[100] The AMRF approved the grant, but Harbutt 'doubted the value of this project and suggested that a close watch be kept on progress'.[101]

In 1964 Harbutt drew attention to a 'marked increase' in the number of A3 smears (CIS) reported during the previous two years, and wondered whether there had been a change in the standard or criteria for reporting. He pointed out that, 'An A3 report indicated that a cone biopsy should be done and sometimes this involved young women. He was doubtful if cone biopsy was justified in all cases.' The committee responded that it was up to the clinician to decide on the management of his cases. Some consultants said that they already referred these cases to McIndoe for examination with the colposcope before deciding on management. Green commented that Coppleson in Australia was using colposcopy as a useful complement to cytology.[102] In 1965, following the resignation of the hospital's medical cytologist Dr Roy Darby, McIndoe with the backing of Green was appointed medical cytologist at the hospital on a 5/10th basis and colposcopist on a 3/10th basis.[103] In 1968, with a grant from the AMRF and the Auckland Division of the Cancer Society of New Zealand, McIndoe visited America and Europe to study current use of the colposcope. He returned fired with enthusiasm, commenting, 'The importance of early examination by colposcopy and the smear test lay in the realisation that detection and treatment of cancer at this stage would result in almost certain cure'.[104] Like some of his contemporaries, he was caught up with enthusiasm for the powers of modern medical technology. Others continued to question whether the condition they were treating was, or ever would be, cancer.

Green had visited Malcolm Coppleson in Sydney in 1964 and reported back to the medical committee that he was impressed by what he saw.[105] Clare Matheson, the patient who was later to be central to the Cervical Cancer Inquiry, visited National Women's for the first time that year with a positive smear test. When she returned, expecting to have a cone biopsy, she was given a colposcopy instead:

> Sitting on the end of the bed, he [Green] spoke in a friendly and reassuring manner. He said he had just returned from Australia, and from what he had learned there, now believed that a biopsy was not necessary at all. He explained that he and others once believed that abnormal cells led to cancer and that hysterectomies used to be carried out on women in my condition. Lots of women, he said, had had hysterectomies needlessly. He did not believe now that they were necessary. Green assured me that I would be kept under observation, however, and that I should come into the clinics regularly for check-ups.[106]

This explanation was in keeping with Coppleson's views as later expressed in his 1967 publication.[107]

When Green submitted a proposal for a new treatment protocol for CIS to the Hospital Medical Committee in 1966, he drew attention to the new colposcopy as a safeguard. His memo suggested a change in official hospital policy relating to the treatment of CIS. The Hospital Medical Committee agreed to a protocol for the treatment of CIS in 1966, just as they had in 1958 and as they were to do again in 1978, collectively adopting what they believed was current best practice. Green almost certainly intended to publish the results of the treatment practised, and in that sense his memo could be regarded as a research proposal. At that time it was generally assumed that the therapeutic relationship would automatically predominate over the scientist-subject relationship.[108] Later asked at the 1987 Inquiry about his 'research programme', Green explained that he was engaged in diagnosing and treating patients, and 'if I could use the data afterwards for some clinical cartography or research I would use it, but the primary business was to treat, to diagnose, treat and cure the patients'.[109] His 'research' was to consist of reviewing the records. Green was, as Professor Barbara Heslop later described him, 'a clinical collector'. Reflecting on Green's research, Heslop explained, 'The investigation was basically a collection of clinical cases whose attributes were to be reported retrospectively. The 1966 proposal was a request for more referrals to try a new treatment. He explained why he wanted the referrals – but this was not a formal project application as we write them today.'[110]

Green's 1966 proposal read:

> It is considered that the time has come to diagnose and treat by lesser procedures than hitherto a selected group of patients with positive (A3-A5) smears . . . . It is suggested that this should be extended to include all cases in women under the age of 35 with positive smears in which there is no clinical or colposcopic evidence of invasive cancer. i.e.
>
> a. The cervix shows nothing more than an everson (erosion) on clinical inspection.
> b. There is no undue bleeding on probing the cervical canal.
> c. Colposcopic findings are consistent with carcinoma-in-situ only.
>
> In the interests of continuity of supervision and patient-confidence it is suggested all such cases should be passed to the care of Professor Green whose conscience is clear and who could therefore accept complete responsibility for whatever happens, with Dr McIndoe to assist on the colposcopic and cytological aspects.
>
> Following a discussion, during which Green answered many questions, the Committee approved the proposal. [111]

In recording that Green's conscience was clear and that he accepted responsibility for whatever happened, the committee confirmed its trust in Green to keep the welfare of his patients paramount, and accepted that women's lives would not be endangered by the treatment protocol he suggested. This protocol would have met with the approval of Professor Per Kolstad from the Norwegian Radium Institute, an international authority on CIS and cervical cancer who was later to be an expert witness at the Inquiry. He wrote in 1970:

> . . . the histologic diagnosis of carcinoma *in situ* may, at times, be difficult and, even when the diagnosis is unequivocal, the clinical outcome cannot be predicted. It, therefore, seems reasonable to advocate conservative diagnostic and therapeutic procedures when facing precancerous lesions of the cervix, especially in young women. By the complementary use of cytology, colposcopy and histopathology, the risk of overlooking invasive carcinoma of the cervix is minimal.[112]

Coppleson also later referred to a number of studies (citing a 1971 article, for example), 'confirming the safety of this approach' which was 'colposcopically directed punch biopsy'.[113]

The minutes for the 1966 meeting recorded that 'Professor Green said that its aim was to attempt to prove that carcinoma-in-situ is not a pre-malignant disease.'[114] Green later explained at the 1987 Inquiry that these minutes were taken by a lay person, who had omitted the word 'invariably' before a 'pre-malignant disease'. He told the Inquiry, 'Had I stated such a bold hypothesis, I am sure that my colleagues would not have approved the proposal I put forward. I was not setting out to prove anything, but I did question the view that CIS invariably progressed to invasive cancer.'[115] A review of Green's publications and lectures supports his claim. Green explained that there was considerable debate relating to what proportion of cases of CIS would advance to cancer.[116] His 1969 article in the *International Journal of Obstetrics and Gynaecology* pointed to current uncertainty that 'the invasive potential of in-situ cancer is as high as had been claimed'.[117] Reviewing the international debates in the *Journal of Obstetrics and Gynaecology of the British Empire* in 1970, he noted that recent estimates about the probability of CIS becoming invasive ranged from 100 per cent to less than 10 per cent, his own estimate.[118] The important point, he believed, was careful diagnosis and follow-up.

As will be further discussed in chapter 3, Green's approach was based on balancing the risks and benefits of treatment. He considered that the clinical approach had to be tailored to the individual woman's needs to avoid doing more harm than good. Yet he did not believe that CIS was necessarily benign. In 1973 he told Dr Algar Warren, the hospital's medical superintendent, that, 'It was always a calculated risk that invasive cancer could be overlooked, although it was hoped that colposcopy, clinical examination, and repeated directed biopsies would minimise, if not actually avoid, this. This had been advocated repeatedly by Dr. J. V. M. Coppleson in Australia and is now urged by American authorities (eg Stafl and Mattingly, Ostergard and Gondos, Selim et al, all 1973).'[119]

Green also justified his conservative approach to CIS in 1966 on the basis of population statistics. As Barron Lerner commented in relation to the history of breast cancer, by the 1960s the disciplines of biostatistics and epidemiology were increasingly used to influence clinical decisions.[120] As noted earlier Green had a particular interest in statistics and epidemiological evidence. When he put forward his proposal for treating CIS conservatively to the Hospital Medical Committee in 1966, he quoted epidemiological findings. He urged a cautious approach, given that 'Cytology and the diagnosis of large numbers of carcinoma-in-situ have not accelerated the declining incidence of cervical cancer or favourably influenced mortality rates.' In a lecture the following year, he explained that although cytology had been used in New Zealand since 1955 the statistics showed no noticeable improvement in incidence or mortality rates. Again, he was far from alone in this. Others were reaching similar conclusions, and Green was well aware of that literature (for a further discussion, see chapter 6). The internationally renowned epidemiologist, Sir Richard Doll, attended a Medical Research Council Symposium at Auckland's Green Lane Hospital in 1973. When asked to sum up at the end of the proceedings, he said:

> There have been many highlights and it is invidious to mention any for special reference . . . however . . . . First, the paper of Professor Green, which he modestly describes as 'stirring the pot' but which I would describe as an example of the most useful thing

that a scientist can do, namely the issue of a challenge to generally accepted ideas . . . . he has succeeded in making us realise once again the disastrous power of words to constrain thought. We label a lesion carcinoma-in-situ and regard it automatically as cancer, whereas we ought to ask ourselves continually, what is the reality behind the words and what it is that has actually been observed.[121]

In her report of the Inquiry, Silvia Cartwright was dismissive of Doll's support. She cited Green's claim that he had the backing of Sir Richard Doll, and quoted McIndoe's statement that 'it is difficult to follow in the above what Doll has recently supported', thus questioning Green's claim. Cartwright agreed with McIndoe.[122] In fact, reading Doll's contribution to the *New Zealand Medical Journal*, it is not at all difficult to understand the point he was making. He was clearly support-ive of Green. The latter's contribution to the symposium was also printed in the *New Zealand Medical Journal*. This was a study of cervical cancer incidence and mortality figures for the Auckland provincial area 1946–72 and for New Zealand 1948–70, comparing them with similar data for British Columbia.[123]

Following the advent of colposcopy, Green urged doctors to pass their CIS patients to his and McIndoe's care. Some CIS patients who met certain medical criteria would be treated by less than cone biopsy. He clearly hoped that accumu-lated statistics would show this approach to be correct. To justify this approach, Green referred to his own clinical experience, overseas clinical studies and popu-lation statistics. As noted above, Knox had stated that same year that population and pathological evidence could be used to suggest not one but two diseases – a benign one and some other hitherto unidentified lesion.[124] They simply did not know.

The Ongoing International Debates, 1967–1990s

In her report Judge Cartwright stated that the '1966 Proposal' depended on a belief that CIS was rarely if ever a cancer precursor. She wrote, 'In 1966 Dr Green and his colleagues on the Hospital Medical Committee were well aware of the world view that CIS was a precancerous condition.'[125] Yet the international litera-ture showed immense uncertainty and debate in the period leading up to 1966. Twenty years later a New Zealand doctor was still citing pre–1966 articles to support the argument that CIS was a pre-cancerous condition. Dr A. R. Chang, senior lecturer in pathology at the University of Otago Medical School and head of the Cytology Department and colposcopist for Dunedin Hospital, wrote in the *New Zealand Medical Journal* in 1985 that, 'A more universally accepted view is that the majority of high grade CIN lesions progress to invasive cancer and that few revert to normality or to a lower grade lesion.' His two references to support these 'universally accepted' views in 1985 were American journals dated 1959 and 1964; one of the authors of the 1964 article, Dr Joseph (Joe) Jordan, was subsequently invited to be an expert witness to the Inquiry.[126] Yet Chang could easily have found counter-arguments in the early 1960s, as he could in the subsequent period. There was no consensus in the 1960s or by the 1980s, and the debates were often heated. The editor of the *British Medical Journal* declared in 1976 that the debate about the 'natural progression or regression' of early cervical lesions had become a 'fierce controversy'. [127]

In *Effectiveness and Efficiency: Random Reflections on Health Services* (1972), Professor Archie Cochrane, director of the British Medical Research Council's Epidemiology Research Unit, discussed cervical smears. In a section on the British National Health Service's 'sins of omission and commission', he claimed that introducing cervical smears in the hope of preventing carcinoma of the cervix was 'the saddest' example of the latter. He thought the original idea which had reached Britain from America was a good one, but considered it unfortunate that it had not been accompanied by a randomised controlled trial. He believed it was very difficult to test the hypothesis by observational evidence as the death-rate from carcinoma of the cervix had been falling before smears were introduced and continued to fall at roughly the same rate in most areas.

Cochrane also suggested that the way the media (as well as many doctors) presented screening to the public was particularly 'sad'. He claimed, 'Never has there been less appeal to evidence and more to opinion.' He recalled a particular television programme which 'sold' cervical smears in a 'frighteningly effective way' (promising certain cure). He also recalled a lecture he gave on screening in 1967, when he made what he thought to be an innocuous statement that, 'I know of no hard evidence at present that cervical smears are effective.' To his surprise, he said, he was pilloried in the local press, which quoted many anonymous colleagues who considered him a 'dangerous heretic', and received many abusive letters, some from colleagues. He added, 'One very distinguished colleague wrote (rather irrelevantly) accusing me of "causing misery to thousands by telling lay people that there was no cure for carcinoma of the cervix". I wrote and asked him what his evidence was that carcinoma of the cervix could be cured but he did not answer. It was a pity as I have always wanted to know.'[128]

In a 1971 article, Cochrane questioned the assumption that CIS invariably progressed to cancer. As in his subsequent text, he commented on the 'strong emotions' associated with cervical cancer. He explained, 'the hypothesis is that on the basis of cervical smears women can be identified who are at great risk of developing carcinoma of the cervix, and that this risk can be diminished or abolished by subsequent therapy'. Assessing this claim, he maintained that in the absence of controlled trials, 'the next best approach is to compare the death-rate from carcinoma of the cervix in an area where screening has been practised for many years with another similar area'. He compared British Columbia, where screening programmes had been pursued, with the rest of Canada but found no differences in mortality rates. This case, he said, 'therefore provided no positive evidence in favour of the hypothesis'. He also referred to 'model-building', comparing the number of cases of CIS with those of cervical cancer. This evidence 'supports the idea of a very high [natural] regression rate'.[129] This was repeated in Malcolm Coppleson's 1981 textbook on gynaecological oncology, which noted that the discrepancy between the cumulative incidence of preinvasive lesions, even if only CIS was considered, and the expected cumulative incidence of clinical invasive carcinoma had been christened the 'yawning gap' by Cochrane.[130]

A 1972 writer in the *American Journal of Obstetrics and Gynecology* on 'current developments' began his article, 'Both the definition and the management of microinvasive carcinoma of the cervix [A5 smears] remain controversial.'[131] In 1975 a leading article on cervical epithelial dysplasia in the *British Medical Journal* questioned the unitary view of lesions, from mild dysplasia to invasive cancer, as postulated by Richart, arguing that there was some reason to think that dysplasia, or what Richart labelled CIN1, and CIS (or CIN3) were alternative rather than successive states.[132] British gynaecologist Albert Singer sided with Richart, responding that it was 'dangerous and potentially lethal to regard dysplasia as a purely benign condition'.[133] In his view all should be treated as precancerous.

The 'Walton Report' advocating population-based screening was published in Canada in 1976. It stated that the significance of CIS as a precursor of invasive disease had been recognised for more than three decades.[134] The reference for this was a 1952 article by two Americans, Drs Arthur Hertig and Paul Younge. What the authors of the Walton Report failed to state was that Hertig and Younge's talk at the annual meeting of the American Gynecological Society, entitled, 'A Debate: What is Cancer in Situ of the Cervix? Is it a Pre-invasive Form of True Carcinoma?' was followed by a contribution by Dr John McKelvey from the University of Minnesota Medical School, which put forward the opposing view and sparked a lively debate.[135] There was no consensus about the invasive potential of carcinoma in situ, either in 1952 or in 1976.

The issue of the *Canadian Medical Journal* that carried the Walton Report contained a presentation by Dr David Popkin of a 'gynecologist's viewpoint'. Popkin cited Dr Leopold Koss who argued that 75 per cent of dysplasia and CIS lesions would eventually become invasive.[136] As an indication of the continuing uncertainties, Koss revised his estimate in 1989, concluding that, 'at the most, one in ten precancerous lesions is likely to progress to invasive cancer if left untreated'.[137]

In Malcolm Coppleson's 1981 textbook on gynaecological cancer, two contributors based in the United States, Warwick Coppleson and Bryan Watson Brown, discussed a mathematical model for assessing control of carcinoma of the cervix. They pointed out that it was still widely believed that CIS was a true cancer and consequently that it rarely regressed spontaneously and eventually became invasive or metastasised. They claimed that this theory was only now being adequately tested and it 'has not stood up well to close examination'. In their view, 'The availability of a fine diagnostic tool, the Pap smear, compelled belief in its infallibility . . . . The distinction of carcinoma in situ from histologically similar conditions, some of which are entirely benign, is also much less reliable than is generally acknowledged and further confuses the issue.'[138] Similarly, Dr James McCormick from the Department of Community Health, Trinity College Dublin, argued in 1989 that, 'the natural history of these lesions is poorly understood, and, furthermore, invasive cancer may arise without evidence of progression through a series of precancerous stages'.[139]

Shortly before the Cartwright Inquiry, a *Lancet* editorial noted that, 'Virological work has lately confirmed the heterogeneity of the natural history [of cervical cancer], showing that almost all invasive cancers are caused by human papilloma-virus (HPV) types 16 and 18, while the great majority of CIN lesions (probably over 95%) are caused by HPV types 6 and 11, which rarely progress.'[140] In other words, most CIN lesions did not progress to cancer.

Green therefore was a participant in an international debate. At the time of the Inquiry, he estimated the rate of invasion of dysplasia and carcinoma in situ to be 5–10 per cent.[141] The Norwegian gynaecologist Professor Per Kolstad, brought to the Inquiry by Cartwright as an expert witness, put it at about 20 per cent. He had argued in 1976 that, 'It is necessary to keep in mind that the majority of *in situ* lesions will not progress to frank invasive carcinoma.' He noted that the intro-duction of the smear test had led to a large number of 'suspect' cases. As many of these occurred in young women of childbearing age, for whom hysterectomy and even cone biopsy could mean considerable trauma to the reproductive tract, he advised caution.[142]

American pathologist Ralph Richart had a very different perception of the invasive potential of CIS. He believed almost all cases were invasive.[143] Richart told the Inquiry that Green's view differed from virtually everyone else in the world who was a student of cervical neoplasia, and that 'most or all patients with untreated CIS will develop invasive cancer'.[144] In her account of the Inquiry, Sandra Coney claimed that Richart brought 'fresh air and a crystal-clear view' and was 'the perfect expert witness for the Cancer Inquiry'.[145] In this international debate, Green was clearly at one end of the spectrum and Richart at the other.

The international debates continued following the Cartwright Inquiry. In 1988 Dr Joe Jordan from Birmingham, England, another of the Inquiry's expert wit-nesses, wrote that he believed it was time to reopen the debate about whether patients with grade I and grade II CIN 'truly reflect a preinvasive disease or whether some are no more than changes caused by infection with human papil-lomavirus without the potential to progress'. CIN1 and CIN2 were indicative of milder dysplasia than CIN3; however, the diagnoses were far from precise, as will be discussed in chapter 5. Jordan explained that an attempt to resolve whether these mild forms of dysplasia were invasive had been carried out by medical researchers at Belfast City Hospital. They had prospectively studied the clinical course of patients with mild dysplasia who had not been treated. The patients were followed up by cytology alone, a test being carried out every six months. Referral to a colposcopy clinic was made if the dysplasia persisted for 18–24 months, if the smear showed more dysplasia, or if the patient had abnormal bleeding or a cervix that looked abnormal on clinical examination. He reported that, 'In just under half the patients the cytological changes reverted to normal without treatment. In only one instance did cytological surveillance fail: the patient was later found to have invasive carcinoma.'[146]

In 1993 an Australian researcher reviewed all the studies of the progression of CIS that he could find since 1955. He commented on how controversial it was. He explained that some people believed dysplasia was essentially benign and rarely progressed to CIS or invasive cancer, while others were convinced that dysplasia was an early form of cancer and that, although some spontaneous regressions did occur, the vast majority ultimately evolved into CIS and invasive cancer. He noted that this was 'the raison d'etre for the massive financial investment aimed at its [dysplasia] prevention'. He concluded that the studies showed the invasive potential of CIN1 to be 1 per cent, of CIN2 to be 5 per cent, and of CIN3 or CIS to be 12 per cent or possibly more.[147]

In 1988 Dr David Slater from Rotherham District Hospital, England, commented on Judge Silvia Cartwright's 'decree' (his word) that CIS should be legally construed as a potentially premalignant condition and all patients given active treatment rather than simple follow-up. He repeated the question asked by Dr Charlotte Paul, an epidemiologist at Otago University and medical adviser to the Cartwright Inquiry, in her *British Medical Journal* article about the 'unfortunate experiment': 'Could it happen again?'.[148] He referred to Jordan's recent article declaring the need for a multicentre comparative study of cytology and colposcopy in the management of minor grades of cervical dysplasia. He also pointed out that Drs David Jenkins and M. H. Jones had given preliminary information on a multicentre comparative study of patients who had presented with mild dysplasia at the Whittington Hospital, London. Slater expressed surprise that one of Jenkins and Jones's groups was managed by cytological surveillance alone, and commented, 'It would appear to be happening again.'[149] It happened again because the jury was still out on the status of CIN. While these studies were of 'mild dysplasia', the categorisation of the dysplasia itself was subject to much debate; as Ralph Richart said in 1981, 'It is a dictum among pathologists that one man's dysplasia is another's carcinoma in situ.'[150]

Conclusion

In her report Cartwright made differing statements on what she accepted as Green's position on the invasiveness of CIS. Discussing the 1966 proposal she wrote, 'I cannot believe that Dr Green would be so cynical as to attempt to prove an hypothesis that carcinoma in situ is not **invariably** [her emphasis] a premalignant disease' as that would be accepting that some cases would advance to cancer.[151] Elsewhere she wrote that he believed that CIS was 'almost always a harmless disease', and that, 'Although Green was tireless in his efforts to diagnose and treat invasive cancer, he believed that CIS was likely to be a benign lesion in the great majority of cases. Therefore, he did not treat that lesion in some cases.'[152] It should be added that Green's decision whether or not to treat these cases was a clinical one, based on risk assessment, as will be discussed in the following chapter.

As noted earlier Green disputed the wording of the 1966 meeting's minutes, which stated that his aim was to prove that CIS was not a premalignant disease.[153] Moreover, he queried the *Metro* article description of his views as controversial, commenting that many overseas authorities were questioning whether there was an automatic progression from CIS to invasive cancer.[154] A reading of the international literature supports that view. Those debates show clearly that Green's was not a lone voice, and that he was not attempting to prove a 'personal belief' that CIS was not invasive, in the face of 'world opinion' to the contrary. He was part of an international community of medical researchers who were questioning the significance and status of a diagnosis of CIS. This questioning was not, they believed, at the expense of the patient. Rather, as will be discussed, they argued that sometimes the treatment was worse than the disease and that less treatment could therefore be in the interests of patient welfare. As British medical sociologist Tina Posner wrote in the 1980s, 'The "medical dilemma" [in relation to CIS] was . . . to know when to treat the abnormality and when to leave it alone because no harm would result from doing so, whereas intervention could lead to a variety of unintended negative consequences.'[155]

3.

Management of Patients
with Carcinoma in Situ

In her prize-winning book about her experiences at National Women's, published following the Inquiry, Clare Matheson (known as 'Ruth' in the *Metro* article) referred to Herb Green's 'unorthodox and life-threatening withholding of proper treatment'.[1] By contrast, Dr James Gwynne, pathologist at National Women's Hospital from 1957 to 1958, suggested after the Cartwright Inquiry, 'It could be that a conclusion might be reached that Professor Green deserves a medal for his contributions to the widespread acceptance of the modern conservative management of self limiting proliferative disorders of the cervix.'[2]

This chapter discusses the debates around management of patients with carcinoma in situ, once again locating Green's work in an international context, and questioning some of the assertions made in the Cartwright Report and the subsequent publicity. It specifically addresses the claim that 'normal' or 'conventional' or 'accepted' treatment was withheld from some patients with CIS. Further, it addresses some of the assertions made at the time – notably that Green was alone in the treatments he advocated, and that he was motivated by a scientific interest in the natural history of cervical cancer, or by a concern to preserve the fertility of young women rather than to treat individual patients to the best of his ability.

Denying Treatment to 'Selected' Patients

In their *Metro* exposé, Sandra Coney and Phillida Bunkle outlined the 1984 article by Bill McIndoe, which he wrote with Malcolm (Jock) McLean (a pathologist), Ron Jones (a gynaecologist) and Peter Mullins (a statistician).[3] Coney and Bunkle explained that the authors had surveyed the data of 948 women diagnosed with CIS at the hospital from 1955 to 1976, making it 'the largest study of its type in the world'.[4] Explaining the Auckland 'study', Coney and Bunkle wrote:

> The study divided the women into two groups – 817 who had normal smears after treatment by 'conventional techniques' and a second group of 131 who had continued to produce persistent abnormal smears. This second group is called in the study the 'conservative' treatment group. Some had only biopsies to establish the presence of

disease and no further treatment. Others had abnormal smears after initial treatment, and were not treated further.

In Group 1, 12 women (1.2 per cent) developed cancer, and in the other – the so-called 'conservative treatment group' – 29 women (22 per cent) developed cancer.[5]

Cartwright too suggested differential treatment. In her report she quoted Coney and Bunkle's statement that: 'Twelve of the total number of women died from invasive carcinoma. Four (0.5%) of the Group-one women, and eight (6%) of the Group-two women who had limited or no treatment [sic]. Thus the women in the limited treatment group were twelve times more likely to die as the fully treated group.' Cartwright accepted that this accurately reflected the findings of the 1984 McIndoe paper.[6] Elsewhere in the report, she wrote that 'those women whose lesion had not been eradicated . . . had at least a ten times higher chance of developing invasive cancer than those who had been treated by generally accepted standards.'[7] Further, she wrote that McIndoe's paper 'distinguished between two groups of women treated for carcinoma in situ at National Women's Hospital. It found that . . . 22 per cent of those whose abnormalities were untreated [McIndoe's Group 2] ultimately developed invasive cancer of the cervix or vaginal vault.'[8]

Bill McIndoe made it clear in 1985, in a reply to an inquiry from Fertility Action, that their study was not based on any premise that the original researcher (Green) had divided the patients into two groups: 'In our study we compared results in women whose only difference is in follow-up cytology findings – in one group follow-up cytology was normal and in the other group abnormal. The detailed management of patients is not under consideration in this paper . . . . there was no choosing of women, nor did they choose to be in any particular group. The two groups we discuss result from a method we have applied to analyse data.'[9] One of the other authors of the 1984 paper, statistician Peter Mullins, later confirmed that, 'The implication that the abnormalities were untreated is, on the information presented in our 1984 paper, quite false: the group was defined as "continuing to produce abnormal cytology", not as having been untreated.' Furthermore, 'the 1984 paper was in terms of a second group of patients who "continued to produce abnormal cytology", not a group that was "conservatively treated".'[10]

It was the authors of the 1984 study who divided the patients retrospectively into two groups – Group 1 whose cytology returned to normal after two years and Group 2 who continued to have positive smears. The point of the paper was to show that those who continued after two years from initial diagnosis to have positive (abnormal) smears were more likely to develop invasive cancer than those who reverted to negative (normal) smears, and that therefore an all-out attempt should be made to ensure the smear returned to normal. However, the same range of treatments had been offered to the women in each group. Of the total of 948 cases analysed in the 1984 paper, 71 per cent had cone biopsy and 26 per cent hysterectomies. Of those in Group 2 (the supposedly untreated or conservatively treated group), 67 per cent had cone biopsies and 25 per cent hysterectomies.

In 1988 Drs Charlotte Paul and Linda Holloway responded to a claim by a senior consultant at National Women's, Dr Graeme Overton, that there were no two groups based on those treated conventionally and those treated conservatively ('conservative treatment' comprised those who had less than a cone biopsy, i.e. 'wedge' or 'punch biopsy' or who received no treatment). Paul and Holloway admitted that women in Group 2 received treatment, noting 'Most of these women did eventually receive treatment', and that 'Many women in group 2 did in fact receive treatment at some point in their follow up, which would be expected to have altered the natural history of progression' (which was presumably the intention). However, Paul and Holloway further argued, 'The assertion that both groups had the same treatment was based on the fact that both groups had a similar range of final treatment types. But the proponents of this view failed to distinguish initial from final treatments. Clearly the initial treatments differed markedly in their success in eradicating the disease.'[11]

The 1984 paper did not differentiate 'initial' from subsequent treatment, except for sample groups. Table 4, for example, included the heading 'Cone biopsy', under which the authors divided the patients into (1) 'Punch and/or wedge, later cone biopsy', and (2) 'Cone biopsy'. Of 53 of the Group 2 women in the first category (i.e. in which initial treatment was less than cone biopsy), four were said to have developed cancer, while ten of the 35 in the second category (cone biopsy) did so. Thus, fewer of those Group 2 women whose initial treatment was less than cone biopsy had developed cancer in this period of study than those who had been given cone biopsies. A similar pattern occurred in the samples given for hysterectomy. The heading 'Total hysterectomy TH' included categories (1) 'Punch and/or wedge biopsy, later TH' and (2) 'Cone later TH'. Of the four Group 2 patients in the first category, none developed cancer, and of the 29 in the second, six were said to have advanced to cancer. Again, those whose initial treatment was 'punch and/or wedge biopsy' were less likely to develop cancer than those initially given a cone biopsy in this period of study. The paper also included three figures (Figures 1, 2 and 3) recording treatment over time for twelve women from Group 1, 23 women from Group 2 and six women from Group 2 respectively. (Three of those in the sample figure for Group 2 [Figure 2] had negative smears at two years, so in fact were wrongly placed and should have been in Group 1.) From these small samples, one cannot make assumptions about overall treatment of the 948 women. However, in the given samples, six from Group 2 had a hysterectomy within the first two years, and only one from Group 1, again showing that those who ended up in Group 2 (positive smears after two years) did not necessarily receive less treatment than those in Group 1 (who ended up with negative smears).[12]

Thus, the 1984 paper provided no suggestion that there were two groups differentiated by treatment, either initial or final. Moreover, a focus on 'initial treatments' could not be construed as 'experimenting' on 'selected' patients, as McIndoe's two groups were not based on initial diagnosis but were classified by their status two years later; there were, in other words, no two initially different groups. This idea of 'selection' was repeated almost 20 years later, in a *Lancet* article of which one of the authors was Charlotte Paul and which noted that 'no records exist of which women were chosen for that study'; it also referred to 'recruitment to the study'.[13] No records existed because, as Green explained at the Inquiry, there was no 'selection' of patients: rather, they selected themselves by their medical condition – they were selected on the basis of the fact that clinical, colposcopic, cytological and histological investigations had shown that they did not have invasive cancer.[14]

Coney admitted in her book (published at the same time as the Cartwright Report) that she and Bunkle had been mistaken in assuming that the women had been divided into two groups on the basis of the treatment they received.[15] Cartwright accepted too that it was the authors of the 1984 paper who divided the patients into two groups[16] but, as noted above, elsewhere in the report she reintroduced the idea of differential treatment.[17] The idea of two groups continued to flourish in the popular press during and following the Inquiry. The *Dominion* reported that there were two groups of women; that 817 in one group had normal smears after treatment by 'conventional techniques' and a second group of 131 had 'limited treatment'. The article stated that 'those who had had normal treatment rarely developed invasive cancer'.[18] Similarly, the *New Zealand Woman's Weekly*, just a few weeks after publication of the Cartwright Report, described the 1984 paper as an outline of a study on patients at National Women's: '814 who had been treated "conventionally" and 131 who'd received "conservative" management'.[19]

Cartwright stated that 'some patients were not treated by generally accepted standards and some not treated at all'.[20] Along with Coney and Bunkle, she claimed that Green denied some patients 'normal treatment'.[21] Explaining the Inquiry to an international audience through the medium of the *British Medical Journal*, Dr Charlotte Paul also asserted that there was 'a deviation from accepted treatment', with patients 'not being offered the generally accepted treatment', and that 'all should have been offered accepted treatment'.[22] In order to test the validity of these claims, we need to determine what exactly was 'accepted', 'proper' or 'normal' treatment at that time.

Hysterectomy – An 'Accepted' Treatment?

Judge Cartwright appeared to suggest that Green was out of touch because he thought that hysterectomy was an all-too-common response to CIS in 1966. She wrote, 'Green in his evidence said that he expected some senior staff to oppose the [1966] Proposal "because some of them believed that hysterectomy was still the correct treatment".'[23] According to Cartwright, hysterectomy was not the 'accepted treatment' for CIS by 1966: 'for some years there had been a move from the more radical treatment of hysterectomy to cone biopsy. The 1966 Proposal was not part of this trend.'[24] She also stated that, 'National Women's Hospital policy was in line with much of the rest of world. It was already treating by cone biopsy in the 1960s. The 1966 trial was not a move from hysterectomy to cone biopsy.'[25] And again: 'From the 1960s, cone biopsy was the preferred method of treatment for CIS both at National Women's Hospital and in many other parts of the world. Hysterectomy following a diagnostic cone biopsy was no longer the norm except in parts of the United States of America.'[26]

In seeking to establish National Women's 'standard' treatment, Cartwright cited a 1958 Hospital Medical Committee motion stating that biopsy was to be the standard treatment. However, she did not specify that this was proposed and seconded by Green and Harbutt, or that two other members of the committee's nine senior consultants (the committee included a paediatrician) voted against the motion.[27] They had some support for this stance: that same year British gynaecologist Dr Aleck Bourne stated in the new edition of *Recent Advances in Obstetrics and Gynaecology*, 'Should we discover a case of pre-invasive carcinoma . . . we would unhesitantly perform total hysterectomy.'[28] Despite Cartwright's attempt to distance Green from a move in medicine which she saw as positive, he undoubtedly helped to lead the push, in Auckland at least, from hysterectomy to cone biopsy as treatment for CIS.

When Malcolm Coppleson discussed the 'correct' treatment of CIS in 1967, he claimed, 'Most authorities believe this to be total hysterectomy . . . .'[29] Coppleson himself clearly had reservations about hysterectomy in relation to CIS, asking, 'Is this operation itself more radical than necessary?' Like Green, he thought it was an all-too-common response.[30]

In 1963 Dr Hugh McLaren, professor of obstetrics and gynaecology at the University of Birmingham, England, claimed that in Baltimore, USA, 95 per cent of CIS cases were treated by hysterectomy.[31] However, in the course of a 1967 discussion in America on treatment of CIS, Drs James Krieger and Lawrence McCormack suggested treating CIS with cone biopsy rather than hysterectomy; this was considered 'controversial and challenging' not only to 'parts of America' but also to the whole of the Association of Obstetricians and Gynecologists. A commentator claimed that the 'dogma' of treating CIS by hysterectomy was 'now well disseminated'. During the ensuing discussion, Dr John Boyd of Cleveland, Ohio, commented on this 'controversial' paper: 'If conization alone will success-fully control carcinoma in situ in 90 per cent of cases, one would reasonably assume that the commonly advocated hysterectomy is serious overtreatment.' Dr Nicholas Thompson, also of Ohio, opined, 'For those of us who believe that total hysterectomy is the treatment for in situ carcinoma and use conization only in the poor-risk surgical patient or the one who is desirous of a pregnancy, it is disturbing to learn that 90 per cent of our hysterectomies may have been unnecessary.' Yet there is no indication that practices changed as a result of this 'disturbing' revelation. Thompson still believed that hysterectomy was the most appropriate treatment.[32]

Hysterectomy continued to be popular into the 1970s. A 1972 discussion in America noted that 'most areas' in the USA employed hysterectomy as routine treatment for abnormal smears.[33] A 1973 editorial in the *British Medical Journal* pointed to 'the increasing tendency for total hysterectomy to replace cone-biopsies as therapy' for cases of positive smears.[34] In his contribution to a 1977 obstetrics and gynaecological textbook, Coppleson noted that the use of cone biopsy for preinvasive carcinoma of the cervix was 'generally unacceptable to many gynae-cologists so that presently with carcinoma in situ, total hysterectomy is still the most used procedure'. He cited Dr Clayton Beecham from the Gynecologic Tumor Clinic, Temple University Medical Center, Philadelphia, USA, who wrote in 1969 that, 'when the patient reaches 30 a hysterectomy should be done'.[35] The 1979 edition of *Recent Advances in Obstetrics and Gynaecology* noted a move towards conservatism in the treatment of CIS, adding, 'A few years ago the only accepted definite treatment of carcinoma in situ was hysterectomy . . . . total hysterectomy is still commonly used in the treatment of patients with CIN who have completed their childbearing.'[36]

Ralph Richart certainly believed hysterectomy to be the appropriate treat-ment. He maintained in his contribution to a 1981 textbook on gynaecological cancer that, 'In the young patient with carcinoma in situ who had not borne children or who wanted additional children, conization was sometimes offered as a temporary therapeutic measure, but the patient commonly had a hysterec-tomy subsequently even if there was no evidence of residual disease, . . . to insure against subsequent recurrence and the development of invasive cancer.'[37] Richart pointed out that many physicians thought hysterectomy the treatment of choice for the older woman with an abnormal smear who had completed her child-bearing, adding that 'hysterectomy may be viewed as an appropriate sterilization procedure for women who wish to terminate use of contraceptives'.[38] He also told the Inquiry that women who had had CIS were often given a hysterectomy once their childbearing was complete, despite negative smears and the absence of any signs of neoplasia.[39]

As noted in chapter 2, Richart introduced the terminology of CIN with various grades to replace terms such as dysplasia, CIS and microinvasion, believing that they were all part of the same process and that the earlier they were treated the better. In 1980 he noted that pathologists 'varied greatly in their categorizing of presumed precursor lesions. Thus, whether a woman had a hysterectomy could depend on which pathologist assessed the slides of the cone-biopsy specimen.'[40]

It is therefore clear that, contrary to Cartwright's claim that the worldwide trend of the 1960s was away from hysterectomy to cone biopsy, hysterectomy was still a common response to positive smears, and continued to be popular into the 1970s. In his 1975 textbook, Sir Norman Jeffcoate, professor of obstetrics and gynaecology at the University of Liverpool, England, and former president of the Royal College of Obstetricians and Gynaecologists, expressed reservations about the continuing widespread use of hysterectomy in relation to treatment of CIS: 'Some gynaecologists advise hysterectomy in all cases. This may be justified if they are dealing with women likely to default from follow-up, but not otherwise.' He added a footnote: 'regular "follow-up" cytology is nearly as important after total hysterectomy as it is after conservative surgery. The woman who suffers from carcinoma in situ of the cervix is the one most likely to develop similar lesions in the vagina.' He also felt compelled to add, 'Hysterectomy is *never* indicated merely on the finding of a positive cervical smear [his emphasis].'[41]

In 1983 Goran Larsson from Sweden reviewed the approach of different countries to the treatment of CIS over the years. He noted that in the United States hysterectomy had been the most common form of treatment since the early 1950s. He referred to British Columbia, Canada, where 75 per cent of women with CIS were reported as having had hysterectomies. In Britain the 'standard treatment' had been hysterectomy until the late 1970s, when conisation became the preferred method. In Scandinavia, 'Around 1973 a debate started as to whether [even] conization was over-treatment for CIS.' New Zealand and Australia, he said, had 'early developed conservative treatment', and on the European continent 'only a minority of authors have recommended hysterectomy or extended hysterectomy as treatment'.[42] Ralph Richart told the Cervical Cancer Inquiry in 1987 that 'in many countries, simple hysterectomy is still recommended'.[43]

Green was wary of Richart's views, which he claimed could lead to a 'golden age of hysterectomy'.[44] Green warned that hysterectomy, 'so very "easy to resort to" often seems to carry the implication that the problem is solved once and for all'. He cited a 1962 study by R. R. Margulis and his colleagues from the Department of Obstetrics and Gynecology of William Beaumont Hospital in Michigan, USA, as proof that this was not the case; the study reported that some women developed cancer even after a hysterectomy had been performed.[45] Like Jeffcoate, Green noted that cytology confirmed that the chances of a recurrence of CIS after hysterectomy were real and that cases managed in this way needed just as careful follow-up as those treated conservatively.[46] Hysterectomy was alive and well during Green's working life; while Green did perform hysterectomies, he and many others regarded the procedure as serious over-treatment in response to a positive cervical smear.

Withholding Hysterectomy and Cone Biopsy

Cartwright claimed that Green's 1966 proposal was not about a move to conservatism – hysterectomy to cone biopsy – as this had already taken place; it was about offering lesser treatment or no treatment at all. She cited the evidence of Green's National Women's Hospital colleague, Dr Bruce Grieve, who declared that Green had been the only clinician at National Women's Hospital treating patients with CIS by less than cone biopsy before 1966. All the others, he said, were performing hysterectomies or cone biopsies.[47] Grieve had opposed the suggestion of passing such patients to Green's care as early as 1958.[48]

Dr Jefcoate Harbutt, Green's senior at National Women's, had expressed doubts in 1964 as to whether cone biopsy was 'justified in all cases' of CIS.[49] The international literature showed that others had expressed similar doubts and offered minimal treatment. Malcolm Coppleson, for example, reported in 1967 on a study he had conducted of cervical lesions in over 300 women. Discovery of the lesions had been 'followed by colposcopy (and . . . cytology) for 2 to 9 years. Nearly all women had a preliminary small punch biopsy for histological diagnosis. This was sometimes repeated at subsequent examinations but care was taken not to remove the whole of the atypical [dysplastic] area. The vast majority of these women have had no treatment to the cervical lesions'; he observed no adverse effects.[50]

In his sabbatical leave report for 1969, Professor Dennis Bonham commented on a visit to a cancer clinic attached to McGill University in Montreal, Canada, run by Dr Paul Latour, with whom he discussed the conservative management of cases with positive smears. He reported that, 'They are working on rather similar lines to us in this connection, although they do not appear to have quite the same number of safety factors, vis. colposcopy and punch biopsies . . . . This clinic had to be set up because excessive numbers of conizations were being done on rather mild positive smear reports.'[51]

Malcolm Coppleson was invited to contribute a chapter to a standard text-book on obstetrics and gynaecology that was published in 1977. He noted that since the late 1960s there had been 'a general disquiet with the large number of cone biopsies and hysterectomies which have been the legacy of many years of loose interpretation of abnormal Papanicolaou smears and cervical histology'.[52] Concerned that they might miss cases, some gynaecologists opted for radical treatment, and many healthy asymptomatic women with a normal-looking cervix but with an abnormal smear were given a cone biopsy followed by hysterectomy based on the cytology. It was only now that 'many thinking gynaecologists . . . question[ed] their true indispensability [i.e. cone biopsies and hysterectomies] in the light of the increasing numbers of follow-up studies reporting excellent results from minimal interference in patients with abnormal smears and pathol-ogy'.[53] The confidence provided by the colposcope had been the major factor in the increasingly conservative approach to management in Sydney's King George V Hospital. Staff there had conducted a trial of 'even more conservative methods of treatment for preinvasive epithelial lesions such as dysplasia, carcinoma in situ and even microinvasive carcinoma':

> Preliminary results from several sources indicate that many such lesions, which would have previously been treated by the more expensive and more hazardous sequence of diagnostic conisation followed by hysterectomy or by therapeutic conisation, have been thoroughly evaluated and safely treated by either cryosurgery, electrocautery or electrodiathermy, or multiple punch biopsy, in the office, in the outpatient department or by a single admission to hospital for a few hours.

Coppleson acknowledged that such treatment was still controversial, but argued that the advantages of avoiding unnecessary cone biopsies were obvious; these included a reduction of the significant surgical complication rate associated with cone biopsy and the minimising of medical expenses. He explained that he would perform a cone biopsy in a case where there was 'repeated abnormal smear in the absence of a colposcopic lesion',[54] though he did not say how many abnormal smears were required before this would occur.

The Cartwright Report stated that, 'One outstanding fact ought to have been clear to him [Green] and to others – following (without treating) patients with positive smears, whether after punch or cone biopsy, or after hysterectomy, was unsafe, as a proportion of those women would subsequently be shown to have invasive cancer.'[55] Yet she did not consider the alternatives, or that perhaps the treatment could be worse than the disease, particularly for those asymptomatic women whose only sign of 'disease' was a positive smear. Once the patients had been subjected to hysterectomy, options for further intervention included radiotherapy or additional surgery. So what were the risks attached to the various treatment options available?

Dr Joe Jordan from Birmingham, England, a medical witness to the Inquiry, viewed Green's case notes and noted the 'dilemmas' faced by Green.[56] Regarding one case of CIS of the lower genital tract he wrote:

> At this stage Professor Green discussed treatment with the radiotherapist who felt that radiotherapy was not indicated because there was no invasive carcinoma. The alternative was a vaginectomy with vulvectomy, in the absence of symptoms, it was decided to await events with the hope and expectation that the lesion would not become malignant . . . . Again one sympathises with the dilemma faced by Professor Green knowing that the treatment of this was either removal of the upper vagina or radiotherapy. Both procedures carrying a high morbidity and almost certainly, removing or interfering seriously with sexual function.

He concluded his commentary on this case by stating that the patient would probably have had a squamous cell carcinoma of the vulva and urethra, even if she had had radiotherapy or vaginectomy to remove the vaginal vault lesion.[57]

As Coppleson noted, hysterectomy and cone biopsy were invasive procedures with potentially serious side effects. In 1969 Green compared his experience at

National Women's with that of Dr Stanley Way, a leading British authority on cervical cancer with whom Green had worked in Newcastle upon Tyne in 1954–55. Green revealed that his follow-up of 539 CIS patients at National Women's, 32 per cent of whom were given hysterectomies, showed the same result as Way's follow-up of 551 patients, where 74.6 per cent had been treated by radical hysterectomy. In both cases one instance of invasive cancer had occurred, but Way's results also included five damaged ureters and three postoperative deaths.[58]

In 1972 Drs Vernon Hollyock and William Chanen from the Dysplasia Clinic at the Royal Women's Hospital in Melbourne reported on complications following cone biopsy. These complications included haemorrhage, cervical stenosis and pelvic infection. They believed cone biopsy and hysterectomy should be abandoned as routine procedures in the management of abnormal smears.[59] A group of American researchers similarly commented that conisation which required a general anaesthetic was sometimes 'complicated by postoperative bleeding or late cervical dysfunction'.[60] Another American research team noted the inherent risks of conisation, including bleeding, cervical canal stenosis, infections, cervical dystocia and occasionally infertility.[61]

Thus, there were considerable risks attached to cone biopsy. The commonest was haemorrhage, which usually occurred from the fifth to the twelfth day after the operation and sometimes required further intervention such as hysterectomy. Other complications included uterine and cervical perforation, infection or damage to the cervix causing painful periods, dysmenorrhoea, amenorrhoea and infertility. Future pregnancies could also be affected, with the possibility of miscarriage or spontaneous abortion, premature labour, low-birth-weight babies and a greater likelihood of requiring a caesarean section.[62]

Jeffcoate advised in his textbook that, 'When cervical smears *repeatedly* contain cells indicative of malignancy or *severe* dyskaryosis [dysplasia], the next step is to carry out cervical [cone] biopsy . . . . This operation is not free from immediate and late hazards and fatalities are reported; so it is wise to be sure that it is really necessary before proceeding to it.' He pointed out that cone biopsy was especially dangerous during pregnancy, when possible complications included serious haemorrhage and abortion. Because of that, '*unless the patient also has symptoms or signs* which suggest she may have invasive cancer', the pregnant woman should not be given a biopsy but kept under observation [his emphases].[63]

Jeffcoate advised more limited biopsies if colposcopy were available. A 1980 American article also supported colposcopy as a way to avoid cone biopsy and enable lesser and more targeted biopsies to be performed. The author once again cited the considerable complications attached to cone biopsy as the reason for this recent trend at his clinic in New Orleans. His article, advocating conservative treatment, was distributed with the New Zealand Women's Health Network *Newsletter.*[64]

In 1981 Coppleson reported rates of complications from cone biopsy to be as high as 25 per cent (and 33 per cent for pregnant women), and noted that symptoms might be immediate or occur years after the operation. Having listed these complications, Coppleson added that these were 'outcomes which some gynecologists seem content to face in order to exclude suspected invasive disease'.[65]

A British feminist sociologist, Peggy Foster, gave an example of the costs of biopsy:

> The extent to which routine treatment following a positive smear test may 'go wrong' is illustrated by an American case . . . . A woman who was given a cone biopsy following a positive smear bled so excessively following the operation that she was given an emergency hysterectomy during which she almost died from the anaesthesia; all this following the results of a test which is by no means a particularly accurate predictor of potential cancer in the first place.[66]

Cone biopsy in pregnancy appeared to be particularly hazardous, and yet positive smears were often discovered among those attending antenatal clinics. In 1966 Green had assessed the adverse effects of cone biopsy on subsequent pregnancies. Sixteen of the 30 patients on whom cone biopsy was performed had complications of pregnancy as compared to only two of the seventeen who had been given the less invasive ring biopsy. The former included eight miscarriages, five premature labours and three term labours with cervical dystocia.[67]

There was also the question of whether radical treatment made a real difference to the outcome. From his studies at the Norwegian Radium Hospital, where he followed up 1121 cases of carcinoma in situ, Per Kolstad concluded in 1976 that radical treatment did not make a difference. He questioned the statement that patients who were 'adequately treated for carcinoma in situ' had a 100 per cent chance of survival. He considered it possible that the 'number of recurrences observed is dependent not so much on the degree of radical treatment as on the length and completeness of the followup'. Kolstad admitted that his study could not substantiate the repeated claims that there was no risk to patients with CIS 'if treated properly'.[68]

Green had shown that his 'conservative' treatment yielded similar results to Way's more radical (hysterectomy) treatment. McIndoe and his colleagues' 1984 paper showed that some patients who underwent a hysterectomy and had subsequent negative smears still ended up with cancer, while others with minimal treatment reverted to negative smears. The authors questioned the claim that 'adequately treated CIS is a totally curable lesion'. They pointed out that 'the reoccurrence of CIS and the development of invasive carcinoma in adequately treated cases is reported by other authors' and that the evidence in their study 'strongly supported' that possibility.[69] In the same year, the *British Medical Journal* cited a case where a woman had been treated for CIS by cone biopsy and, despite six subsequent negative smears, developed cancer.[70]

There were no straightforward answers to the treatment of women with positive smears. A 2008 retrospective study of CIS patients at National Women's concluded that those whose initial treatment was less than cone biopsy, and who continued to have a positive smear, had a 50 per cent chance of developing cancer in 30 years (31 women in this category had developed cancer).[71] While an aggressive approach to CIS could be shown after 30 years to be a better approach than the more conservative one, this knowledge was not available at the time. Additionally, the study itself did not include those women whose smear returned to normal within two years. (The authors wrote that their 'approach might overestimate the invasive potential, because it ignores the possibility of spontaneous regression within the first 24 months'.) Under a more aggressive interventionist regime, these women might have been treated unnecessarily with possible side effects. Angela Raffle and Muir Gray wrote in 2007 that most positive Pap smears were transient minor cell changes which would never advance to cancer, and they referred to the unnecessary treatment which resulted in loss of fertility and even death from operative complications on occasion.[72] The 2008 *Lancet* article included the comment that the dangers of untreated CIS might even be higher than their study indicated as 'even a small diagnostic biopsy might be curative for some women'. The reference given for this was a 1962 article by the American CIS expert Dr Leopold Koss, which was available to Green and which showed that minimal treatment could be beneficial.[73]

Other studies continued to show the benefits of 'conservative' treatment. In 2004 the *New Zealand Herald* reported a study that had been undertaken at National Women's Hospital. This study, involving 652 treated and 426 untreated women with cervical dysplasia visiting the colposcopy clinic at National Women's Hospital from 1988 to 2000, showed 'benefits for "conservative" management of some cases'. Lead researcher Dr Lynn Sadler explained that,

> We found that women who had the most extensive cervical biopsies were at the greatest risk in pregnancy . . . . Women in their reproductive years who have been shown with biopsies to have pre-cancerous cervical changes – known as cervical dysplasia – should be managed conservatively to avoid premature births for later babies. Significant numbers of women treated for cervical dysplasia later had their waters break early, triggering pre-term births, which has implications for child health.

The study advocated regular monitoring of low-grade disease, with treatment only if the condition progressed, and the researchers noted that, 'Most of these cases will spontaneously regress back to normal without therapy.'[74] Similar results were reported in the *British Medical Journal* in 2008, by researchers using citation tracking from 1960 to 2007, and from a Norwegian population-based cohort study, leading the editor to advise, 'To protect against serious adverse pregnancy outcomes, the minimum amount of tissue should be removed or destroyed.'[75]

In their evidence to the Inquiry, two National Women's consultants, Andrew Mackintosh and Murray Jamieson, commented on the problem of over-treatment. Mackintosh averred, 'it does seem unfortunate, if I may use that word, to over-treat if you don't need to . . . and if you unthinkingly and unquestionably do extensive procedures, you will think you're marvellous, the patients will all think you're marvellous, except of course the ones who have complications. . . . and I am sure we have overtreated, yes, we have overtreated a lot and in this series . . . there are many many people who didn't need to have a degree of overtreatment . . . .'[76] Jamieson agreed, 'It is very easy to over-treat and feel good about it, but you are moving into areas of more radical treatment and the side effects can be severe.'[77] The consultants at National Women's Hospital were not alone in recognising these dilemmas, nor were the dilemmas unique to this particular condition in the practice of medicine.

Follow-up Studies

Professor Ralph Richart told the Inquiry that, 'I have known of no one except Dr Green who would recommend prospectively following patients with persistent CIS of the cervix, vulva or vagina rather than treating them and eradicating the lesions.'[78] Dr Joe Jordan told them, 'From 1966 onwards, I know no unit other than Professor Green's which was prepared to allow patients with carcinoma in situ to continue without treatment.'[79] Yet Professor Jeffcoate's advice in his 1975 textbook (and 1980 reprint) that in some cases it was preferable to 'observe' patients rather than treat them has already been noted.[80]

The international medical press reported on many studies in which the doctors followed up cases of CIS without treatment. In 1962 American gynaecologist Associate Professor John Graham surveyed follow-up studies of CIS to that time. He referred to a Swedish study by Dr Hans-Ludwig Kottmeier who had followed 850 cases of CIS without treating them; three had developed cancer within a year and twelve had cancer seventeen years later. He also referred to American gynaecologist Paul Younge's prospective series of 58 patients with CIS followed without treatment for five years or more, of whom only one developed cancer (which occurred within a year). Graham claimed the most informative was O. Petersen's Copenhagen study of CIS cases followed from 1930 to 1950 which showed an approximate 33 per cent invasion rate, though he added that the status of the original diagnosis was uncertain.[81]

Dr Leopold Koss tracked several hundred CIS patients without treatment between 1950 and 1963. The follow-up was based on cytology, Schiller's test[82] and punch biopsies of the cervix. At that time, colposcopy had not reached America. Koss explained, 'We were particularly interested in determining whether there were any intraepithelial lesions that were more likely than others to disappear, remain stationary, or progress.' Results published in 1963 showed that in

25.4 per cent of cases diagnosed as CIS, the lesion disappeared, in 61.2 per cent it persisted, in 7.5 per cent it progressed to questionable invasion and in 5.9 per cent it progressed to invasive carcinoma (four patients). Of those lesions which were 'borderline', 38.5 per cent disappeared, 15.4 per cent persisted, 42.3 per cent advanced to CIS and 3.8 per cent (one patient) advanced to cancer.[83] The conclusions of Koss's study, which Cartwright repeated in her report, were strangely inconsistent with the statistics presented. Koss concluded that CIS 'beyond doubt is a precursor of invasive cancer', and that 'Spontaneous disappearance of CIS apparently does occur, but it is an extraordinarily rare event.'[84]

The 'largest and longest follow-up experience available with this disease', according to the 1979 edition of *Recent Advances in Obstetrics and Gynaecology,* was a series of 1121 patients with CIS reported by Per Kolstad and Valborg Klem from the Department of Gynecology at the Norwegian Radium Hospital in 1976. These patients were followed from five to 25 years. [85] This study, unlike McIndoe's 1984 study, was an attempt to correlate treatment with results, and led to the rather sobering conclusion for those advocating radical treatment that it was possible that the 'number of recurrences observed was dependent not so much on the degree of radical treatment as on the length and completeness of the follow-up'.[86]

Another study reported in *The Lancet* in 1978 covered 9000 women aged 20 years and under who had cervical smears in the period 1967 to 1976. Abnormal smears were found in 145, and follow-up of these 145 women for ten years showed that in over half the cases subsequent smears had reverted to normal without treatment, and nineteen of the 145 patients progressed to have smears which were suggestive of malignancy.[87]

Some follow-up studies were designed to follow mild lesions only, though the categorisation of smears was far from a precise science, as will be discussed in chapter 5, making these studies not so different from other follow-up studies. Richart himself conducted a follow-up study of 557 patients with cervical dysplasia without treatment at the Medical College of Virginia and the Columbia-Presbyterian Medical Center in the 1960s. While he insisted that all were treated if they were diagnosed with CIS and it was 'unethical' to proceed further,[88] 'three patients developed invasive carcinoma while under surveillance'. [89]

Clinicians at Belfast City Hospital, Northern Ireland, undertook a study of 1781 cases of mild dysplasia discovered through a cervical smear (CIN1 and 2) between 1965 and 1984 and reported the results in the *British Medical Journal* in 1988. During the first two years, the patients were given five smears. In 46 per cent of those observed, the lesion regressed within two years without biopsy or other treatment. Invasive cancer occurred in seven of the 1347 cases followed and in three of the 434 patients lost to follow-up.[90] Dr David Slater referred in the same year to a follow-up study of 'mild dysplasia' at a London hospital, in which one of the groups was 'managed' by smears only.[91]

Thus, there were many follow-up studies of patients diagnosed with CIN1, CIN2 and CIN3 (CIS) as researchers and specialists in the field sought to determine the appropriate management of such cases. In 1993 an Australian researcher attempted a literature review of follow-up studies conducted and reported on since 1950, and found a large number of such studies.[92] The dilemma for the clinician was at what point to intervene and whether the intervention would be beneficial or would do more harm than good. There was no shortage of studies but there were no definitive answers. Richart and Jordan were misleading when they suggested in 1987 that no-one apart from Green followed up CIS cases rather than eradicate the lesion.

Green's Follow-up Studies

Like others working in the field, Green reported his follow-up studies in international medical journals. He wrote provocatively in the *International Journal of Gynaecology and Obstetrics* in 1969 that by commonly accepted standards he had 'almost disdainfully undertreated' many in-situ lesions. For example, he explained, the proportion of patients treated by hysterectomy was now 'almost confined to those with gross uterine pathology in addition to carcinoma in situ', and some women with positive smears alone had been followed up for periods up to ten years.[93]

Yet how 'disdainful' was his treatment in reality? Of the 576 patients treated 'conservatively' (for whom the initial treatment included cone biopsies in most cases) at National Women's Hospital, one developed cancer. This woman had been given a cone biopsy for carcinoma in situ in 1958. By July 1966 she had had a total of 15 negative clinical and cytological examinations. She returned to the hospital in November 1966 following two months of postmenopausal bleeding. Her notes recorded, 'Although the cervix still looked normal and the cytology was negative, biopsy showed unequivocal invasive cancer, which was treated radically.' She was alive and well the following year.[94] Green later stated that the importance of this case was that cytology by itself could be misleading.[95] This view was vindicated many years later, when a British study including more than 60,000 women concluded that Pap smears were only 53 per cent sensitive for detecting CIS.[96]

Reporting in 1970 on a group of 27 patients treated by punch biopsy only, like Kolstad's 25 'observed' patients in his 1976 study,[97] Green wrote:

> Except for a special series of 27 patients diagnosed and treated by punch biopsy alone, all those with carcinoma in situ have had at least a cone-type biopsy for diagnosis . . . . The 27 treated by punch biopsy form part of a special series, commenced in 1965 relative to patients whose only abnormality at presentation was a positive smear and in whom invasive cancer has been excluded as far as is possible by clinical, colposcopic

and cytological examinations; the histological diagnosis rests on punch biopsy alone, and no further therapy is being undertaken.[98]

This was very similar to many overseas studies and was considered, according to Kolstad's 1970 statement, a 'safe' approach. Urging conservative treatment, Kolstad claimed that the use of cytology, colposcopy and histopathology made the risk of overlooking invasive cancer minimal.[99] As Green explained in another article (his third on the topic in 1969–70, showing that he was not afraid to put his views up for peer assessment), 'clearly patients treated in this manner must be assessed and followed carefully, and if clinical, cytological or colposcopic evidence requires it, be subjected to more radical diagnosis and treatment'. This had been necessary for three patients – two were then given ring biopsies and one a hysterectomy.[100] He stressed the importance of careful monitoring of cases, explaining that if a patient failed to keep an appointment or did not reply to an enquiry she was visited by one of the hospital's medical social workers. Local and national home-nursing, social security and legal agencies were used to locate defaulting patients. With its conscientious staff, the hospital had achieved 100 per cent follow-up.[101]

By 1974 Green was reporting conservative management of 750 cases, among whom ten or 1.3 per cent had 'apparently progressed to invasion'. He added a rider, however, that in only two of those cases was there 'no clinical or histological doubt about the progression'.[102] Here he was alluding to disputes about diagnoses which will be explored later. As with similar studies elsewhere, he found a small minority of cases slipping through the net. Like others, he did not believe that radical treatment for the vast majority was justified, given the attendant side effects, both physical and psychological. As he made clear at the Inquiry (see below), he treated each case on an individual basis, trying to avoid over-treatment.

Cartwright appeared convinced that Green was more interested in researching the natural history of CIS than in treatment and patient welfare. She explained, 'the use [by McIndoe] of the Group 2 category corresponded with Dr Green's category of patients with positive follow-up cytology. Patients had been categorised according to the likelihood that they had persistent disease. This was appropriate for an analysis of the natural history of CIS, while it would not have been appropriate for an analysis of the effectiveness of different forms of treatment'.[103] Yet Green's 1974 *New Zealand Medical Journal* article which discussed those with continuing positive smears did so in the context of the treatments given.[104] Cartwright concluded that 'follow-up was not conducted for the purpose of monitoring and treating the patient where necessary, but to ensure no data were lost which could be included in the study results'.[105]

Green firmly believed that he was primarily engaged in diagnosing and treating patients and, as noted in chapter 2, explained at the Inquiry, 'if I could use the data afterwards for some clinical cartography or research I would use it, but the primary business was to treat, to diagnose, treat and cure the patients'.[106] His responses to questioning about individual case files at the Inquiry support that statement. His attempt to avoid over-treatment must be viewed in the light of the known side effects of treatment discussed above. Green told the Inquiry that the word 'carcinoma' frightened some doctors into responding unnecessarily.[107]

He also explained to the Inquiry that with colposcopically directed punch biopsy, as advocated by Coppleson, it was hoped that the lesion would regress with the removal of the most significant part.[108] If there was any question of progression, then more extensive surgery was advocated. Punch biopsy was adopted to avoid the morbidity associated with cone biopsy.[109] Coney told the Inquiry, 'Well I don't believe Professor Green could have known when he started using punch biopsies only on people that that was going to be sufficient. I would say that he was trying it out to see what happened.'[110] This conjured up an image of an unscrupulous experimenter, and yet, as noted, Jeffcoate for one advised this approach in his gynaecological textbook.[111]

Cartwright pointed out that the women often found the follow-up inconvenient, particularly as many had young children.[112] The repeated visits to the hospital caused constant disruption to their lives, including marital problems, family difficulties with arranging childcare or time off work.[113] Yet elsewhere in her report, she acknowledged the importance of 'monitoring their [CIS patients'] condition for life'.[114] Green and others saw follow-up as not just a matter of gathering statistics, but as something that was in the patient's best interest and the hallmark of conscientious medical practice. Describing the follow-up system to the Inquiry, Green claimed that, 'It was because of our concern for the wellbeing of our patients that we went to those lengths.'[115]

Sometimes patients did not understand the need for follow-up. One patient told the Inquiry that because she had been advised that the cells could become malignant at any time, the visits to the hospital for colposcopy and biopsies had an adverse effect on her. She wondered at each visit whether this time it would be malignant. She explained her concerns to Green who told her that he did not think she needed a hysterectomy but that he would do one if that was what she wanted. She told him that was the case, and he performed the hysterectomy. She added, he 'also told me that a hysterectomy didn't mean that I would be able to stop visiting the hospital'. Following Green's retirement, a new doctor told her that she would have to come back for the rest of her life because in 5 per cent of cases like hers there was a possibility of cancer forming elsewhere. She remembered coming away thinking, 'My God, I had always felt the hysterectomy would be the end of my problems and the idea that cancer could still form, got to me.'[116]

Green noted there had been some criticism about the financial burden imposed on some patients by the follow-up system, and explained how he had instituted a Patients' Welfare Fund at the hospital from a bequest left by a patient. Approximately $2000 a year was spent on assisting patients with expenses such as travel costs.[117] Reporting on allegations that university medical staff were making a profit from seeing private patients, Auckland Medical School dean David Cole commented that this did not apply to Green, who did not collect fees; if he did receive any, he placed the money in the patients' welfare fund.[118]

In her evidence to the Inquiry, Mrs W. commented on Green's concerns about patient welfare. She had a hysterectomy in 1957 and attended the hospital for check-ups until 1982. Although she remained well, nine times out of ten her smears were positive. She clearly enjoyed coming for check-ups, saying that she got to know Professor Green really well over the years. She readily agreed to be the patient in a demonstration by Bill McIndoe of his new colposcope ('It was really funny that day', she told the Inquiry). On one occasion when she had been made to wait, she went home and wrote Green a 'shirty' letter. She related:

> Professor Green rang me up himself and apologised and said there had been an emergency in one of the wards. When I went to the hospital the following week for my appointment he rang my boss, while I was there and made apologies for keeping me so long the first time and for having to get me back a second time, after that, I never used to lose any money at work when I had to go to the hospital.[119]

Green's Defence of his Clinical Decisions at the Inquiry

At the Inquiry, Fertility Action counsel Rodney Harrison questioned Green on why he did not adopt a more aggressive approach to the treatment of CIS. Harrison referred to Mrs W., asked why Green did not advise further intervention and suggested that he did not perform a vaginectomy in order to keep her for research purposes. Green replied that he discouraged her from having a vaginectomy –

> . . . because in a woman like Mrs W., a vaginectomy would have been surgically very dangerous if not impossible, because it is impossible to excise the whole vagina. The urethral orifice and urethra itself is derived from the same tissue as the vagina. If the vagina is removed, [CIS] is bound to recur in the urethra or sub-urethra tissues, for the simple reasons that you cannot excise the whole vagina and even if you tried it and skin grafted it, the skin would adopt the same lesion. She had carcinoma in situ of the vagina which I believe is stationary. I spoke several times to Dr R, the radiotherapist about giving her x-ray treatment as the only thing that would cover her vagina. He declined because of her obesity and secondly because it was only carcinoma in situ and he wasn't going to irradiate people with carcinoma in situ. I certainly did advise her against . . . a vaginectomy. It would have been, I am sure it would have been the end of her.[120]

Richart agreed with Green on this, pointing out that vaginectomy was a difficult operation, and that radiotherapy was occasionally used for 'in situ' which could not be treated by other means, but it was extremely uncommon: 'I agree that it would be inappropriate to radiate a patient for a vaginal carcinoma in situ because the complications of radiation itself are significant.'[121]

Harrison also questioned Green on his treatment of Mrs H., whom Green remembered as 'a thin rather delicate woman'. Harrison asked him, 'If you only did a couple of vulvovaginectomies how could you possibly say that is not a treatment which is worth trying on the patient?', to which Green replied, 'You've only got to do one to realise that.' Harrison retorted, 'Perhaps . . . the problem was that they were too hard Professor Green?' Not recognising this as a possible slur on his professionalism, Green replied, 'Yes. Too difficult and too dangerous.' Harrison continued, '. . . you preferred to leave these women to the possibility of death from invasive cancer, rather than subject them to the vulvovaginectomy, is that right?' Green responded, 'Of course not. I couldn't possibly inflict vulvovaginectomy on a woman like this particular patient, and be able to promise that she would be free from disease. All I would do would be to inflict a lot of misery and physical mutilation on her and this is why in the end . . . .' Harrison interrupted, 'Didn't she ultimately develop invasive cancer and have a lot of misery?', to which Green responded that she 'was thought to develop invasive – she didn't have a lot of misery, only with the xray treatment of the cancer. . . . Fortunately because the biopsy had been labelled invasive cancer, the radiotherapist agreed to radiotherapy and the whole thing melted away by the time she was halfway through the course and she is alive and well, but she has the complications of radiotherapy.'[122] Thus, at the Inquiry, Green was being pushed to defend decisions against ever more radical therapy – to eradicate the lesion at all costs – whereas the realities of medicine were about balancing the disadvantages and the benefits of different treatments.

In 1961 a 40-year-old woman, Mrs M., went to National Women's following a positive smear and was treated with a cone biopsy. This was in line with what Cartwright claimed to be the preferred method of treatment of CIS at that time both at National Women's and in many other parts of the world.[123] With persisting CIS (A3 smear) in 1963, another cone biopsy was performed, after which her smears were graded A2 (suspicious but less than CIS) during the following three years. In 1966 when the smear test was A4, she was given a colposcopic examination, which did not indicate further treatment. With persistent positive smears, however, a hysterectomy was suggested around 1970. Dr Florence Fraser recorded in the clinic notes in 1971 that Mrs M. was not keen to have a hysterectomy and Green also noted this the following year. Following another cone biopsy, the smear test again reverted to A2. Another cone biopsy was performed in 1975 and, in 1977, with smears at A5, Green suggested radium treatment. The doctor responsible for administering radium treatment was unwilling to treat Mrs M. with radiation. (The Radiotherapy Department at Auckland Hospital was reluctant to treat pre-cancerous cases because of the known side effects of radiation treatment and the experience that it could cause further CIS or cancer elsewhere in the genital tract.[124]) The only other option was 'total hysterectomy and total colpectomy'. The clinical notes stated that she was 'most unwilling to have anything done'. She was admitted for cryosurgery and biopsy and later in 1978 for a full assessment for radium treatment. The 1978 clinical notes indicated considerable discussion between four doctors (including Green) about treatment. Radiation treatment was again not considered justified 'unless there was definite histological diagnosis of invasive cancer'. The other suggested treatment was total hysterectomy and vaginectomy. The notes recorded, 'Possibly this could be regarded as even more radical than radiation at the present stage, especially in an asymptomatic patient.' In other words, she was still physically well. Finally, when she returned in 1979, Green 'noted that her condition was malignant and would have to be treated by external radiation'. This was done, and her smears returned to A2. Thereafter her condition deteriorated, however. She had a colostomy (for metastases in the rectum) and further treatment for metastases in the lungs. There was a recurrence of the pelvic tumour in 1981, and Mrs M. died that year.[125]

Coney wrote that Mrs M. became 'a real presence for us as we prepared our case and not a few tears were shed over her'. Mrs M.'s daughter told Coney her mother had been a real battler. Coney said they felt alternately sad and angry about what she had endured. By the time she died, she had visited the hospital 65 times. She had had at least 50 vaginal examinations and 24 major procedures, including twelve general anaesthetics, but she 'never had the hysterectomy she needed. She had died with her uterus intact, though with no bowel, and no function in her bladder and her vagina scarred beyond recognition.'[126]

To Coney, a hysterectomy would have solved the problem once and for all. Green regarded that as too simplistic. He told the Inquiry that, 'One of the most difficult things in medicine is to avoid over-treatment even in asymptomatic patients and certainly when you have no guarantee that your treatment is going to effect a cure, it is not much good for the individual patient.' Harrison noted that Green had also mentioned Mrs M.'s unwillingness to have major surgery. He asked, 'Was it ever put to her in terms "it is essential for you to have a hysterectomy Mrs M. this is life-threatening"?' Green replied, 'I can't remember that, but a hysterectomy wasn't going to cure her.'[127] Elsewhere he spoke of those cases in which CIS of the vagina occurred following hysterectomy.[128] Jordan too, reviewing case notes, referred to the 'dilemmas' Green must have felt in treating asymptomatic patients radically. Mrs M. was not in the sole charge of Green, in any case; as Coney noted, at least ten doctors were involved with her treatment.

At the Inquiry Green appeared to be upset by allegations relating to his treatment of Mrs M. He described her as 'one of a handful of patients who caused me considerable distress and concern as a clinician', and declared that he believed that he had treated Mrs M. with his utmost ability, and that 'It is a complete anathema to suggest that I was not acting in the best interests of the patient, or that I had no intention to cure her.' He claimed to be 'particularly distressed that members of Mrs M.'s family should have been told by Ms Coney that Mrs M. died as a result of experimenting on her or that I failed to discharge my medical responsibilities to that patient'.[129]

Under questioning Green discussed diagnostic and treatment difficulties, such as with Mrs P. who developed cancer of the genital tract following treatment for CIS by hysterectomy. While admitting that in hindsight his clinical decisions were not always necessarily the right ones, he insisted that he attempted to act in the best interests of the patients.[130] He told the Inquiry, 'these prolonged cases of CIS of the vagina following a total hysterectomy, these are the cases that have caused me more unease than any other type'.[131] The cross-examination focusing on individual case files gives a fair indication of what Green thought he was doing. He had an excellent recall of the patients he had treated, had consulted widely, particularly with radiotherapists, and was clearly much exercised by the 'dilemmas' referred to by Jordan. He told Harrison, 'They are all individuals and every cancer is as different as every thumb print'.[132] When Harrison asked Green whether he agreed that there were 'no deaths of women with CIS which would have been preventable, had you been willing to adopt a more radical method of treatment', Green replied, 'No person who treats the cancers, can make such a claim.' He said that no surgeon could predict the outcome; he could only hope he was doing the correct thing. Harrison retorted, 'And such a surgeon should also, as a minimum, ensure that he does not shut his eyes to the obvious of what is going on around him.' Green replied, 'Some may.' 'Are you one of those Dr Green?' Harrison asked. 'My conscience tells me no', was Green's reply.[133]

Green and Fertility

While Cartwright did not suggest that Green was motivated in his advocacy of conservative treatment by a desire to retain women's fertility, Coney and Bunkle put forward this idea and others adopted it. For instance, Pamela Hyde claimed in her PhD thesis on the history of cervical cancer in New Zealand that, 'Green regarded women's fertility as their most precious asset and opted for conservative treatment'.[134] Coney and Bunkle wrote in the *Metro* article that Green was 'concerned at any reduction in women's fertility'.[135] They claimed that, along with his colleagues William Liley and Pat Dunn, he played a leading role in the anti-feminist anti-abortion Society for the Protection of the Unborn Child (SPUC), and that he opposed sterilisation and refused to work with doctors who performed abortions. They stated, 'it was experience with young childless women which started Herb Green asking questions about the diagnosis and treatment of cervical cancer', implying that the life of the potential baby was more important to Green than the life of the woman.[136] Green denied that he ever questioned the treatment of invasive cervical cancer, but added, 'I did question treatment of carcinoma in situ, because I questioned the need for unnecessary hysterectomies which could cause unnecessary trauma for women of all ages.'[137] He also pointed out that he had never been a member of SPUC, had never refused to work with doctors performing abortions and had sterilised a number of women in circumstances where sterilisation was justified.[138]

In 1965 Green recorded that while some authors who favoured conservative treatment for women were keen to preserve fertility, others – among whom he included himself – 'urged local excision without regard to age, parity, or the desire for more children.'[139] His 1966 proposal 'extended' the conservative management to all women with CIS under the age of 35 but it was clearly not his intention to limit it to that age group.

Coney referred to a paper Green had written on sterilisation in support of her claim that he opposed fertility control. She cited his paper: 'Only medical problems would be considered and projected sterilisations for social or contraceptive reasons viewed with great suspicion. Such cases constitute a subtle threat to the public welfare and are therefore illegal and unethical – in the widest sense of the terms.'[140] This paper was written in 1958 at a time when the New Zealand branch of the British Medical Association considered it unethical even for doctors to work in family planning clinics. Green was reflecting the dominant medical view of the time.[141] Significantly, Green's article stressed the importance of patient consent: 'He [the doctor] should make sure that the patient is mentally able to give consent and that such consent is fully and fairly given without influence of others.'[142]

One of the eleven patients who gave evidence publicly at the Inquiry said that she had asked Green for a tubal ligation or hysterectomy for contraceptive purposes. She explained, 'He said he did not do tubal ligations because it was against his religion, so he referred me to Professor Seddon.' She was under the impression he was Catholic (in fact he was Anglican). At the Inquiry, Green denied that he ever said to this patient that he did not do tubal ligations because of his religion: 'I certainly sought a second opinion from Professor Seddon on whether or not tubal ligation was necessary, but I did not refuse to do the operation on religious, moral, ethical or any other grounds.'[143] The patient notes penned by Green during consultation stated that the patient requested tubal ligation 'and she probably has a case for this in view of her unreliability with the pill. Would Professor Seddon please opine.'[144]

Mrs Elsie Barnes who ran Green's gynaecological clinic from 1957 to 1961 was a witness at the Inquiry. She described him as 'a very compassionate man concerned about women as individual people with problems', and she addressed his attitude to sterilisation and abortion. Until the Contraception, Sterilisation and Abortion Act 1977, abortion in New Zealand was illegal unless the life of the mother was under threat. Barnes explained the situation at National Women's:

At the time I was working at National Women's Hospital, abortion for other than medical reasons was absolutely and totally unthinkable. Associate Professor Green's opposition to abortion was not the kind of Roman Catholic doctors for example. It was not on that kind of ideological ground, and in fact, I remember more than one instance of his being very angry at the effect the hierarchy of the Church had on a woman's right to manage her own fertility. He was certainly not opposed to the routine measures of contraception that were available at the time, nor to tubal ligation where a need was indicated for social or family reasons, or for medical reasons. Consequently, it wasn't only where women's physical health was at risk that he would do a sterilising operation. As for Professor Green's being chauvinistic or paternalistic towards his patients, I remember his once saying to me 'that the more he had to deal with the problems of women, the more he came to despise his own sex'. [She added that Green was deeply moved when he said this] . . . . He was certainly the only senior member of the staff at National Women's Hospital who ever expressed any such feeling to me. I also remember a patient who, domestically, socially and personally would have bene-fited greatly from some measure to remove the fear or actuality of further additions to her family. However, because of her very deep and very sincere religious convictions, no form of contraception or sterilisation was acceptable to her. I will never forget Professor Green's impish and triumphant delight, conveyed to me on examining this woman, that she had a uterine fibroid that justified hysterectomy.[145]

Reflecting on her time in the hospital, Barnes said, 'Oh we all knew which doctors we hoped that patients who needed a tubal ligation wouldn't end up in front of. I mean we had Roman Catholic doctors on the staff and if a woman really needed a tubal ligation for social reasons, we would hope that they would get someone who was sympathetic to their problems and Professor Green was such a person.'[146] She reiterated, 'he was very moved by the plight of some of the women, particularly those from the lower socio-economic groups who were very ill-used by their husbands.'[147]

The suggestion that Green might wish to help women retain their fertility was given as evidence of his chauvinism. And yet the desire to retain fertility was not necessarily anti-feminist. A 2008 study by a University of Auckland nursing student suggested that the emotional toll of fertility damage from cancer treat-ment in younger women could be 'huge, with many saying it is like rubbing salt in the wound . . . . Some even say that the fertility effects are worse than the cancer itself. This can affect a woman's self-esteem, confidence, spirituality, sexuality and psychological recovery from cancer, and have lasting effects on current or poten-tial partners and relationships.'[148]

One of Green's colleagues told a *Woman's Weekly* journalist following the Inquiry, 'In a way I think Herb Green was New Zealand's first feminist. He was really the great protector of people against unnecessary surgery and discrimina-tion of all kinds. The one thing I know about Herb Green is that he was fairer and stronger than anyone I've ever met in medicine.' Another said, 'He was very strong on human and individual rights.' They admitted that Green could often be 'a bit gruff', but that he was 'straight and frank. And now, he's being made to sound like an ogre.'[149]

Conclusion

What then was the 'conventional treatment' that the patients at National Women's Hospital were apparently denied by Herb Green? According to Cartwright it was not hysterectomy, which had already been rejected throughout the world as a routine response to CIS in favour of cone biopsy or local excision by the 1960s. Yet many gynaecologists still believed that hysterectomy was the appropriate response to the problem, including star witness to the Inquiry Ralph Richart. A significant minority of gynaecologists was questioning the appropriateness of hysterectomy and cone biopsy, both of which were far from benign procedures. Kolstad might have queried Green's clinical decisions (see chapter 11), but he was the first to admit that there were no clearcut answers. Jordan might also have been critical of Green's approach, but he did acknowledge the 'dilemmas' in deciding appropriate treatment for asymptomatic women when the treatment options themselves carried a 'high morbidity'. Jeffcoate recommended cone biopsy only when smears repeatedly contained cells indicative of malignancy.

Discussing treatment of preclinical invasive carcinoma of the cervix in 1981, Coppleson commented on a 'bewildering array of opinion in the world literature. Descriptions from a selection of the more authoritative authors is warranted [he wrote] because from this array gynaecologists the world over have tended to choose an opinion more often on no firmer grounds than expediency and the personality of the exponent.'[150]

McIndoe and his colleagues' 1984 paper was used to show that CIS was a preinvasive condition.[151] However, the article itself also pointed to uncertainties; the authors referred to some patients 'who had normal cytology after initial management . . . [who] later developed invasive carcinoma despite complete removal of the original lesion', and that 'whether or not the lesion is completely excised does not appear to influence the possibility of invasion occurring subsequently'.[152] Intervention could be defensive medicine, but not necessarily in the best interests of the patient. As Dr Angela Raffle and colleagues in Bristol, England, were to declare in 1995 in relation to a screening programme launched there in 1966 and encompassing 225,000 women: 'Despite good organisation of the service, much of our effort in Bristol is devoted to limiting the harm done to healthy women and to protecting our staff from litigation as cases of serious disease continue to occur.'[153] Although there was a high uptake of cervical smear invitations, the expected eradication of deaths from cervical cancer had not occurred. There was pressure to screen more frequently, to substitute or add different screening methods, to read every smear test more than once, and to lower the thresholds for progression to colposcopy, biopsy and treatment:

> During each screening round in Bristol, over 15,000 healthy women are being incorrectly told they are 'at risk'; over 5500 women are being investigated, with many also treated for a disease that would never have troubled them, and are being left with problems that include lasting worries about cancer, difficulties in obtaining life insurance and worries concerning the effect of their treatment on their subsequent reproductive ability.[154]

The authors pointed out that if a smear showed anything that an expert witness could later claim to be a missed abnormality, then the safest thing was to err on the side of caution. The desire to avoid over-diagnosis was outweighed by the need to avoid any possibility of being held responsible for missing a case. The authors explained that even with treatment available, some women would go on to develop cancer.[155] Barron Lerner also found that in relation to breast cancer treatment, radical treatment was safest for the physicians, who could then claim they were doing everything possible to eradicate the disease and, in America, to avoid possible future litigation.[156] During the Inquiry, Richart was asked whether, if he missed a case of invasion at his institution, he could expect to hear about it. 'I would expect to hear from her . . . or from her lawyer in the US', was his reply.[157] Green's decisions had not been dependent on fear of legal proceedings, but on what he believed to be best practice. Nor was there any evidence that he was motivated by the desire to study the natural history of the disease or to retain women's fertility at all costs, as suggested by some commentators.

4.

The Therapeutic Relationship
and Patient Consent

'The real issues are trust and abuse of it', Sandra Coney declared. 'The patients didn't know.'[1] Cartwright agreed: 'The great majority of patients did not know, except intuitively, that they were participants in the 1966 trial.'[2] In its submission to the Inquiry, the Ministry of Women's Affairs Te Ohu Whakatupu (Maori Women's Secretariat) stated, 'Dr Green certainly was not accountable to his patients who appeared to be totally ignorant of the fact that they were being used as subjects in his studies and that they were being denied full treatment of their condition.'[3]

The '1966 trial' was not a trial in the sense of a controlled trial or the trialling of an unknown drug. It was not 'controlled' in the usual understanding of the term; there was no control group of any kind, let alone a randomised control group.[4] But it did involve the introduction of a form of management of CIS which was controversial, albeit supported by a significant minority of 'thinking gynaecologists', as Malcolm Coppleson called them.[5] This chapter explores what patients were told about their condition and treatment options and what the norm was during the period of the 'unfortunate experiment' of the 1960s and 1970s. The relationship between doctors and their patients was more formal then than it would later become as the result of a change in the general social milieu. Paradoxically, Green probably had a better and closer relationship with his patients than did many of his medical colleagues.

Treating the Whole Patient

In her submission to the Inquiry, Sandra Coney claimed that, 'the question isn't whether CIS was adequately treated, but whether women were adequately treated. A doctor may cure a patient's condition while mistreating the patient. Thus, the clinical management of CIS at NWH is not the only question. The concern of the Committee [of Inquiry] should be how patients were treated, not how a certain collection of cervical cells were treated.'[6] Green himself told the inquiry: 'I would agree with Ms Coney that the objective of a clinician in treating a patient should be to treat the entire patient, not simply the clinical disorder that she may have. I hope that is the approach I have always taken with patients and it is the way I have taught students.'[7] Green put much store on the treatment of 'the whole patient'. Ironically, the focus of McIndoe and his colleagues' paper on positive or negative cytology was concerned solely with the status of cells, an approach which, according to British medical sociologist Tina Posner, disregarded women's experiences and feelings.[8]

In his evidence to the Inquiry, Richard Seddon, professor of obstetrics and gynaecology at the University of Otago, confirmed that Green stressed treatment of the total person in his lectures to students. Seddon had been a registrar at National Women's 27 years previously, in 1960, and he told the Inquiry,

> . . . the impression I gained in 1960 was sustained in my subsequent years at National Women's namely that I believe that Professor Green taught me and others to manage the total person when dealing with cancer or suspected cancer and not just manage the lesion. I think he taught us that you could be objective and analytical, scientific if you like, but at the same time compassionate. That's a firm impression that I have vivid recollections of.[9]

Several of Green's colleagues said that he was more conscientious about giving explanations to his patients than were most other consultants. In his evidence, Dennis Bonham addressed the question of whether patients were told there were differing medical views about the methods of treating CIS. Bonham declared that he had 'frequently heard Green explain to patients, in considerable detail, the nature of their condition, the treatment that he recommended, the options available for treatment and the likely consequences of treatment'. He believed that Green 'spent more time than most clinicians with his patients ensuring they fully understood what was happening'. Indeed, Bonham claimed that Green's manner of dealing with patients was 'extremely commendable', so much so that he was chosen to be the clinician to teach students how to explain to patients with cancer their condition and treatment. According to Bonham, he had 'a very honest but humane approach which ensured that patients understood what was wrong and consented fully in their treatment'.[10]

Those who helped run Green's clinics were equally adamant that he explained things to patients. Mrs Elsie Barnes, who ran his gynaecological clinic from 1957 to 1961, told the Inquiry that Green was very concerned that patients should always have full knowledge of their own health condition. According to Barnes, he deplored the fact that sometimes women would have scars from abdominal operations but would not know what had been done to them. In her view, 'He was not someone whose policy it was to be secretive; quite the contrary, he felt that patients should know very well what their medical history and future course of treatment was.' [11]

Among the letters published in the local press following the Inquiry was one from Olga Carr, who first saw Green in 1964. She confirmed that she and her husband were fully aware that CIS was 'normally' treated with hysterectomy which would stop her having children. She stated, 'During my association with these doctors at National Women's, any questions my husband and I asked were always answered completely to our understanding. We both felt we knew what was going on, as we were well-informed.'[12] Mrs L., the first witness to appear before the Inquiry, similarly declared that all her questions had been answered: 'He [Green] always consulted me and told me how, why I was being treated. . . . I never felt that I was being deliberately being kept in the dark about anything.'[13]

Mrs G., upon whom Dr Alastair Macfarlane performed a hysterectomy at National Women's Hospital in 1964, told the Inquiry that she first saw Green in 1976 and found him totally supportive. She commented on his understanding of the whole patient:

> I had very many family health problems to cope with at that time and Dr Green was caring and concerned to the extent that he assisted me over obtaining a postponement of the special return airfare [to Sydney] I had purchased. I was then able to stay in New Zealand for three months, to be with . . . my mother. He was always articulate in his explanations of any condition to me and always took the time to ensure that I fully understood his prognosis and he took the time to answer any questions . . . . Dr Green's total honesty has always impressed me and I am convinced that he would treat each and every patient with the same respect.

She added that she came forward as a result of reading the article in the *Metro*: 'I came forward because of my disgust.' Harrison asked if she had been contacted by anyone in particular, to which she answered, 'I certainly wasn't.'[14]

Mrs L. told the Inquiry that she had always felt secure with Green, who gave her the impression that he really cared about her wellbeing. She added, 'He never at any stage made me feel, as some doctors do, that he was carrying out treatment regardless of my feelings.'[15] Health Minister Michael Bassett similarly received letters in support of Green, including one from 'Mary', who had been Green's patient from 1970 to 1981. She told Bassett that Green was 'a very dedicated doctor' and that she was 'fully aware of his type of treatment'. Mary confirmed that he willingly answered all her questions, and explained in detail why it was preferable to delay major surgery, adding, 'If for any reason I had not been happy with my treatment, I would certainly have gone elsewhere for another opinion.'[16] Gabrielle Collison, medical superintendent at National Women's Hospital, similarly received many letters from patients in support of Green; one example was Mrs E. L. L. who was treated by Green in the 1960s and had negative smears in 1983 and 1987, and had had 'complete trust' in Green.[17]

At the Inquiry, Rodney Harrison accused Green of withholding information about the seriousness of a patient's health status. He suggested to Green,

'You could have written to her saying that she was in real peril of developing some serious condition if she did not come in for examination.' Green replied, 'I wouldn't write to a patient and say she was in serious peril unless she came to the hospital . . . that's pure blackmail.' He said that he was not about to 'bludgeon' them into coming back to the hospital.[18]

Green also told Harrison that he believed it important to give information to ensure the patient's cooperation. When Harrison asked him, 'Did you tell the patients that they had something like two to three per cent risk of developing invasive cancer and dying?', Green replied:

> Every patient knows that if she has some form of cancer, there is that chance of dying. I told these people that if the smears were positive, there was a possibility they might develop cancer of the cervix, and possibly die from it. I told them that with conservative treatment, it would be possible to prevent this in the great majority of cases, but how can you explain to a patient the uncertainties of cancer. I did my best to explain things to the patient because otherwise I couldn't have expected anything from her.[19]

Green told Harrison that he thought it 'quite unnecessary to frighten people with the . . . unnecessary mention of the word "cancer"'.[20] Harrison replied, 'And you adopted the practice of not telling people that they had cancer.' 'No, I didn't', Green said. 'No patient with cancer admitted to National Women's Hospital under my care has failed to be told the correct diagnosis. Even despite recommendations from other people that they shouldn't be told.' At the Inquiry, the daughter of a patient who had died of cervical cancer gave evidence. Though the woman apparently knew she had cancer, according to her daughter, she did not know the seriousness of the disease. In his reflections on the treatment of cancer patients, American historian and physician Barron Lerner commented on patient under-standing. Despite having informed a patient of a poor prognosis, this patient insisted on being optimistic. He added, 'As a doctor I knew that hope and denial were two powerful motivating forces for patients undergoing cancer treatment.'[21] This could equally have applied to the Inquiry witness's mother; she might not have been misled. Patients' responses were variable and highly subjective, for those with cancer as well as for those with CIS.

Cancerphobia

The Inquiry found that, 'Almost all patients knew that they had an abnormality in a cervical smear test.' Patients recalled phrases such as 'funny cells', 'inconclusive smears', 'cell changes' and 'cancerous cells'.[22] Coney also noted that, 'most of the women believed they had something like "suspicious" or "abnormal" or "changed cells", but were unaware of the significance of their condition'.[23] This raises the question, 'what exactly was the significance of their condition?' And what did other doctors tell patients at that time?

If Green had sought guidelines on how to communicate with patients in the mid–1960s when he began his 'trial', he might have turned to a recent book by Hugh McLaren, four years his senior and professor of obstetrics and gynaecology at the University of Birmingham, England. The book included a chapter entitled, 'Explanations to the Patient and her Husband', in which McLaren stated that he avoided the term 'carcinoma in situ' and preferred to talk about 'restless skin'. He warned of the dangers of contributing to 'cancerphobia', citing the case of a 31-year-old woman with a Grade IIIA cervical smear (CIS): 'After at least 18 subsequent visits to the clinic and endless explanations and interviews . . . nothing but hysterectomy would satisfy her so great was her fear of cancer. But this I would not do and in the end she had to be admitted for psychiatric treatment under a diagnostic label of "acute phobic anxiety"'. McLaren conceded that this case was exceptional, 'but salutary and an obvious pointer to the need for most careful handling of the patient with a positive smear'.[24] Coppleson wrote in 1967 that hysterectomy was indicated 'when the woman is so fearful of the disease and cannot be reassured as to the safety of conservative measures'.[25] He warned that, 'With the recall of the woman, even if the smear has only been "doubtful", the fear of cancer becomes a reality. Often, despite all reassurances as to her future safety, the woman may be subjected to a traumatic emotional experience'.[26] In 1981 he repeated that hysterectomy might be resorted to in cases of 'fear of conservative methods', though he added that this rarely happened when patients received confident and soundly based gynaecological advice.[27]

As early as the 1950s, those seeing women with positive smears showed an awareness of the problem of 'cancerphobia'. In a paper on CIS presented at the 1952 meeting of the American Gynecological Society, Dr John McKelvey told his audience:

> There is another reason for avoiding the term carcinoma in situ for those lesions in which proof of malignancy is lacking. The patient is, under such circumstances, presented with a lifelong fear which may well be completely unjustified and unnecessary. Cancerophobia without something to fix it in the patient's mind is bad enough. When a term such as carcinoma in situ is used, the patient may well be permanently haunted.

He thought it would be wise to 'drop the term carcinoma in situ as it applies to the squamous epithelial lesions of the cervix. We do not use the term.' He reminded his audience that the 'vast majority of which [lesions] clearly do not lead on to clinical malignancy'.[28]

At an American conference in 1967, when Dr James Krieger was asked about patients' concerns relating to his proposed conservative treatment, he replied,

'I believe that the degree of patient concern is in direct proportion to physician concern.' He added, 'It is too bad that the word carcinoma was ever applied to the condition under discussion. Had it been almost anything else, much concern on the part of both physicians and patients might have been avoided.'[29]

There is a great deal of evidence to suggest that patients did indeed fear cancer and interpret a positive smear as a likely death sentence. Clare Matheson was not told she had a suspicious smear result by her practice nurse when the latter phoned her about her first appointment at National Women's. When Matheson subsequently asked the nurse why she did not tell her, the nurse replied that 'some women got so upset by such information and made such a fuss' that it was better not to tell them.[30] Commenting on cancerphobia, Jeffcoate suggested, 'One tactic which can be employed . . . is to describe the lesion in terms of inflammation which, if neglected, might possibly develop into a growth. This gives the excuse necessary to ensure continued observation and leaves the patient happy at the thought that she has avoided the dreaded disease.'[31]

While Green might therefore have felt justified in avoiding the use of the term 'cancer', he claimed it could be used provided the doctor displayed confidence. In 1965 he outlined his philosophy regarding 'the approach to the patient', something to which he had clearly given some thought. He explained that the doctor should always explain what a positive (abnormal) smear meant – 'something which is probably not true cancer but *may* become so if left untreated, that it can probably be cured by local treatment, that regular and not too frequent follow-up smears will be necessary, and that with her co-operation in this way she will not develop true cancer'. This approach, he explained, 'without being afraid to mention the word "cancer", has hardly ever failed', and in his opinion was less psychologically harmful than vague generalities about 'growths' or 'ulcers' or advising hysterectomy 'lest the condition become serious'.[32] Like others, Green stressed a confident approach: 'Close follow-up was not mentally traumatic to patients – if the physician does not worry too much about the disease then neither will the patient!'[33]

When Clare Matheson first visited the hospital in 1964, she was clearly frightened. Her account of this time demonstrates how Green put his beliefs into practice. He told her, as Matheson later explained:

> There were a few cells that were not normal, but he did not think they were anything to worry about. A close watch would be kept on my condition and he recommended I have a biopsy under anaesthetics within a few months. A biopsy, he explained, involved taking a small piece of tissue to be examined under a microscope . . . . His confidence was reassuring. After all the waiting and shuffling around the various parts of the clinic, the examination and discussion were over in a matter of minutes. I relaxed. The relief was enormous.[34]

When she went back for the biopsy a few months later, Green had just returned from a visit to Coppleson's clinic. He now believed that a cone biopsy was an over-reaction and booked her in for colposcopy. He was not afraid to use the word 'cancer'. Matheson described the encounter, in which she found Green 'friendly and reassuring'. He explained that a hysterectomy was not considered necessary and that she was to be kept under observation and attend the clinic regularly for check-ups.[35]

'Ruth' – Clare Matheson's Story

In his interview with Coney and Bunkle prior to their writing the *Metro* article, Green commented on the unnecessary anxiety women faced through diagnostic tests following a positive smear: 'They're told they've got cancer cells and they think they've got cancer. You ask a woman who's been through it.'[36] Coney and Bunkle asked Green, 'What would you tell women?' 'I would tell the patient: I don't think it's serious. That it may be enough to take all abnormal cells. Unnecessary diagnostic work creates psychological morbidity. You mightn't worry, you're an educated, intelligent woman. Would depend on who I was speaking to.'[37]

One such intelligent woman, according to Green, was Clare Matheson. 'Ruth', as she was called in the *Metro* article, first went to National Women's Hospital with a suspicious smear in 1964 and returned regularly to the hospital until she was discharged following five negative smears in 1979. Green referred to a note on her file from 1968 when he had written, 'She is a very sensible patient and knows all the ins and outs of the situation.'[38] Matheson later disputed this claim, commenting that her 'consent to be part of a research programme or to receiving less than proper treatment had never been sought'.[39] She now believed she had been one of Green's 'guinea pigs'.[40] Asked at the Inquiry what he meant when he said she 'knew', Green declared that he believed she understood the difference between carcinoma in situ and cervical cancer.[41] He also pointed out that she had signed the consent form for cone biopsy under anaesthesia, acknowledging that the nature and effects of the operation had been fully explained, on each of the four occasions she had a biopsy.[42] Despite Matheson's sympathetic account of her 1964 encounter with Green in her book, Judith Medlicott's review of the book claimed that, 'In 1964 . . . [Matheson] submitted without question to the humiliating processes forced on her at National Women's.'[43]

During the period 1964 to 1979, Matheson had, Cartwright wrote, 'been subjected to 14 years of regular monitoring at National Women's Hospital, repeated colposcopic examinations and four procedures under anaesthetic before being discharged from the clinic to the care of her general practitioner who was not told to ensure she had regular follow-up examinations'.[44] The Medical Practitioners' Disciplinary Committee subsequently investigated the practice of the general practitioner, at the instigation of Matheson and her lawyer, Rodney Harrison. The doctor had apparently replied to regular inquiries from National Women's

Hospital about the state of Matheson's health without taking a smear test. The committee charged him with 'professional misconduct in the management of this patient and the completion of the reports he forwarded to the hospital'. The committee considered the 'over-optimistic report of the specialist in the letter of discharge from National Women's Hospital' to be a 'mitigating factor'.[45]

In his evidence to the Inquiry, Green emphasised that 'I can fully understand Ruth's frustration and to some extent she was let down by a member of the medical profession who falsely informed the hospital of smear results which had never taken place.'[46] Cartwright clearly did not agree with the disciplinary committee's verdict that it was the general practitioner's ultimate responsibility. She argued that Green had not stressed the urgency of the situation at Matheson's discharge. By way of explanation, Green pointed out that Matheson had returned five negative smears over a period of three years and a negative clinical examination before she was discharged. Richart told the Inquiry that the end point of treatment for him was three negative smears at monthly intervals and no clinical signs of disease.[47] He stated, 'After they have had three negative smears, then the risk in general drops to that of the population as a whole, the high risk population as a whole.'[48] Green adopted a more cautious approach than Richart, explaining that he did not think three negative smears within this timeframe were adequate.[49] He pointed out that as Matheson was fed up with attending the hospital on a regular basis it seemed reasonable at that time to discharge her to the care of her general practitioner.[50]

Commenting on her discharge in 1979, Matheson said, 'I and my family were delighted. We believed that Professor Green had monitored my condition closely and that I had never approached any condition that looked like cancer.'[51] In other words she was aware that Green had been monitoring her for any possible development of cancer, which is what Green too believed he was doing. But should Green have revealed to her the medical disputes on the significance of a positive smear or even microinvasion?[52] This would have been highly unusual pre–1980, when social relations in medicine were reliant on trust and the belief that doctors were acting in the best interests of patients. Green told the Inquiry, 'we believed that we had excluded invasive cancer and that the lesion might still regress, with or without further treatment'.[53] In her book, Matheson wrote that she should have had a cone biopsy in 1964 which would have saved all the subsequent visits to the hospital.[54] Yet Kolstad concluded in 1976 that it was possible that the 'number of recurrences observed is dependent not so much on the degree of radical treatment as on the length and completeness of the follow-up'. He admitted that his study of 1121 cases of CIS could not substantiate the repeated claims that there was no risk to patients with CIS 'if treated properly'.[55] There is no evidence to suggest that if Matheson had been given a cone biopsy in 1964 the outcome would have been different, and, according to medical convention, she would still have required follow-ups. Questioned by Harrison as to why he did not perform a cone biopsy on Matheson in 1965, Green replied that she was two months pregnant at the time. Harrison retorted, 'In other cases you were prepared to do cone biopsies on pregnant women weren't you?', to which Green replied, 'Yes, I have done them and regretted them.'[56]

In a critique of the findings of the Inquiry, journalist Jan Corbett argued that Matheson's treatment at National Women's had not been as bad as she had suggested. Corbett pointed out that if she had been given a hysterectomy when she first had a suspicious smear in 1964, aged 27, she would not have had the child she bore the following year (this was her fourth pregnancy). Corbett could also have added the fertility problems associated with cone biopsy.[57] Corbett claimed that even if Matheson had forgone pregnancy and had a hysterectomy with complete excision, that would have been no guarantee against the later development of cancer somewhere else in the genital tract, as happened with 31 other women recorded in McIndoe's paper as having had hysterectomy with complete excision. Corbett noted that in 1971 when Matheson's smear was suggestive of malignancy, Green performed a cone biopsy. In 1976 she had a ring biopsy, which showed CIS but no malignancy: 'By today's standards a ring biopsy is the standard treatment of a grade 3 smear.' From 1976 to 1979, her smears were normal, 'which is no doubt why in 1979 she was discharged to the care of her GP'.[58] In 1985, 20 years after she first attended National Women's, Matheson was found to have cancer and had radiation treatment followed by a hysterectomy. The tumour was so small and localised that it was completely obliterated by the radiation. The pathology report following her hysterectomy showed no residual tumour, nor was there evidence of malignancy in any other part of the uterus. The 1987 notes read, 'Fit and well.'[59] Twenty-one years later Matheson attended an anniversary of the Cartwright Report.

Green adopted a paternal approach to his patients. Matheson recounted a comment he made around 1974. Green had been so reassuring to her about the nature of her condition that she was beginning to wonder why she was recalled to National Women's Hospital so many times. When she suggested to him that she was one of his 'guinea pigs', he apparently replied, 'You will do as you're told.'[60] At the Inquiry, Green's lawyer, David Collins, suggested to Matheson that he may have snapped in a situation in which he believed he was carefully monitoring her condition but felt, in making this comment, that she was totally unappreciative.[61] There are no other examples in the public evidence, or in submissions or letters by former patients, that suggest such an abrasive approach (at least to patients – colleagues were another matter). Most patients found Green supportive and friendly, as noted by Cartwright.[62] Matheson wrote that one woman 'aptly' commented following the revelations of the Inquiry that, 'Suddenly the father figure turns into a dirty old man.'[63] This patient had clearly viewed him as a father figure. She had now been persuaded otherwise, not by his behaviour but by the Inquiry.

Blurring the Distinction between CIS and Cancer

In 1988 Tina Posner and Martin Vessey from the University of Oxford's Department of Community Medicine conducted a study of patient responses to a positive cervical smear.[64] This showed just how 'unnecessarily dangerous to patient wellbeing' it was to equate CIS with cancer. The study reported patients' responses to a positive smear: 'The feelings of alarm, extreme anxiety and horror related to beliefs that the abnormal (positive) smear implied cancer, necessitating a hysterectomy, an end to childbearing, and possible death.' [65] 'The GP put me in quite a state. I walked out with jelly legs. It left you with a horrible dread . . . . I kept looking at the children. We had a terrible weekend. Cancer's a dreaded word. Known friends with cancer – always fatal.'[66] The authors noted, 'Whatever the doctor says about future prospects will carry great weight and affect the way the patient thinks about her health status.'[67] They concluded that the challenge to health education in this field was 'to make the black and white convincingly grey' and to convey the knowledge that abnormal cells did not necessarily progress to cancer.[68]

During the Inquiry, National Women's Hospital medical superintendent Gabrielle Collison forwarded to Cartwright a letter in the *British Medical Journal* which highlighted this very point. The authors, from a hospital in England, complained that the way in which the disease was publicised had left many women with the impression that it was 'common, rapidly progressive and invariably fatal'. They explained that, in the desire to simplify the explanation of the role of cervical cytology, many used the term 'a test for cervical cancer'. They were concerned that this led to unnecessary anxiety, and advised the medical profession to do its best to ensure that all publicity made clear that cervical cancer was a relatively uncommon cause of death, and that any precancerous abnormality detected was 'harmless and likely to remain so for many years'.[69] This was not a message women would have got from the Inquiry and the publicity surrounding it, both of which presented the situation in black and white terms.

Coney stated categorically in her book that 'the 1984 paper showed that untreated CIS will progress to invasion', and that 'Overseas medical experts to the commission would testify that by the early sixties it was "scientifically established" that CIS progressed to invasion.'[70] When Bunkle wrote to Bonham asking 'if the untreated women chose not to have their cancer treated', he correctly replied that none of the patients had cancer.[71] According to Coney, this was simply playing semantic games,[72] a statement which clearly reveals the easy slippage between 'carcinoma in situ' and 'cancer', a hallmark of the subsequent public debates.[73] The submission to the Inquiry from the Christchurch women's group The Health Alternatives for Women (THAW) declared, 'As well as being denied their right to informed consent, the women experimented on by Professor Green, were not informed of other methods of treating cervical cancer.'[74] The confusion was perpetuated by the Cartwright Inquiry itself which was an investigation of the treatment of carcinoma in situ at the hospital, but was billed, 'The Report of the Cervical Cancer Inquiry'.

This blurring of the distinction between CIS and cancer was perpetuated by some members of the media. Cathy Campbell, designated a health reporter when she joined TVNZ in 1986, covered the Inquiry for the 6.30 news. She later explained to a researcher that as a lay person she had to come to grips with the medical terminology and received some coaching from Sandra Coney and Dr George Hitchcock, cytopathologist at Auckland Hospital. In her submission to the Inquiry, Coney cited Hitchcock's claim that, 'The concept that "cancer in-situ is not cancer" is quite indefensible'.[75] Campbell explained to the researcher that she saw her role as that of an educator, getting the material across to ordinary people who had little knowledge or understanding of either the condition of cervical cancer or of the issues to be addressed by the Inquiry. For the purposes of audience understanding, she changed specialist terms like carcinoma-in-situ to 'signs of cancer' and colposcopy to 'a way of detecting cancer'.[76] This was exactly the scaremongering that the 1988 letter in the *British Medical Journal* warned against when it regretted that smears were popularly known as a 'cancer test' with all the unnecessary anxiety which accompanied this.[77]

This equation of CIS with cancer was accepted by the general public. Following the Inquiry, a letter to the editor of the *New Zealand Listener* proclaimed:

> The National Women's Auschwitz-like experiment brings disgrace to the medical profession and causes one to lose faith in the universities. In the light of clear evidence that carcinoma in situ leads to invasive cancer, how is it that the academic doctors remained silent for 20 years while the gynaecological department of Auckland University's School of Medicine used women as 'lesions to be studied' and failed to treat their cervical cancer?[78]

This equation of CIS with cancer contributed to the popular belief that patients' welfare had been placed in jeopardy by treatment at National Women's without their knowledge.

Informed Consent and the 'Therapeutic Relationship'

A recent history of medical ethics since 1947 noted that the 1964 Declaration of Helsinki, which was the authoritative source for guiding research protocols internationally at the time of Green's 1966 proposal, did not mention the rule of informed consent among the basic principles of research ethics.[79] Though the 1964 Declaration provided for informed consent of human subjects before starting a study with no direct therapeutic benefit, for clinical research it only required consent 'if at all possible, consistent with patient psychology'.[80] As American professor of bioethics Robert Baker has pointed out, it 'presupposed that the therapeutic relationship would automatically predominate over the scientist–subject relationship; it also presupposed that the conscience and sense of personal honor and integrity of the decent (that is non-Nazi) therapist–researcher provided a reasonable safeguard against abuse.'[81]

Green was of the generation that believed in the 'therapeutic relationship'. He commented in 1965, 'The leaving of the decision about hysterectomy to the patient is most undesirable; if we are uncertain about the natural history of the disease which cytology has revealed in her how can we possibly expect her to make what is really our decision?'[82] His belief that decisions about treatment lay with the doctor was consistent with medical views of the 1960s. In 1963 Sir Austin Bradford Hill, doyen of controlled trials and professor emeritus of medical statistics, University of London, wrote about doctor–patient relationships in response to a ruling of the World Medical Association's Ethical Committee. The committee decreed that the nature, reason and risks of experiments should be fully explained to patients, who should have complete freedom to decide whether or not to participate. Hill considered this sometimes unrealistic, pointing out that it was 'often quite impossible to tell the ill-educated and sick persons the pros and cons of a new and unknown treatment versus the orthodox and known. And, in fact, of course one does not know the pros and cons.' He asked, moreover, if the doctor could describe that situation so that the patient did not lose confidence, which was in his view 'the essence of the doctor/patient relationship', and in such a way that the patient fully understood and could therefore give an understanding (or informed) consent:

> If the patient cannot really grasp the whole situation, or without upsetting his faith in your judgment cannot be made to grasp it, then in my opinion the ethical decision still lies with the doctor, whether or not it is proper to exhibit, or withhold, a treatment. He cannot divest himself of it simply by means of an illusory or uncomprehending consent.[83]

In 1963 the British Medical Association and the British Medical Research Council issued research guidelines exempting 'therapeutic experiments' from the stringent constraints of informed consent, provided the medical attendant was satisfied that a new procedure benefited the patient, in which case 'he may assume the patient's consent to the same extent as he would were the procedure entirely established practice'.[84]

The World Medical Association clearly took Bradford Hill's concerns on board when in 1964 it drew up the Declaration of Helsinki, which was then adopted by that association and the World Health Organization. Coney wrote that Green's research breached the 1947 Nuremberg Code with its insistence on informed consent.[85] However, historians of bioethics have pointed out that both national and international medical organisations, including the World Medical Association, practically ignored the Nuremberg Code from 1947 to 1975.[86] They considered the Nuremberg Code to be applicable only to the oppressive and unscrupulous Nazi regime and not to Western medicine generally; this viewpoint prevailed until the patients' rights movement of the 1970s revived the code. The Declaration of Helsinki, which was revised in 1975, continued to give doctors greater freedom from consent requirements than the Nuremberg Code.[87] Only in 1979 did the American Belmont Report of the National Commission for the Protection of Human Subjects of Biomedical and Behavioral Research underwrite the 'universality' of the Nuremberg Code.[88]

As late as 1985, Lord Scarman, a British Law Lord and former head of the Law Commission, declared in the Medical Protection Society's annual report that the doctor had a 'therapeutic privilege enabling him to withhold information . . . if disclosing would pose a serious threat of psychological detriment to the patient'.[89] Epidemiologist and professor of medicine at Oxford, Sir Richard Doll, agreed:

> So long as physicians limit trials to situations in which they genuinely do not know what is the best way to treat patients, weighing potential risks against benefits, it is, I believe, frequently undesirable to be explicit about the nature of the trial, just as the doctor who is not carrying out a trial is normally not explicit about all the uncertainties associated with the treatment he prescribes.

He gave the example of not revealing uncertainties in the case of myocardial infarction because of the acute anxiety this would engender; he could equally well have used the example of the dreaded disease, cancer, with the undoubted anxiety attached to it.[90]

On patient consent Green was no different from any other medical specialist from this era. Green did indeed, as Robert Baker has suggested of American physicians, believe the therapeutic relationship took precedence over the scientist–patient relationship, as he told the Inquiry.[91] Green did not breach the 1964 Declaration of Helsinki, which included a strong exemption for patient consent in therapeutic research. The American-based National Institutes of Health, which funded most medical research there, issued guidelines for clinical research in 1966. These guidelines stressed peer review and expressed reservations about being able to convey all the information necessary for patients to make an informed decision.[92] Green was therefore in accord with these latest American recommendations when he took his proposal for conservative treatment to the Hospital Medical Committee for approval. Later, in 1972, he submitted another proposal to the committee, to conduct a randomised trial comparing radical surgery and external radiation in the treatment of the early stages I and II of cervical cancer.[93] This was a time, as Barron Lerner also noted in America in relation to breast cancer, in which medical decisions were in the hands of the clinicians.[94]

The Medical Research Council of New Zealand (MRCNZ) endorsed the Declaration of Helsinki as an appropriate guideline in 1968, published its own guidelines in 1969 and set up an ethics committee in 1972. The revised MRCNZ application form from 1973 required a signed statement to the effect that clinical research protocols had been examined and approved by a properly constituted ethical review committee in a hospital or university. The emphasis, as elsewhere, was on peer review.[95]

Dr Jim Hodge, then director of the MRCNZ and involved with the Medical Council of New Zealand from 1968,[96] wrote in his submission to the Inquiry that correct procedures had 'certainly' been followed by research grant applicants based at National Women's Hospital since formalised ethical review procedures had been instituted by the MRCNZ and by Auckland Hospital (1973). He maintained that the general standard of ethical review in Auckland was known to be high, and that, 'At no stage prior to 1987 has the Council had reason to suspect that ethical review procedures were unsatisfactory in any of the teaching hospitals in Auckland.'[97]

National Women's Hospital set up an ethics committee in 1973, although this was not the start of monitoring medical research, which had previously been done by the Hospital Medical Committee. Sometimes the Hospital Medical Committee considered patient consent appropriate, as in Dr Barton MacArthur's 1969 proposal to conduct a follow-up study of premature and low-birth-weight babies. The committee approved this, with the proviso, 'that the consent of the parents should be obtained before Mr MacArthur was given access to the records'.[98] Verbal consent was still considered adequate in 1973 when Drs Graham Liggins and Ross Howie submitted a proposal for a trial for corticosteroids to induce labour in prolonged (past term) pregnancy. They told the committee, 'Patients will be asked for their co-operation in a trial and will know that they may receive placebo. They will be given a written outline of the project to read . . . and their consent obtained verbally after a discussion of the benefits and risks involved.'[99] That they relied on verbal agreement was an indication of the two-way trust which existed at that time.

When it was set up in 1973, the Hospital Ethics Committee was initially concerned with animal experimentation, and a veterinary adviser was appointed. Nevertheless, it quickly came to deal with human research as well; the minutes reveal that the committee was assessing clinical trials in 1973.[100] The committee appointed a lay member in 1975, at the request of the Auckland Hospital Board. Harry Israel, a former pharmacist and coroner, assumed this role, and he was still the lay member at the time of the Cartwright Inquiry.[101] Research grants considered by the committee included one by Graham Liggins in 1976 for Sandoz Ltd on the evaluation of bromocriptine in the treatment of premenstrual tension. The minutes recorded, 'Green said that he assumed that the informed consent of patients to be obtained by the investigators meant that the patients will be fully informed of known possible side effects. Dr Baird said that this was the case.' That application was approved on the understanding that the investigators were to obtain the informed consent of each patient to participate in the trial.[102] There was, therefore, monitoring of research and awareness of the issues long before 1977. What changed in 1977 (the date given by Coney and others for the establishment of the ethics committee at National Women's Hospital)[103] was that the ethics committee became a sub-committee of the newly reconstituted Hospital Medical Committee.[104]

Peer review formed an important part of medical research in the 1960s and 1970s. However, the climate was changing, with patients demanding greater input into medical decision-making as a consequence of consumerism, civil rights and the women's movement. Yet when Green was practising there was nothing unusual about his approach and behaviour. Had he adopted a more interventionist approach, this would still have been his decision. Joe Jordan was no less (and possibly more) authoritarian in his approach than Green. He told the Inquiry, 'It would seem important . . . that when a patient is known to have premalignant disease, she should be told the exact nature of the disease, the planned treatment to eradicate the disease, and that she should be reassured that the disease has gone following her treatment.'[105] Such unqualified reassurance was not, however, possible, as Jordan himself admitted: 'Most authorities would like to think that carcinoma in situ is 100 per cent curable and for all practical purposes this is true. However, even following cone biopsy and hysterectomy invasive carcinoma can and does develop.'[106] There were no certainties but the uncertainties were clearly not to be conveyed to the patient.

Yet even in this new climate, with patients demanding more information, reassurance appeared to be important. At the Inquiry, Mrs B. described her visits to National Women's Hospital: 'I used to see the same nurses and that was very reassuring. I would have preferred it if I could always see the same doctor, as when you see a different doctor they tell you a different thing, and that is a bit worrying.'[107]

Informed Consent and a Clinical Trial, 1972–1982

Another group of patients who came under the scrutiny of the Cartwright Inquiry and its discussion of patient consent were those known as the 'R' series, part of a randomised controlled trial conducted by Green and colleagues from 1972 on the treatment of cervical cancer. Cartwright wrote that Green 'decided not to seek patient consent'.[108] Green was interrogated about this at the Inquiry and it was reported in the press:

'Coin tossed over cancer says Prof'. The trial was to solve the 40-year old argument in medical circles whether radiation treatment alone got better results than radiation and surgery. Green explained both treatments were considered equally acceptable. Once the treatment was proposed the patient could object and opt for alternative treatment. If they objected they were immediately dropped from the trial and treated as they chose, he said. Dr Rodney Harrison QC argued that this was not real choice.[109]

The problem of ensuring 'real choice' had been the subject of much debate, as discussed above. In the world of bioethics, a precondition for randomised controlled clinical trials was 'clinical equipoise', defined as 'the state of genuine uncertainty within the expert medical community about the preferred treatment'.[110] Green told the Inquiry, 'It was considered by myself and all the staff, since there was genuine doubt as to the best treatment and that each method was fully acceptable by appropriate authorities, that such a randomised trial of treatment was ethically and clinically justifiable.'[111] Harrison asked Green if the patients were asked for their consent 'to be used as experimental subjects', to which Green replied, 'They were not experimental subjects.' Harrison then asked Green how his decision to treat some patients radically fitted in with his conservative approach. Green pointed to the difference between CIS and cervical cancer, and stated, 'I will be as radical as anybody if I am dealing with invasive cancer.'[112]

In explaining the rise of the use of radiotherapy as an alternative to hysterectomy for the treatment of cervical cancer in the early twentieth century, historian Ornella Moscucci argued that the former was promoted as a feminist issue by women doctors, who argued that radiotherapy was less invasive and 'mutilating' than surgery.[113] Yet the correct treatment continued to be controversial, as noted by Sir John Stallworthy, professor of obstetrics and gynaecology at the University of Oxford, in 1981, which he attributed in part to the 'stultifying and selfish rivalry which existed between the two disciplines' of radiotherapy and surgery.[114] The gulf between the two sides was highlighted by another contributor to the same textbook who declared that, 'Hundreds of thousands of women have been cured of cervical carcinoma by radiotherapy.'[115] Coney stated that the results of the 'R' series trial showed that survival rates were the same, but that those treated by radiation alone had more complications. As a result, she explained, the hospital subsequently treated all early stage cancers with radiation and surgery, rather than radiation alone.[116] In a later discussion of patient experiences, Linda Kaye (who had acted as counsel for the Ministry of Women's Affairs at the Inquiry and later acted for the women seeking compensation) quoted a woman who complained that she had not been given the option of a hysterectomy and that radium treatment had caused her long-term debility.[117] The implication was that Green was denying women hysterectomy for the sake of the study, causing unnecessary suffering. Yet Green told the Inquiry that most of those who objected and were removed from the trial were those who objected to having a hysterectomy.[118] The National Women's Hospital trial ran from 1972 to 1982, and as late as 1981 Stallworthy still saw it as a controversy. The 'R' series conformed to international standards for conducting randomised controlled trials in the 1970s.

Conclusion

Green's relationship with his patients was little different from that of any other medical professional of his era. If anything, he had a better relationship than did most clinicians. Green's patients rallied to his support to such an extent that his solicitor David Collins bound their letters into a folder as an appendix to his evidence; Harrison disparagingly described this as Green's 'fan mail'.[119] This folder, however, did not comprise the only source of support by patients. Ex-patients also wrote to Gabrielle Collison as superintendent of the hospital, to Michael Bassett as Minister of Health, and sent submissions to the Inquiry itself. Phillida Bunkle wrote that the so-called 'experiment' at National Women's Hospital that led to the Inquiry could occur only 'in a climate in which patients were perceived as non-humans' and that it was 'an extreme example of a culture in which women become things, and in which patients become objects'.[120] Yet without a doubt Green treated 'the whole patient' and not just a collection of cells. His research was secondary to the therapeutic relationship.

5.

A Profession Divided

Judge Cartwright made damning statements about Herb Green's professional competence. She wrote, 'An analysis of Dr Green's papers points to misinterpretation or misunderstandings of some data on his part, and on occasion, manipulation of his own data.'[1] She believed that the evidence of invasiveness 'appears to have been disregarded by [Green] or not fully understood'.[2] At best she portrays him as stupid, at worst devious. She extended these assessments to some of his colleagues. Green and others at the hospital were accused of making 'confused statements'.[3] Commenting on Professor Dennis Bonham's explanation to Coney that the women had abnormal cytology but not cancer, Cartwright wrote that this was 'misleading' and a 'misinterpretation [of] factual matters'.[4]

What Cartwright labelled 'misunderstandings', 'manipulation' and 'misinterpretation' were actually reflections of the uncertain state of medical science and of medical disagreements. At the time of the Cartwright Inquiry, medical practitioners widely debated the interpretation of histopathology, the management of patients and the value of screening programmes. The Cartwright Inquiry inadvertently revealed a profession divided. This chapter considers medical disputes that arose between Green and his colleagues McIndoe and McLean, while chapter 6 discusses disagreements between Green and those who were developing a national cervical screening programme.

Interpreting Histopathology

Green's supposed misunderstandings or manipulations related to diagnoses based on histopathology. Diagnosis of CIS was widely recognised as problematic. In 1962 American gynaecologist John Graham noted the lack of unanimity about the identity of these lesions: 'Epithelium that is called carcinoma in situ by one observer may be called invasive carcinoma by another and mildly anaplastic by a third.' Graham referred to an experiment in which 20 slides of borderline lesions were submitted to 25 eminent pathologists, who were asked to allot one of ten diagnoses ranging from normal to invasive cancer to each of the histologic sections. He explained that no two pathologists agreed on all the diagnoses, and that, 'Each of the slides was called carcinoma in situ by at least one observer

and negative by at least nine observers. As an example of the disparity of interpretation, one slide was called negative by one observer, various intermediate
diagnoses by twenty-three and invasive cancer by the remaining observer.[5]

Others conducted similar studies. In 1965 Britain's Royal College of
Obstetricians and Gynaecologists organised a panel of five gynaecological pathologists to assess the pathological reports of cases of dysplasia, carcinoma in situ
and microinvasive carcinoma treated between 1955 and 1965. In 234 cases (32 per
cent), the panel disagreed with the diagnosis of the submitting pathologist.[6] The
1966 edition of *Recent Advances in Obstetrics and Gynaecology* commented on
the 'conflict [around] the pathological significance of the lesion', and referred to a
diagnosis as being 'not cancer in the clinically accepted sense of the word'.[7]

Recognition of the subjectivity of pathological and cytological reports continued into the 1970s. American pathologist Leopold Koss commented in 1979 that
there were many surveys on record which demonstrated great individual differences even among experienced pathologists in the assessment of intraepithelial
lesions of the cervix. He also reported that if the same set of slides was shown
twice to the same observer or groups of observers, there were usually major differences between the first and second diagnoses. He wrote:

> Truly it can be repeated that one man's dysplasia is another man's carcinoma in situ
> . . . . Precancerous lesions of the uterine cervices have many faces, every one of which
> may or may not lead to invasive cancer . . . . There is no publication on this subject
> where one could not reshuffle the photographs and substitute pictures labeled as dys-
> plasia for those labeled carcinoma in situ and vice versa. This unfortunately includes
> some of the official pronouncements on this subject that are intended to guide others
> in the diagnosis of these lesions.[8]

A 1979 article in the *American Journal of Obstetrics and Gynecology* reported a
study in which pathologists reviewed 265 cases of microinvasive carcinoma of the
uterine cervix and reclassified 50 per cent of them as belonging to a category less
than microinvasion. The author wrote of the tendency to overdiagnose for fear
that the patient might not receive adequate treatment.[9] There was also the question of legal action for failure to diagnose accurately, particularly in America, as
noted by Ralph Richart.[10] This contributed to the 'risk-aversion' approach which
Barron Lerner identified as a cause of overtreatment (by radical mastectomy) of
suspected breast cancer in America.[11]

Commenting on a study showing that 30 per cent of those diagnosed with
CIS did not have 'proven carcinoma in situ', Professor Jeffcoate of Liverpool wondered how many errors there were worldwide, and commented that the toll of
unnecessary surgery and patient worry was inestimable.[12] In 1981 Dr Malcolm
Coppleson noted that the opinion of the histopathologist had always been considered definitive in determining treatment, and that the decision to remove a

woman's uterus had frequently been made on differences in the appearances of a few surface cells. He believed there was now sufficient evidence that faith of this sort was unwarranted, given that 'the same slide proffered among cytopathologists of great expertise was rated from benign to invasive cancer'.[13]

Disputes at National Women's

With experts of international standing in the United States, Britain and Australia offering diagnoses which ranged from mild dysplasia to invasive cancer, it is little wonder that disputes arose in New Zealand. However, they appeared to reach an abnormally high pitch at National Women's. In her 2000 discussion of the 'unfortunate experiment', Dr Charlotte Paul wrote, 'There were heroes as well as villains in this story. What Coney and Bunkle laid bare was not unitary professional power that set doctors' interests against patients' interests. On the contrary, it was a profession torn by dissension, both privately and publicly.'[14]

Paul followed Coney in identifying only two heroes in the story, Bill McIndoe and Jock McLean. She explained that they 'took the appropriate steps, according to the traditional obligation to use their power to help the sick', and wrote that their actions could be described as 'models of good practice in medicine'. The villains were 'the powerful medical leaders in the hospital, with rigid hierarchies and strong personalities', by whom she meant Dennis Bonham and Herb Green.[15] This division into 'heroes' and 'villains' is, however, too simplistic.

Green's relationship with pathologist Jock McLean can only be understood in the context of Green's long-term interest in pathology. As noted earlier Green had worked in the 1950s with the British gynaecologist Stanley Way, who ran his own histopathology laboratory in Newcastle upon Tyne. Like Way, Green developed a particular interest in histopathology. He told National Women's Hospital medical superintendent Dr Algar Warren in 1973 that, 'Ever since 1949, when I wrote my MRCOG commentary on "The Early Diagnosis of Cervical Cancer" and discussed carcinoma in situ with the National Women's Hospital pathologist (Dr Lindsay Brown), I have been studying the histology of in situ and invasive cervical cancer. Since September 1956 I have been personally involved in the histology of approximately 1050 cases of invasive cervical cancer.'[16] Green also told the Inquiry that the pathologist John Sullivan 'discussed his pathological findings with me and encouraged me to make my own pathological examinations'. He also recounted that he and Sullivan discovered that there was no pathological evidence of CIS in the uteri of many women who had been treated for this condition by hysterectomy.[17]

In 1962 John Sullivan moved from National Women's into private practice and Malcolm (Jock) McLean was appointed pathologist-in-charge, a position he held until his retirement in 1989. McLean had graduated from Otago Medical School in 1948 and gained his MD in 1957, after which he spent a year working in the pathology department at University College Hospital, London, and then four years as pathologist to the Hutt Hospital, Wellington.[18] Dr James Gwynne, who, as pathologist at National Women's from 1957 to 1958, had applauded Green's commitment to learn about pathology, speculated after the Inquiry about the poor relationship between Green and McLean, 'it seems that Green was doubtful of his [McLean's] opinions when he came to National Women's Hospital because McLean was not a specialist in gynaecological histology. On the other hand, McLean probably was not aware of Green's expertise in cervical histology and apparently obstructed his access to histological material because he was not a pathologist'. Gwynne said of Green, 'His high ethical standards and professional integrity are in my view beyond question.'[19]

Pathologist Stephen Williams, chairman of the New Zealand Society of Cytology, which was set up in 1968, commented in 1971 on Green's involvement in histopathology. Williams believed that Green was 'sincere, although perhaps bigoted, on this subject'. He wrote of his concerns to American pathologist Dr George Wied, one of the founders of cytopathology, who had worked in Papanicolaou's laboratory at the Cornell Medical Center in New York and who had been a guest speaker at a conference on cancer at National Women's in 1959.[20] Williams told Wied that it appeared that in a number of cases where invasion had clearly followed the original in situ diagnosis, Green reviewed the histology himself although he was not a trained histopathologist, and removed the cases from his series on the grounds that they were invasive carcinomas from the outset. Williams considered this 'an embarrassing and awkward development and one which does not appear to be susceptible to reasoned argument and discussion'.[21] Green was not alone, however, in conducting retrospective reviews of the histology of invasive cancer. Reporting his follow-up of 1121 cases of carcinoma in situ, Per Kolstad wrote that in three cases of invasive cancer 'the primary lesion must have been a microcarcinoma which was overlooked during the initial histopathologic examination. This was confirmed by re-examination of the operation specimens in two cases. It should be emphasized that such cases must occur in large series of patients with carcinoma in situ.'[22]

At the Inquiry, Harrison asked McLean if Green 'crossed out his diagnosis on the report form and put in his own diagnosis'. McLean recalled that this happened in about six cases, explaining, 'He would see the slides and if he disagreed with my diagnosis, he would substitute or cross out my diagnosis and put down his diagnosis and this was quite open in actual fact because the report would then go back to the CC notes and it was available for anybody to see.'[23] Jamieson also told the Inquiry, 'Some of the cases as I recall were downgraded by Dr McLean, some by other pathologists, some by Professor Green and that applies also to the cases downgraded which did not develop invasive cancer.' He continued, 'The whole point about these cases was that there were doubts and the doubt often existed in the mind of either the pathologist or the clinician involved. I don't think that unequivocal decisions were available often in these cases on the basis of their initial biopsy.'[24] Neither Jamieson nor McLean suggested any underhand or dishonest manipulation of the data.

In 1965 Green had asked five Auckland pathologists, including McLean, Sullivan and Williams, to view the slides of 22 cases previously treated for invasive cancer at National Women's by radiotherapy or hysterectomy; only in three cases did they agree there was evidence of invasive cancer.[25] This exercise by Green might have been interpreted as a vote of no confidence in McLean (and indeed this was suggested to Green during his cross-examination at the Inquiry),[26] but it simply followed conventions used in similar overseas studies.

Green told medical superintendent Algar Warren in 1973 that the profession generally recognised that 'what constitutes histological invasion or not, particularly just the possibility of it, is a very personal and subjective opinion that may vary greatly from one pathologist to another'.[27] Green wrote, 'Dr McLean is entitled to his opinion, even on clinical matters if he so chooses (and he has so chosen), but similarly I must be allowed my opinion on histological matters.' He pointed out that McLean hardly ever saw the patients before treatment and never subsequently, whereas he (Green) was intimately concerned with their history, including histological sections, their treatment and follow-up; as a result he believed he was in a better position to say what might or might not happen to a given patient.[28] In 1981 Coppleson also wrote that in the past too much power had been given to the pathologist in determining treatment and that only recently had 'the art of the clinician [been] reinstated to a traditional role as captain of the management team'.[29]

While some of his colleagues perceived Green as arrogant and dogmatic, McLean could be equally determined. As a colleague, Dr Andrew Mackintosh, later said of McLean, 'he was very proud of his work and didn't have the kind of personality that took kindly to it being questioned'.[30] Nor did McLean appear to appreciate Green consulting Sullivan, who was now working privately, seeing this as an insult to his professionalism. Green wrote in 1973, 'I have never been able to get Dr McLean to ask for a second opinion from Dr J. J. Sullivan, a pathologist whose opinion I and many others value highly . . . and I have been forced to consult almost surreptitiously with Dr Sullivan on occasions about cases in which I have had histological doubts.' Green was also frustrated by the difficulty of accessing McLean's records; he explained, 'Dr McLean keeps the best diagnostic slides from many cases in his own private collection under a disease index only; these are not always made available to me and if Dr McLean changes his classification later these special slides become as good as lost for that patient.'[31] Sir Graham Liggins, a colleague at National Women's Hospital and professor of obstetric and gynaecological endocrinology, later observed that arguing with a pathologist about his diagnoses was a red rag to a bull.[32]

In the early 1970s, McLean did indeed challenge Green's authority in the area. He told Warren that for many years Green had 'entered in an authoritative manner into the field of histopathology . . . in which he may have seen many cases, but scarcely a field in which he has had adequate training and background, a field in which he is not an acknowledged expert'. McLean pointed out that Green had challenged his diagnoses: 'Challenging of this nature is a serious matter as it raises the question of my competency to undertake my duties as a histopathologist in this hospital.' He wrote disparagingly of Green's 'themes of retrospective rediagnosis to suit the situation and that delays in diagnosis and treatment had not altered the outcome'; and he believed that patients were at risk as a result of Green's mode of management.[33]

McLean referred to the delayed treatment of one CIS patient, adding in response to Green's defence, 'The fact that this patient is alive and well is beside the point.'[34] Revisiting this at the Inquiry, Green claimed that, 'the fact that this patient is alive and well is NOT beside the point'.[35] On the delayed treatment, had McLean consulted American pathologist Leopold Koss, he would have learned that Koss put CIS in the 'non-urgent' category for treatment. Koss wrote, 'carcinoma in situ and the lesser but related lesions appear to have an extraordinarily slow evolution, continuing over periods of many years. For this reason they should not be considered in the emergency category and may be treated less urgently.'[36] Similarly Coppleson wrote in 1981 that, 'Once invasive cancer has been excluded by the complementary use of histology and colposcopy, the handling of the cervical intraepithelial lesion becomes a more academic and leisurely matter.'[37] Green responded to McLean that, 'A survey at this hospital (Green 1971) shows that the length of disease time (as judged by symptoms) before treatment affects the survival rates not at all (except that possibly the longer the symptoms in Stage I the better the survival rate) . . . . It could reasonably be anticipated that even if some "early" case were overlooked and treatment delayed, or if some progressed to Stage 1A invasion, that such patients would not be put at a disadvantage.'[38]

Green reviewed the case notes of fourteen patients identified by McLean as delays in diagnosis and treatment and stated in regard to one case that, 'The only "delay" has been in treating a young, asymptomatic woman on the basis of a very disputable histology report alone. I have since discussed the case and all the reports, National Women's Hospital and Australian, with the patient's GP; he supports my management and in view of what he knows of the patient suggests that further temporisation is now justified – with which I agree.'[39] Green added that delays might have cut both ways: 'Dr McLean and his laboratory administrative staff [had] . . . delayed reports. I mention the latter advisedly for I have known the delay to be as long as 4 years in one case of malignancy, despite my repeated requests.'[40] Referring to another delay in the case of a woman who died of cervical cancer, Green later told the Inquiry that this was a colposcopic failure though it did not vitiate the colposcopic method. He explained, 'this patient had an extremely aggressive tumour beneath the epithelial surface, and it is very unlikely that making the correct diagnosis almost a year [earlier] would have made any difference to the ultimate survival time of two years, eight months'.[41]

It was important for clinicians and pathologists to work together in this area. The *British Medical Journal* stated in 1974 that its evidence of diagnostic disagreement 'demand[ed] a critical review of diagnostic criteria for cervical lesions in every gynaecological department'.[42] While Richart told the Inquiry that it was the clinician who should take ultimate responsibility, he believed that in the past pathologists had been the principal and sometimes the sole arbiters of therapy.[43] However, with Green's own long experience and definite views, this was not going to happen in Auckland.

While there is evidence of a poor relationship between Green and McLean, Green's relationship with the hospital colposcopist Bill McIndoe appeared to be for some time collegial. A memorandum by McIndoe, included in the Cartwright Report, showed that in 1969 he still shared similar views with Green. McIndoe told Green, 'From my reasoning and findings during my study tour last year, I am convinced that some of the patients treated here successfully by local excision and classified non invasive are treated in many places radically and classified INVASIVE. I know you have the same view.'[44] In the same year, McIndoe published with Green.[45] It was this article that formed the basis of McIndoe's invitation to the First World Congress of Colposcopy and Uterine Cervical Cytology and Pathology in Argentina in 1972.[46]

Relations between Green and McIndoe deteriorated in the 1970s, however. In 1971 McIndoe suggested in a letter to a gynaecologist in Wellington, Graeme Duncan, that there was a 'slight difference in our [his and Green's] views'.[47] In a 1973 memorandum to Warren, McIndoe commented on Green's 'belligerent response . . . to comment and criticism', and claimed that he (McIndoe) 'had endeavoured by all means possible in a mature and dignified manner to make my feelings plain'.[48] He wrote that the previous year Liggins had interrupted one of these occasions with the comment, 'Are you two still at it?'[49] Professor Per Kolstad visited from Norway in 1973 at McIndoe's invitation and funded by the Cancer Society of New Zealand. During that visit Kolstad examined six patients and disagreed with Green's management, a point that was raised at the Inquiry, though Green added that this had included one case of a benign tumour that Kolstad had 'unhesitatingly diagnosed as invasive cancer'.[50] It was Kolstad's close friendship with McIndoe that persuaded him to come to the Inquiry in 1987 'to defend the memory of my dear friend McIndoe' (who had died in 1986).[51] In 1974 Green wrote to Warren about McIndoe's 'extraordinary attempt to discredit me by the allegations he made behind my back to staff members . . .' and pointed out that their approaches to CIS remained similar: 'I now have details of some ten cases which Dr McIndoe has handled in exactly the same manner that he apparently found so reprehensible in me and will let you and the staff have details of these cases whenever you like.'[52]

In her book on the Inquiry, Coney reported Bonham's view that there were 'personality problems' between McIndoe and Green, and recalled how McIndoe's daughter told the Inquiry that her father refused to attend any social functions at which Green was present. Coney rightly stated that everyone agreed McIndoe possessed great personal integrity. Bonham agreed that McIndoe was 'a charming chap', but pointed out that he did have a 'vicious streak' which could make him difficult to work with.[53] Dr John France, associate professor in steroid biochemistry, also said of McIndoe, 'the trouble was he was passive, and he was also a bit undermining in that he would go behind the back of people to say it'.[54]

The version of events given by McIndoe to Phillida Bunkle when she was preparing the 1987 Metro article appears to be somewhat distorted. He told her that Green had been 'treating people in all sorts of different ways and observing the results . . . . About the mid–sixties when the results were discouraging, he just stopped publishing his results'.[55] In fact Green continued to publish on the subject until after his retirement in 1982, though his last paper on the National Women's Hospital CIS data was published in 1974 and was referred to in McIndoe's 1984 paper.[56] Green's final publication was a joint one with Dr Minoru Ueki, who had worked at National Women's before becoming head of the Cancer Detection Clinic of the Osaka Medical College, Japan. Published in 1988, this was a study of the conservative management of 211 women with CIS.[57]

In their 1984 paper, McIndoe and colleagues discussed the conservative treatment practised at the hospital, pointing out that, 'The few clinicians who initially performed punch or wedge biopsy alone had abandoned the practice by 1970.'[58] Yet McIndoe suggested that some hospital staff were still practising Green's 'conservative' treatment regimen in 1985. He told Bunkle, 'the management is being repeated by doctors who are Green's followers. That's the worst thing, they are still doing it. It is still affecting the way they teach GPs. It didn't end when Green retired [in 1982]; that's why I wanted to do something, to bring it out.'[59] If 'conservative treatment' meant less than cone biopsy (i.e. punch or wedge biopsy), as stated by Cartwright, these two statements by McIndoe appear to be at odds with one another.

At the time of the Inquiry, Dr Sadamu Noda was in charge of the cytopathology department of the Cancer Prevention and Detection Center and the Department of Gynaecology at the Osaka City University Medical School, Japan. He had visited National Women's in 1967, and was so impressed by what he saw that he engineered a return to the hospital to work with Green for a year in 1972. In his submission he told the Inquiry that during that visit he performed over 100 investigations at National Women's using a colposcope. 'Sometimes', he said, 'I found it difficult to agree with Dr McIndoe's findings. As a guest in New Zealand I could not take issue with Dr McIndoe but would direct a further cytology smear to indicate his more obvious mistakes.' He told the Inquiry that Green's protocol was of a high standard and one that they adopted in Japan.[60]

It is clear from McIndoe's 1973 memorandum that a territorial dispute was developing between him and Green. Complaining that Green wanted to improve his personal skills in colposcopy, McIndoe expressed concern about 'any technique in inexperienced hands . . . if the person's attitudes to the technique and the way it is being applied are not soundly based'. He added that it 'must be applied with maturity and not elevated in importance out of its due place by over-enthusiastic advocacy'.[61] Not only did McIndoe discourage Green from learning the technique, but from about 1974 he also stopped providing the service for Green who thereafter had to rely on 'visiting experts'.[62] Once he no longer had access to colposcopy, Green generally reverted to performing cone biopsies on new cases.[63]

Green later commented on his 'fierce argument' with McIndoe, and the fact that they did not speak to each other for the last ten years of his tenure at National Women's. He maintained that, 'They [McIndoe and McLean] tried to get the other staff against me'.[64] New Zealand was not alone in these disputes and Richart commented that the 'equivocal smear' often placed cytologists and clinicians in conflict.[65] The clash in Auckland, while representative of an international trend, was exacerbated by Green's legendary obstinacy. Cartwright assessed Green's character as follows: 'Any person reading Green's papers . . . will rapidly gain the impression that Dr Green was a person of strong views, impatient with criticism and with total confidence in his own judgement . . . . [and that he had] lack of patience with any system which implies accountability amongst colleagues'. An example cited by Cartwright was Green's 1974 memo to Warren in which he referred to McIndoe's 'complaints about my handling of certain patients, particularly of his extraordinary attempt to discredit me by the allegations he made behind my back to staff members'.[66] Green's dismay at being undermined 'behind his back' does not, however, imply a lack of patience with accountability.

By contrast to this representation of a very forceful character, McIndoe was portrayed (rightly, according to oral accounts) as gentle and retiring.[67] Yet Cartwright suggests that he felt so strongly about this issue that he 'went so far' as to co-author a letter with Dr Stephen Williams in support of cytology in the *New Zealand Medical Journal* in 1972.[68] However, this was not the only time he published in the *New Zealand Medical Journal*, and it could hardly be described as an extreme measure.[69] He also published an article on cytology in the *Australian and New Zealand Journal of Obstetrics and Gynaecology*.[70] McIndoe took Green's views on cytology very personally and regarded the latter's questioning in the press of the value of cytology as 'really an implied questioning of the integrity of Dr Williams and myself'.[71] There is no evidence that Green himself saw it in this light.

Other staff members were drawn into the disputes. The hospital set up a Tumour Panel to discuss 'controversial and difficult cases openly', as McLean suggested in 1971.[72] The Hospital Medical Committee thought this would work provided 'normal courtesies were observed'.[73] Gynaecologist Ron Jones later commented that the proposal to set up the Tumour Panel was obviously an attempt by McLean to involve a broader medical audience than himself and McIndoe. Dr Bruce Grieve, a senior consultant at the hospital, chaired this meeting for many years, but was unable to resolve the underlying conflict between Drs Green, McLean and McIndoe in the management of the difficult cases of carcinoma in situ of the vulva, vagina and cervix which were presented at the Tumour Panel.[74] Green himself felt the main role of the Tumour Panel was to reveal to students that there were different viewpoints, and he expressed disappointment that more students did not attend.[75]

In 1974 senior consultant obstetrician Bruce Faris suggested that a sub-committee be set up to review the situation.[76] The committee comprised three senior members of the consultant staff and its brief was to review the case notes of thirteen women diagnosed as having CIS who had advanced to cancer. Its report noted that all except one had been successfully treated, and concluded, 'It is the firm opinion of this committee that all staff members involved in the implementation of the policy concerned with the conservative management of carcinoma-in-situ of the cervix have acted with personal and professional integrity.' The committee also considered that the effective continuation of the trial depended upon the following factors:

1. The staff members concerned subjugating personality differences in the interests of scientific enquiry and
2. The initial policy be clarified . . . :
   a. Define whose concern for the safety of the patient is to be acted upon
   b. Assess numerical significance in this trial of histopathology reports stating "?Invasive lesion nearby"
   c. Pathology reporting on cervical biopsy material must be made or confirmed by senior histopathologist
   d. Restate the age limits of the patients for inclusion in the trial (committee members had differing interpretations of the present policy statement)

Signed: Alastair Macfarlane, R. J. Seddon, Bruce Faris. 18.9.75.[77]

Cartwright did not accept the 1975 committee's finding that all the doctors under examination had acted with personal and professional integrity. She claimed that, instead of considering patient safety, they followed 'the time-honoured tradition of confusing etiquette with ethics'.[78] Yet medical historian Roger Cooter has recently argued that the distinction between 'medical etiquette' and 'real' medical ethics was a contrivance of sociologists of the 1970s who narrowly defined 'traditional' medical ethics, and that the latter did not necessarily imply 'neglect of issues such as the sanctity of life, patient confidentiality and autonomy, or "proper" doctor-patient relations'.[79] Cartwright interpreted the 1975 committee's conclusion on setting aside personality differences as meaning that patient welfare had been compromised for the sake of scientific inquiry. However, no less an authority on evidence-based medicine than Professor Archie Cochrane had argued a few years previously (1971) that 'in a situation where there is so much honest doubt', even a randomised trial would be ethically justifiable if technically possible.[80] In demanding clarification of 'whose concern for the safety of the patient should be acted upon', the committee pinpointed genuine differences of opinion about the status and meaning of dysplasia, carcinoma in situ and microinvasion, and appropriate treatment.

The third point of clarification that the committee recommended arose from Green's concern that diagnoses were often made by junior staff and not checked by the senior histopathologist, Jock McLean. In another 1973 memorandum to Algar Warren, Green had referred to 'complaints about pathology reports appearing under the names of the pathology registrars .... the position of a junior, inexperienced pathologist could be quite invidious if some future reviewer (pathological, clinical or even legal) saw fit to dispute his diagnosis'.[81] At Green's urging McLean agreed to 'initial all reports that he sees and has actually commenced doing this'.[82] In her report Cartwright did not comment on this concern, which suggested that Green was interested in high standards.

In December 1976 the Hospital Medical Committee asked Green, McIndoe and McLean to produce 'forward-looking' reports on the management of CIS as a prelude to setting up new hospital protocols.[83] Green's and McIndoe's reports were submitted in July 1977, and McLean's three months later, in October 1977.[84]

Green suggested in his report:

1. Since cytology helps greatly in the detection of cancer in those at risk, National Women's Hospital should continue taking smears of all patients of hospital clinics – including antenatal patients.
2. All patients who returned classes 3 to 5 smears from cervices not held to be suspicious of invasive cancer should have a cone biopsy diagnosis as soon as convenient.
3. Since the most important management of a positive smear is the adequate exclusion of invasive cancer, a cone biopsy should be performed and colposcopic examinations and selective punch biopsies not be relied upon.
4. If smears continue positive after a cone biopsy, diagnosis of carcinoma in situ, but the clinical findings are normal, a further cone biopsy should be considered.
5. Continuing positive smears in post menopausal women, despite normal clinical findings, may indicate occult invasion and warrant very careful repeat investigations and biopsy.
6. Hysterectomy is not immediately necessary following a report of 'incomplete excision' of carcinoma in situ. The National Women's Hospital cases and many other series show that there is no correlation between 'complete' or 'incomplete' excision

and what is found in the excised uterus. Only if follow-up smears are still positive after six months need hysterectomy be considered. And even then a repeat cone biopsy may be all that is required to excise the remaining lesions.

7. Every patient must be followed up at least ten years clinically and thereafter annually by letter to her GP.[85]

These recommendations, made four years before Green retired, do not look 'eccentric'; yet this was how Coney characterised Green's approach at the Inquiry:

> I think that in the [Metro] article, the whole picture that we have portrayed is that Dr Green was a person who had ideas about CIS which were out of step with medical thinking and that those ideas, the longer he held them, became more and more out of step or out on a limb, as I would call it and finally to the extent of becoming somewhat eccentric.[86]

Green's 1977 report suggested a return to more cone biopsies, even for Grade 2R smears (i.e. doubtful smears, or less than CIS). However, he added that this recommendation should be seen in light of his 'personal feeling that the whole subject of cancer is far too emotive to both medical and lay people, and that this emotional outlook has vastly overrated the sociological importance of this disease'.[87] Cochrane, too, had commented on the emotionalism attached to this disease (see chapter 2).

In 1978 new hospital guidelines specified that the aim was to 'obtain an adequate biopsy which completely excises the lesion with no delay'.[88] While Cartwright claimed in 1988 to have no evidence that the 'trial' had ever ended, the Medical Council in 1990 accepted the 1978 guidelines as confirming its conclusion: 'Thus, on 16 March 1978 the Hospital Medical Committee passed a resolution effectively putting an end to the 1966 proposal.'[89] Bonham told the Inquiry that by the mid–1970s Green was regularly doing cone biopsies.[90] Green himself stated that once he no longer had the services of a colposcopist as an added protection, he performed an early cone biopsy on almost every case he had to deal with.[91] Dr Tony Baird, another gynaecologist at the hospital, could not recall any patients in the mid–1970s being managed by less than cone biopsy.[92] McIndoe dated it at 1970.[93] Gynaecologist Dr Ron Jones, who was appointed to the hospital in 1973, said that from the mid–1970s the management of CIS and its results at National Women's had 'by and large followed internationally accepted principles'.[94] Yet, as Coppleson noted in 1981, 'many thinking gynaecologists' were questioning the 'true indispensability' of cone biopsies and hysterectomies, as a result of the excellent results from minimal treatment of these cases.[95] In his review of treatment in different parts of the world over time, Swedish author Goran Larsson later wrote that in Scandinavia, 'Around 1973 a debate started as to whether [even] conization was over-treatment for CIS.'[96]

Thus, by 1977 Green was propounding a more invasive approach using cone biopsy in all CIS cases. This change might have been the result of pressure from his colleagues, or because he no longer had access to colposcopy for diagnosis. Coppleson recommended conservative treatments only with the aid of colposcopy, but even the colposcope could give no guarantees. Green reported in 1974 that one woman whose CIS had advanced to cancer demonstrated that 'rather much reliance had been placed on colposcopically-directed punch biopsies' since her invasive cancer had been revealed only when she had a cone biopsy three years later.[97] His willingness to change his approach and to question previous methods shows that Green was not wedded to his 'personal beliefs', nor was he 'eccentric' and 'out of step' with medical thinking.

Dr Murray Jamieson joined the hospital staff in 1981, shortly before Green retired. Jamieson had previously been a Rhodes Scholar in Oxford researching neuro-endocrinology (1971–74) and community medicine in Edinburgh (1979–80) before working for the Lothian Health Board in Edinburgh as an obstetric and gynaecological specialist in 1980–81. His report to the National Women's Hospital's Gynaecological Cancer Advisory Group in August 1982 stated that hospital staff were increasingly performing cone/ring biopsies for benign disease, which he believed was 'therapeutic over-kill'. He wrote that the management of CIN was at that time 'quixotic, unnecessarily radical, and expensive in bed occupancy, theatre use, and complication terms'.[98]

Meanwhile, McIndoe continued to work on his own statistical analysis. He presented this at various conferences, and was urged by Richard Mattingly, an American gynaecologist and editor of *Obstetrics and Gynecology*, the journal of the American College of Obstetricians and Gynecologists, to publish in that journal. His paper, which formed the basis of Coney and Bunkle's article in *Metro*, appeared two years after Green had retired. Jock McLean, gynaecologist Ron Jones and statistician Peter Mullins were co-authors of the paper. As discussed in chapter 3, this was a statistical study of the records of 948 CIS patients followed from five to 28 years. It divided the patients into two groups, not according to treatment but according to whether they had a positive or negative smear two years after initial diagnosis. The authors showed that those who had a positive smear two years after initial diagnosis were more likely to get cancer. The paper was intended to show the importance of obliterating abnormal cells at all costs. Among the 817 patients with negative smears, twelve (1.5%) developed cancer. Among 131 patients with positive smears, 29 (22%) developed cancer. The latter were, in other words, 25 times more likely to develop cancer than those whose smears had returned to normal. Green later questioned the diagnosis of some of these women: 'An in situ or an invasive cancer is not necessarily so because McIndoe, McLean and Jones say it is so – a general statement proven by the numerous papers in the literature which show that what is one man's in situ cancer can be anything from even normality to invasive cancer for another man.'[99]

Deaths were not, however, subject to misdiagnosis. There were four deaths in the Group 1 women and eight deaths in the Group 2, though it is debatable if eight deaths over a period of 21 years is statistically meaningful. Certainly when these figures were converted into percentages they looked more dramatic; when Coney and others did so they arrived at a death rate sixteen times higher for Group 2 than Group 1. McIndoe had succeeded in bringing his disputes with Green into the public arena.

Conclusion

While McIndoe was not a natural debater, this was one debate he felt passionate about. His daughter called McIndoe's dispute with Green 'the battle of his life'.[100] McIndoe did not to live to see the cudgels taken up so successfully by Coney and Bunkle; he died a year before the Cartwright Inquiry. Another gynaecologist at National Women's Hospital, Ian Ronayne, later commented that the whole affair had contributed to McIndoe's early death in 1986, and to McLean's death in 1993; both were 68 years old when they died.[101]

Green's views were undoubtedly controversial and he enjoyed debating. Dr J. Grimley Evans, later professor of clinical geratology at the University of Oxford, recalled being in New Zealand in the 1970s and joining a medical debating group of which Green was a member.[102] Peter Herdson, who joined the staff of the University of Auckland School of Medicine as foundation professor of pathology in 1969, later commented that some of Green's views and opinions were 'contentious' and that he often disagreed with him.[103] Green was engaged in a debate and he knew it. In 1971 he accepted an invitation to participate in a debate at the British University of Birmingham with Professor Hugh McLaren, an avid supporter of cervical screening.[104] As Sir Richard Doll noted in 1973, challenging accepted ideas was one of a scientist's most useful contributions.[105] Green wrote to Warren that same year:

> In my opinion, and without wishing to appear derogatory, both Dr McIndoe's and Dr McLean's comments demonstrate what Sir Richard Doll (2.11.73) at the MRC Symposium described as 'the evil power of words to constrain thought', whereby the labelling of a case with a name like 'carcinoma in situ' (possibly he might have included "?invasive cancer nearby") is synonymous in our minds with 'cancer' and all the implied necessity of radical treatment.[106]

Even in the United States, the home of radical medical intervention, Green was not totally disregarded. During a visit to America in 1975, Green's colleague at National Women's Hospital, Sir William Liley, wrote him a letter, conveying greetings from his hosts: 'Lots of your old friends here enquiring after you kindly.' He went on to name them and added:

. . . but the comment I liked best came from Harry C McDuff Jnr (Chief of Gynecology; Director Gynecology Tumour Clinic, Rhode Island Hospital Providence Rhode Is; Clinical Professor Ob–Gyn, Tuft's Medical School Boston, Mass.) I'm not sure where – or if – you have met him but I fancied his remarks so much that I promptly wrote them down. 'I think Green's work on the natural history of Ca in situ is just as important as your [i.e. Liley's] work on haemolytic disease or Liggins's work on fetal endocrinology. He has saved a lot of young women from mutilating surgery.' I thought this well discerned, the more so as your dogged long term data collection lacks the instant appeal of some of the other exploits about the place – and prophets are not without honour etc. A framed and signed copy might be appropriately presented to Bill Mac, Jock McLean and Cancer Soc. With kindest regards, Bill [Liley].[107]

Green's love of debating continued beyond his retirement in 1982. Three years later the executive of the New Zealand Colposcopy Society wrote in the *New Zealand Medical Journal* that 'those who questioned the value of cytology screening and the invasive potentiality of intraepithelial neoplasia have now been clearly shown to be wrong', citing Green as a solitary dissenter. The latter responded that they were 'perhaps being provocative and it would be churlish of me not to rise to the bait'.[108] This debate will be addressed in more detail in the next chapter.

When Green retired, Sir Graham Liggins wrote about his 'uncompromising dislike of humbug and authoritarianism. Whether you liked him or not you certainly respected him as a person to whom a principle is something one died for if need be, and on whom the underdog could always count on for support whether deserved or not'.[109] Green hoped to transfer this questioning approach to his students. He told the Inquiry that he had many discussions with students, and encouraged students, postgraduates and GPs to seek the opinion of others: 'one of my favourite expressions at the end of the lecture is to say, "You don't have to believe what I say, go away and read about it and ask Mr Grieve or somebody else what he thinks about it."'[110]

Green was widely read; one nurse later recalled that his office was strewn with international journals and other papers.[111] He was markedly influenced by the wider intellectual movement that challenged modern medical intervention. This extended to caesarean sections as well – as colleague Dr Tony Baird recalled that in order to justify a caesarean to Green, 'You had to make a case'.[112] Colleagues later described him as 'a bit gruff' but 'straight and frank'.[113] To McLean and McIndoe, however, he came across as arrogant, even bigoted and difficult to work with. When he questioned pathological findings or debated population-based screening, they took it as a personal affront. Personality clashes were crucial in the disputes that arose at National Women's Hospital, but different interpretations of science were also important. When McIndoe claimed that Green had 'all sorts of strange ideas',[114] he did not seem to appreciate that they were based on a broader intellectual movement and wide reading.

6.

Population-based Cervical Screening

Diagnosis of CIS and its appropriate management were not the only medical disagreements between Green and McIndoe. Green's views on population-based cervical screening also apparently upset McIndoe. Mary Whaley, McIndoe's daughter, revealed that following Green's public utterances about screening, her father would be 'introspective, sometimes agitated, and alternatively depressed for a day or so'.[1] Green's repeated questioning of the value of population screening also caused strife with the community health professionals who were gearing up for a national programme, led by David Skegg, professor of epidemiology at Otago University. In 1985 Skegg, together with epidemiologist Dr Charlotte Paul and Richard Seddon, professor of obstetrics and gynaecology at Otago University, published recommendations for population-based cervical screening, which became known as the Skegg Report.[2] The following year Skegg declared that 'Cervical screening is one of the very few effective tools that doctors have for preventing cancer.'[3]

It was Skegg who first coined the phrase 'the unfortunate experiment at National Women's Hospital', which was used for the title of Coney and Bunkle's *Metro* article and which became the accepted public description of Green's management of CIS at the hospital.[4] Skegg used this term in a letter to the *New Zealand Medical Journal* in 1986, in response to a letter by Green, who in turn was reacting to a letter from the New Zealand Colposcopy Society. Green had taken issue with the Colposcopy Society's statement that 'those who questioned the value of cytology screening and the invasive potentiality of intraepithelial neoplasia have now been clearly shown to be wrong'. He referred to a 1979 article in *Science,* and also to a 1978 editorial in *Time* magazine describing 'A flap about Pap' which 'was threatening to undermine the cytology empire [the colposcopy empire too, Green added] so that I am in good company'.[5] Skegg responded to Green's letter, pointing out that these articles were questioning *yearly* Pap smears, which was not something his group advocated. Yet, while the articles questioned yearly smears, they also outlined the controversies about the effectiveness of Pap smears and questioned their contribution to declining mortality from cervical cancer. Further, Skegg wrote that in order to show that he was in good company, Green cited 'new articles from Time and Science seven years ago'. In fact, Green did not claim they were 'new' articles, and accurately dated them in the text.[6] There was clearly no love lost between Skegg and Green. When a conference was held by the Health Department and the Cancer Society in 1985 to implement the recommendations of the 'Skegg Report', Green and his colleagues at National Women's Hospital (apart from medical superintendent Gabrielle Collison) were not invited to attend.[7]

The Cartwright Report stated that, despite a worldwide consensus on its effectiveness, the importance of cervical screening 'has been consistently undermined by the confused statements from Dr Green and some of his colleagues at National Women's Hospital . . . . Even in 1987 [they] believed that they were right and the rest of the world was wrong in its assessment of the benefits of a mass screening programme.'[8] Cartwright also asserted that teaching at National Women's Hospital had hindered the implementation of a nationwide screening programme by devaluing the importance of the smear test to students. In her book Coney quoted a conversation she had with American gynaecologist Dr Richard Reid of Detroit who was sponsored by the Cancer Society of New Zealand to attend a symposium at National Women's in 1986 on cervical cancer and its precursors.[9] He told her, 'nowhere else in the world were the benefits of screening doubted'.[10] Dr Ralph Richart from New York (whom Coney contacted on Reid's suggestion) asserted that the evidence that screening prevented cervical cancer deaths was irrefutable: 'While most of the world is trying to reach every segment of their population for cytological screening, there is continuing debate in New Zealand as to whether screening is effective – maybe all the nations are out of step with New Zealand, but the evidence is dead against it.'[11] Coney told *Broadsheet*, 'People see it as a legitimate medical controversy . . . . But anywhere else in the world it is not a controversy.'[12] This chapter will consider the claim that doctors at National Women's were pitted against the rest of the world in their views on population-based cervical screening.

Population-based Cervical Screening – The Debates

The opening sentence of Posner and Vessey's 1988 study on the prevention of cervical cancer in Britain reads: 'The prevention of cervical cancer through the implementation of efficient screening programmes has been the subject of much debate and media coverage.'[13] A 1988 history of the British Imperial Cancer Research Fund concluded that, 'Despite its long history, the organization of a screening programme for cervical cancer remains controversial at the present time. Doubts are frequently expressed about the effectiveness of the existing screening programme in Britain, reinforced by the reported increases in mortality from the disease in younger women.'[14] As the Inquiry itself was sitting, an article entitled, 'Twenty Years' Screening for Cancer of the Uterine Cervix in Great Britain, 1964–84: Further Evidence of its Ineffectiveness' appeared in the *Journal of*

Epidemiology and Community Health. The article concluded, 'The results support the belief that the screening programme has been largely unsuccessful.'[15]

Green made it clear that he was not opposed to the smear test, which he used regularly as a diagnostic tool, 'though my opponents decry otherwise'.[16] While he supported the use of the Pap smear for diagnostic purposes, Green wrote in 1972, 'The wholesale and very expensive screening of asymptomatic populations with the avowed aim of abolishing death from cervical cancer is a very different concept, the validity of which has rightly come under serious question from Auckland and many other centres.'[17] From the mid–1960s, Green had utilised epidemiological evidence to assess cervical screening. He pointed out in 1967 that, 'Although cytology has been used in New Zealand since 1955 the above table demonstrates no noticeable improvement in incidence or mortality rates.'[18] He was also sceptical of the extreme claims made by some proponents of screening. He drew attention in 1974 to a recent claim by Dr Sidney Farber, an American pediatric pathologist, that, 'the Papanicolaou test has already saved thousands of lives and if applied universally could wipe out all deaths from cancer of the cervix'. Green noted that despite the prevalence of such optimistic views, they were increasingly being questioned.[19] In 1979 he argued from a statistical analysis of mortality data and screening programmes in New Zealand that, 'It can be concluded that not only has cytology screening, of an intensity comparable to that in Canada, failed to accelerate an already-present decline in cervical cancer mortality in women 35 years and older, but it has also failed to prevent a significant mortality rise in young women.'[20]

In the United States, annual Pap smears were recommended by the American Cancer Society and had been widely adopted since the 1950s. As sociologist Pamela Hyde put it, American women were 'encouraged to view their bodies as harbouring a potentially deadly disease and to acquire a new health habit as rational and responsible individuals'.[21] By 1974 more than 56 million women over the age of 17 had undergone Pap smears. Its popularity was believed by one observer to stem from the fact that it was heavily promoted by the American Cancer Society and that much of the cost was borne by the individual women themselves.[22]

In 1967 Professor George Knox declared that in Britain there existed 'informed skeptics' of screening programmes, and as Dr Angela Raffle and Sir Muir Gray noted much later, in the late 1960s there had been 'growing concern amongst rigorously minded academics about the sweeping claims for screening that had been made by apparently authoritative bodies, based on no particular evidence'.[23] One British questioner was Dr Leonard Franks of the Imperial Cancer Research Fund, another was Professor Archie Cochrane and a third was Sir Richard Doll. Analysing the situation in Canada in 1968 and again in 1973, Doll found no evidence that the screening programme in British Columbia had reduced cervical cancer death rates.[24] Epidemiological evidence was also used to query the American situation. Inspecting the data of the American Cancer Society in 1972, one doctor found that the death rate from cervical cancer had decreased at a nearly uniform rate from 1930 to the present. He commented that it was not possible 'to see any effect of the mass cytology screening program'. He added, 'Interestingly, during this time interval, the death rate of uterine cancer has nearly paralleled the decline of the stomach cancer death rate for which no one takes any particular credit.'[25] A 1973 editorial in the *British Medical Journal* concluded that, 'even after 20 years of effort there is a lack of convincing evidence that they have reduced mortality from cancer of the cervix anywhere in the world, though mortality has been declining even where there has been little or no screening'. Noting that those most likely to be screened were middle-class women with the lowest risk of contracting cervical cancer, the author suggested that it was conceivable that 'the low rates of clinical disease reported in screened women might not be very different in the absence of screening'.[26] In 1972 a pathologist from Australia's Royal Newcastle Hospital described the smear test as a 'fizzer' because it had not contributed to the falling incidence of cervical cancer. He pointed out that statistics for British Columbia revealed that the mortality rate for cervical cancer had been declining at about the same rate both before and after the introduction of cytology screening.[27]

Professor Jeffcoate of Liverpool questioned the effects of mass screening on rates of cancer of the cervix in the 1975 edition (and 1980 reprint) of his textbook, and concluded that, 'Despite having applied it to great advantage in the care of every patient seen during the last 20 years, it seems to me that the results and value of cervical cytology as at present applied to the mass screening of women still deserve critical study and appraisal.'[28] Again he distinguished between the smear as a diagnostic tool and population-based screening. Sir Richard Doll did not publish any further studies on cervical screening after 1973, but included it in a list of possibly useful preventive measures in 1983.[29] Others, however, continued to question it. An editorial in *The Lancet* in 1985 pointed out that since cytological screening had been introduced in Britain on a large scale around 1964, mortality had declined at 1 per cent a year, which was the rate at which it had been falling for several decades previously. The author called the 40,000 smears and 200 excision biopsies performed for every life saved 'a grievously poor cost/benefit ratio' given the 'unnecessary harm inflicted'.[30] In the letters column, George V. Mann, professor of medicine and biochemistry at Vanderbilt University in the United States, agreed that, 'The slow and steady decline of deaths attributed to cancer of the cervix since 1930 does not provide evidence that cervical smears or any other "programme" has made a difference to cancer mortality.'[31]

One strident critic of population-based cervical screening was Dr Petr Skrabanek from the Department of Community Health, Trinity College, Dublin. Born in the Czech Republic in 1940, Skrabanek had joined the department in 1984, where he was promoted to associate professor and fellow of the college, and gained his fellowship of the Royal College of Physicians of Ireland. At his death in 1994, the *Times* described him as –

> ... an astute critic of medical humbug. From his base at Trinity College, Dublin a stream of scientific papers and articles exposed the claims of public health doctors, epidemiologists, dietary evangelists and others that many diseases were preventable. This was not a popular message and he evoked strong antipathy in certain circles which was more than offset by his stream of admirers around the world. He believed that wrong ideas flourished in medicine because it is an authoritarian institution that has to present a façade of systematic knowledge. Doctors find it difficult to admit their ignorance or that there may not be a solution medical.[32]

A colleague described Skrabanek's 'formidable intellect' and how he had a wide circle of friends all over the world, partly as a result of his habit of writing to those whose work he admired.[33] David Skegg, however, dismissed Skrabanek as a 'unique individual who had been active in opposing all forms of screening, and indeed, most forms of preventative medicine . . . . He certainly does not have any credibility as an epidemiologist.' Coney cited this assessment in *The Unfortunate Experiment* (1988).[34]

In 1987 Skrabanek critiqued a study of screening progammes in Scandinavia which purported to show that 'organised screening programmes have had a major impact on the reduction in mortality from cervical cancer.'[35] The study reported that Iceland, with a nationwide screening programme and the widest target age range, showed a fall in mortality from 1965 to 1982 of 80 per cent. Finland and Sweden also had nationwide programmes, with a fall in mortality of 50 per cent and 34 per cent respectively. In Denmark, where about 40 per cent of the population was covered by organised programmes, mortality fell by 25 per cent, but in Norway, with only 5 per cent of the population covered by organised screening, the mortality fell by only 10 per cent.[36]

Skrabanek responded that the authors did not 'attempt to disentangle the effects of a spontaneous decline in incidence and mortality'. Norway, he pointed out, had always been regarded as 'the black sheep among the Nordic countries' in regard to public health, though the published graphs showed death rates from cervical cancer falling on a similar slope to that of Sweden from 1955. He also drew attention to a local programme in Norway from 1960 to 1974, in which no reduction in the incidence of cancer was observed. For Iceland, 'the annual mortality in the most screened group (ages 25–59) before screening was 11.7/105 (1955–59); a decade after screening it was 12.2/105 (1970–74).' He added that estimates of the benefits of screening ignored the fact that in many countries such as Japan, France and Italy, which had no nationwide screening programmes, mortality from cervical cancer was much lower than in Sweden and had been falling for the past 20 years at rates similar to those seen in Sweden.[37]

Dr James Le Fanu of London also addressed cervical screening pro-grammes.[38] The *New Zealand Herald* reported his views in 1988. He described cervical screening as a 'blunderbuss approach' to the prevention of cervical cancer. Not only did the programmes have little discernible impact on the disease, but in his opinion they also played havoc with women's sex lives. He pointed out that it was widely appreciated that most abnormalities detected in this way did not progress and would get better without treatment. But, as it was not possible to tell which would progress and which would not, all abnormalities had to be treated. He estimated that 2 million smears would be taken in Britain that year from women under 40. About 5 per cent would be considered abnormal and so 100,000 women would require treatment. But the number of cases of cervical cancer in this age group was 100 a year, so possibly 1000 women were being 'treated', with attendant problems, to 'prevent' one case of the disease. Not all cervical cancers could be prevented for the simple reason that those most likely to get the disease were working-class women over 40, who were least likely to go to their doctors for a cervical smear. He added that increasing the number of cervical smears would reveal more abnormalities and subject even more women to unnecessary treatment.[39] He cited Skrabanek's view that it was 'a mixture of Utopianism and totalitarianism. Utopian because it aspires to achieve what is unobtainable, totalitarian because it encourages women to participate in the screening programme without informing them of its drawbacks as well as its merits.' He urged a greater openness on the part of those running the programmes.[40]

Around that time Professor Alwyn Smith of the Department of Epidemiology and Social Oncology, University of Manchester, England, wrote about the absurdity of conducting a screening programme 'in such a way that nearly 40 women are referred for an expensive and possibly hazardous procedure for every one who is at risk of developing serious disease'. Smith advised, 'We need common sense and a more discriminatory assessment of the results of screening tests.'[41]

Within New Zealand, Dr Murray Jamieson who had joined National Women's Hospital in 1981, after training in the UK at Oxford and Edinburgh, questioned the value of population screening.[42] Commenting on the proceedings of the 1985 screening conference, Jamieson described as 'extraordinary' the remarks by pathologist Dr Stewart Alexander that 'false positives don't really matter very much' (a 'false positive' was a smear which wrongly showed abnormal cells). On the contrary, Jamieson said, 'They matter a very great deal, I believe, to the thousands of women whose smears are falsely positive.'[43] Jamieson's view was supported by the international literature. An article in *Cancer* (1981) declared that, in assessing the overall value of screening, one must consider the psychological costs of the false positive test, and the morbidity associated with diagnostic tests performed on persons with false positive tests.[44] Jeffcoate also believed that a 'high false positive rate' was far more dangerous than a high false negative rate because it meant that many women were exposed to unnecessary surgery and worry, the effects of which could be lifelong. He explained:

> Every year I have referred to me for treatment several women who, on the strength of an alleged positive smear, have been told they have cancer and that they require hysterectomy; yet on repeating the cytology no abnormal cells can be found in one smear after another .. . . Once a woman has been told she has a positive smear or that she has cancer it is almost impossible to remove all fear and doubt from her mind.[45]

Professor Eugene D. Robin from Stanford University School of Medicine wrote in 1985 that the failure to accumulate data on the number and fate of those with false positive smears was critical. He maintained that, 'The patient with a false positive cervical smear is the principal (and usually hidden) victim. All such patients suffer emotional trauma.'[46]

False negatives (i.e. showing up as normal when that was not in fact the case) were another problem. Green commented that false negatives sometimes resulted in delayed treatment.[47] National Women's consultant Liam Wright told Professor Skegg in 1985 that he dealt with 50 to 70 cases of cervical cancer a year and had 'frequently' encountered women with symptoms suggestive of significant disease who had a smear reported as either normal or showing minor abnormality. Because of this the subsequent investigation and perhaps diagnosis of cancer had been delayed.[48] Jamieson stressed the importance of clinical examinations and checking to see if women had any abnormal bleeding or vaginal discharge, in addition to the smear. He too considered false negatives a problem as they gave a false sense of security.[49]

Dr James McCormick from the Department of Community Health, Trinity College, Dublin, and president of the Scientific Committee of the Royal College of General Practitioners in the UK, was another who questioned population-based cervical screening. He claimed in 1989 that cervical screening did not satisfy any of the criteria which would justify its use. He explained that there was no proof that cervical cancer was reduced by screening, but that there was evidence of disease caused by cervical smears. He pointed to the large numbers of false positive smears in healthy women, which caused distress and anxiety that might never be fully allayed. Further investigations, involving a biopsy, were by no means risk-free, and like others he listed the possible side effects.[50]

The Sexual Revolution

A negative feature of screening identified by some doctors and feminists alike was that women with positive smears ran the risk of being regarded by others as promiscuous, since it was now widely held that cervical cancer was sexually transmitted. James McCormick argued that, 'Their character, as well as their cervix is smeared.'[51]

Professor Skegg told the *New Zealand Woman's Weekly* in 1984 that one of the reasons behind a sense of urgency in setting up a cervical screening programme, in New Zealand as elsewhere, was the belief that cervical cancer rates were increasing as a result of the so-called sexual revolution. Earlier that year he had warned a conference of health professionals in Dunedin, 'I'm afraid there will be an epidemic of cancer of the cervix in New Zealand unless we institute appropriate control measures.'[52] It was around this time that researchers identified an association between the human papillomavirus and cervical cancer, proving what had been suspected for some time, namely that cervical cancer was almost always a sexually transmitted disease.

At the 1985 screening programme planning conference, Skegg noted that cervical cancer was increasing among young women in New Zealand, Australia and Britain, and that was entirely predictable given 'changes in sexual behaviour, possibly the effects of smoking, and so on'.[53] Dr Graeme Duncan of Wellington also maintained that New Zealand was facing an epidemic of cervical cancer unless an effective screening programme were introduced. 'The signs are ominous', he said. 'The epidemic that was predicted all around the world certainly has struck Wellington Hospital this year . . . . So I'm sorry to introduce this slight element of urgency.'[54] Skegg did not agree there was a 'huge epidemic' but stated that there was a 'disturbing trend' in younger women while the disease was declining in older women.[55] In contrast to New Zealand, the Americans and the Scandinavians, who led the sexual revolution, had avoided an increase in invasive cancer, which he attributed to screening. He pointed out that the Americans did not have screening programmes as such: 'they've taken a completely different approach, it's entirely private practice oriented. It's presumably extremely expensive but if you encourage every one to be smeared annually it seems to work.'[56]

In 1985 Murray Jamieson collaborated with Petr Skrabanek on an article entitled, 'Eaten by Worms: A Comment on Cervical Screening'. In it they questioned the victim-blaming by modern epidemiologists: 'some southern authorities [by which they presumably meant Professor Skegg] have promised an epidemic of cervical cancer' as 'our young people are "becoming more permissive and promiscuous"', and all needed to be screened: 'The wages of sex is a positive smear.' They believed the threat of a crisis was overdrawn. They asked, 'Should one million New Zealand women over the age of 19 live in daily fear, though only 0.009% of them will enter the final statistics?'[57] While Skegg questioned their mathematics (it was 0.009 per cent per annum and not over a lifetime as implied), there were many others who believed that the costs of screening to the individual woman far outweighed the benefits. British feminist sociologist Peggy Foster, for example, claimed in 1995 that there was a considerable body of evidence documenting the harm suffered by many women who had participated in cancer screening programmes. These included psychological as well as physical effects.[58]

Jamieson and Skrabanek's statement that 'the wages of sex is a positive smear' was an attack on the way the new understanding of cervical cancer as a sexually transmitted disease was being used to convey a sense of urgency in initiating a screening programme. They objected to the inference that women got what they deserved. Others felt equally uncomfortable about this. Some feminist writers believed that highlighting the link between the two made women feel tainted and guilty. During the Inquiry, Te Ohu Whakatupu of the Ministry of Women's Affairs stated, 'An attitude that many doctors are reported to hold which affects their attitude to women in relationship to cervical cancer is that a major causative factor is promiscuity. This belief then reinforces an attitude that women get what they deserve.'[59]

Green told Coney and Bunkle in his 1986 interview with them that this was 'a slander on women'; in his opinion, promiscuity was 'no worse or better than in Victorian times'.[60] In 1979 he had written, 'Since it is known that cervical cancer is associated with early age of onset and frequency of sexual intercourse the most obvious explanation of the increased mortality from cervical cancer is the alleged increase in promiscuity in the last 15 years or so among young women.' He added, 'Quite apart from the disservice this explanation does to many patients, for lack of hard data on past and present rates of promiscuity (but no lack of unsupported assertions), it is almost as difficult to establish as it is to refute.' He maintained that an argument against the 'promiscuity theory' was that the pattern of cervical cancer mortality in young New Zealand women had changed suddenly whereas sexual mores do not. He came up with an alternative explanation: 'In any search for an associated or causal factor introduced around 1960, therefore, one must consider steroidal agents': namely the contraceptive pill which had reached New Zealand that year. He noted that the continued use of estrogens was considered a potential cause of endometrial cancer, pointing out that the question of the effects of steroids on cervical epithelium had been debated at length in recent years and that it seemed clear the matter needed much more study.[61]

British feminist and long-standing patients' advocate Jean Robinson criticised cervical screening programmes, which she claimed were based on a view of women as passive patients. She wrote in 1985, 'More young women are dying of cervix cancer now – in spite of screening – than died 35 years ago before it existed. Many more are losing their uterus before they have had a child. Women's groups responded by saying that screening should be more frequent – not realising that it is the [contraceptive] pill that is blowing cancer up fast.'[62] In her view, medical technology (screening) allowed people to overlook the ill-effects of another medical technology (the contraceptive pill).

Green and Robinson were not alone in suggesting this connection between the pill and cervical cancer. In her history of the contraceptive pill, Lara Marks wrote that by the early 1960s a number of medical practitioners were 'increasingly uneasy about the potential carcinogenic effects of the pill'.[63] Research into the incidence of cervical carcinoma among 34,000 women attending Planned Parenthood Federation clinics in New York City between 1965 and 1969 showed the incidence of cervical abnormalities to be twice as high among women using the pill than among those using the diaphragm.[64] Interestingly, in light of the above discussion concerning different interpretations of the smear, the results of this study were queried because of 'discrepancies in readings of the Pap smears taken and the classification of cancer'. Controversy erupted when the *Journal of the American Medical Association* refused to publish the report and when the *British Medical Journal* published it instead. Reporters in the lay press accused the American Association of succumbing to pressure from pharmaceutical companies. While American public health officials were sufficiently worried by the findings to admit that 'urgent research was needed to settle the question', by the end of the 1960s investigators 'had come no closer to resolving the riddle of the pill and cervical cancer'.[65]

Green was, as ever, suspicious of commercial interests. Even if there was no evidence directly implicating the pill in cervical cancer, he thought that its use might disguise early symptoms. He wrote to the *British Medical Journal* in 1965 about a case of delayed diagnosis of cervical cancer of a woman on the pill – the intermittent bleeding was considered by her doctor to be a side effect of the pill and unimportant. He wrote that once he envisaged the possibility of such cases arising, he suggested to a well-known drug firm that packages of contraceptive tablets should carry a suggestion to patients to have a regular smear test and to report any unusual bleeding, but got nowhere with this request. He believed it unfortunate that when a patient reported inter-menstrual bleeding, it could be casually dismissed as a side effect of the pill.[66]

In 1983 Professor Martin Vessey of Oxford University conducted a ten-year follow-up study of 10,000 women – 6838 taking the pill and 3154 using intra-uterine devices (IUDs) – who had attended family planning clinics in Britain. His findings suggested that the pill might have contributed to the rise in cervical cancer in England and Wales over the previous decade.[67] However, most of the debate focused on the relationship between the pill and breast cancer. Given the New Zealand feminists' interest in the relationship between IUDs and cancer (see chapter 7), it is perhaps surprising that they did not recognise Green as being on their side in this instance.

Screening: Women's or Commercial Interests?

James McCormick believed that some feminists advocated screening as the solution to cervical cancer without recognising all the attendant problems and without acknowledgement of the victim-blaming. He wrote, 'Because this is a terrible disease something must be done; to suggest that [screening] is worship of a false god is to risk being labelled as heretic, chauvinist, or more politely, nihilist.'[68] Skegg described McCormick's views as 'unbalanced', 'emotive' and 'entirely at variance with those of medical experts around the world'.[69] By contrast, British public health officer Dr Angela Raffle described McCormick as 'very kind, honest and principled', and stated that his views were 'not at variance, it [was] just that he was brave enough and informed enough to voice them'.[70] There was nothing apparently 'emotive' in his writings.

Raffle saw screening as a false feminist icon: 'it is very hard to explain that cervical cancer was a rare and diminishing cause of death before we even began screening'.[71] During the Cartwright Inquiry, the Ministry of Women's Affairs declared that 'Cervical screening is a major women's health issue' and maintained that the level of concern among women in the community about the lack of an organised screening programme was 'very high'.[72] One feminist had told Ann Hercus, Minister of Women's Affairs, in 1985 that she believed, 'if Cervical Cancer was a male illness, this Service would be already operational'.[73]

Veteran feminist Germaine Greer disagreed that screening promoted women's interests. She believed it was many times more likely to destroy a woman's peace of mind than to save her life, and she questioned its feminist status. She complained that, 'If the American government now spends $4.5 billion on Pap smears every year, it is not because they are being pushed around by a bunch of noisy feminists but because of the power and priorities of the medical establishment'.[74] Others too mentioned professional and commercial interests.[75] Ralph Richart himself owned a very large private laboratory, which reported an average workload of 35,000 smears per screener in 1986, three times the caseload recommended by the American Cytology Society.[76] Green wrote on screening for cervical cancer in 1985, 'To apply more and increasingly expensive doses . . . of the same medicine will bring lots of business to cytologists, colposcopists, clinicians and clerical workers, but on past New Zealand experience, the million women 20 years and older being put in fear of cancer . . . are most unlikely to benefit to the same degree'.[77]

During his 1971 sabbatical, Green spent three months in the cyto-pathology laboratory of the Toronto General Hospital, Canada, where the head pathologist was Dr D. W. Thompson. Green was surprised that only about 8000 cervical smears were processed each year at the hospital, despite its size. He later told the Inquiry that the bulk of Toronto's smear processing was carried out in a private laboratory run by four private pathologists, including Thompson. This laboratory processed 200,000 smears a year, and was paid $5.25 (Canadian) per slide by the Ontario government. He commented, 'The exercise of population screening was clearly very profitable for some pathologists'.[78]

Following similar overseas analyses, Green estimated in 1986 that in New Zealand, extrapolating from 1980 figures, 350,000 smears would reveal 14,000 abnormal smears requiring further investigation, which would lead to the discovery of about 900 cases of 'pre-cancer', of which some 120 women would ultimately die of their cancer. The 13,000 or more who did not have cancer would neverthe-less 'provide expensive diagnostic work for numerous doctors, nurses, and clerical staff, together with much psychological and physical morbidity for themselves'. He believed that 'although mass screening may benefit society (and doctors) by reducing the annual 120 deaths, the chances of benefit accruing to the individual average woman presenting herself for a smear test are minute – but not so for her chances of diagnostic morbidity', as a result of viewing it as a 'cancer test'.[79] Green repeated these statistics to Coney and Bunkle in his interview with them, and spoke of the 13,000 'being put through the diagnostic hoop unnecessarily. They're told they've got cancer cells and they think they've got cancer. You ask a woman who's been through it.'[80]

The Fertility Action submission to the Inquiry gave an example of the 'debunk-ing' of cervical screening by the medical superintendent of Wairoa Hospital, who questioned the concept of 'an epidemic' of cervical cancer. He noted that there had been around 90 deaths a year over the past ten years in New Zealand, and compared this to cancer of the large intestine (470), breast cancer (530) and lung cancer (290). He estimated that only 3 per cent of all cancer deaths in women were due to cancer of the cervix. In Wairoa, then, he would expect one death from cer-vical cancer every five years. Screening all sexually active women in Wairoa on an annual basis would require at least three full-time workers and a full-time cytolo-gist. In his view the money required for such a programme could save more lives if it were allotted to other health areas such as screening for bowel cancer, diabe-tes or hypertension, or used in an effort to prevent the 'true epidemic' (smoking) which was killing women at that time.[81] Although Fertility Action reported this as an example of not taking cervical cancer seriously, it could also be interpreted as a serious attempt to weigh up the options in preventive medicine given limited resources.

Thus, there were lively local and international debates about the costs and benefits of cervical screening around the time of the Cervical Cancer Inquiry. Green and his colleagues were not pitted against the rest of the world in ques-tioning the value of such a screening programme as claimed by Cartwright. This debate continued beyond 1988. A 2005 World Health Organization publication on cervical cancer listed the 'hazards' of a screening programme. These effectively summarised the concerns voiced by Green and his colleagues, and show that more than 20 years after Green's retirement they were still regarded as valid. They included:

- Psychological consequences of a positive screening result, with increased anxiety and fear among women;
- Misunderstanding by women and health-care providers of the meaning of a positive screening test, such that a positive result is interpreted as a 'cancer diagnosis';

- Misunderstanding by women and health-care providers of the meaning of a negative test as implying no risk rather than low risk for cervical cancer, which may lead to under-investigation of symptoms;
- False positive screening results leading to unnecessary interventions, with both human and financial cost implications;
- False negative screening results giving false reassurance;
- Overtreatment of preinvasive lesions that left alone would neither progress nor cause any clinically significant disease, particularly as there are still no reliable markers to determine which high-grade cervical cancer precursors will progress to cancer or will remain clinically insignificant;
- Complications of treatment such as cervical stenosis, cervical incompetence and infertility, as well as the result of more radical therapies, such as hysterectomy, with a range of potentially negative sequelae related to the surgical intervention;
- Opportunity costs to the health-care system of introducing a screening programme;
- Impact of incidental findings during screening.[82]

In 2007 Raffle and Gray commented that, 'In truth, most pre-cancer is transient, and screening therefore leads to a major problem of overdiagnosis.' Moreover, they pointed out that 'the high rates of positive tests, most of which are transient minor cell changes, cause major anxiety unless you can change the universal belief that every abnormality must mean cancer.'[83] These modern authors shared similar views to Green.

National Women's Hospital and Medical Students

In his submission to the Inquiry, Dr Alan Gray, medical director of the Cancer Society and consultant radiation oncologist to the Wellington Hospital Board, claimed that, 'The Cancer Society's view is that, because of Professor Green's opposition, it has not been possible to provide a comprehensive national screening programme in New Zealand.' He believed that Green had influenced doctors around the country and even internationally, shaping their negative attitudes to screening.[84] The Ministry of Women's Affairs' submission stated that, 'we consider it vital that previous trainees at the Hospital be contacted and re-trained in the proper detection and treatment of cervical cancer.'[85]

As noted earlier Green made it clear that he was not opposed to the smear test, which he used regularly as a diagnostic tool.[86] Professor Richard Seddon told the Inquiry that he had been responsible for organising the undergraduate teaching programme at National Women's Hospital from 1970 to 1975. He explained that it had been 'standard teaching' to promote the taking of smears for diagnostic purposes when gynaecological symptoms were present, but also whenever an appropriate opportunity arose. He maintained that students were taught that CIS must always be taken seriously, since a proportion of cases became invasive and that the affected area should be totally destroyed or excised either by the preferred conservative treatment of cone biopsy or by hysterectomy.[87]

Colin Mantell who was in charge of undergraduate teaching at National Women's Hospital from 1977 similarly insisted that medical students were taught the guidelines put together by Professor Skegg and published in the *New Zealand Medical Journal* in 1985 (and prior to that date the 'Walton Guidelines' from Canada), and that the curriculum stressed the value of the smear test.[88] Cartwright did not give weight to this evidence and her report emphasised the need to combat the negative influence of National Women's Hospital. She cited Professor Skegg's comment that it would be 'impossible to overestimate the effect [their anti-screening views] had had on medical training'.[89] Two doctors trained at the hospital who were interviewed for the Inquiry were both very pro-screening. Cartwright disregarded the submission of the University of Auckland that the training of these two students had clearly not prejudiced them against screening.[90]

Dr Bruce Phillips sent a submission to the Inquiry about his training at National Women's Hospital. He had graduated from the Auckland University School of Medicine in 1976 and did his undergraduate obstetrics training at National Women's. As a general practitioner, he passed the National Women's Diploma in Obstetrics examination in 1978, following a lecture course there. In his submission he wrote:

> I was very disturbed when I read Ms Cooney's [sic] comments in the New Zealand Herald that it is likely that New Zealand doctors trained in Auckland may pay little attention to abnormal cervical smear results. I wish to state that for me this is completely untrue and as my experience must be similar to that of many other Auckland graduates it is surely quite untrue for them also. These sweeping statements (which may of course be incorrectly reported) must needlessly be affecting the security of many New Zealand women and I hope you can take steps to reassure them and at least partly repair the damage done.[91]

Dr David Davidson, specialist obstetrician and gynaecologist to the Hawke's Bay Hospital Board, also forwarded a submission in which he explained that twelve of the GPs who had practised in the Hastings area since 1969 had diplomas in obstetrics from National Women's. He believed that these practitioners had set an example within their practices to their colleagues with regard to screening programmes for cervical cancer, the subsequent management of these patients and referral to the specialist services. In his view, their exposure to different viewpoints, which was what one would expect at a postgraduate training centre, had stimulated their interest and resulted in well-trained practitioners. He himself had an appointment at National Women's in 1969–70 after completing his membership of the RCOG in Britain. He was convinced that the postgraduate training with regard to cervical neoplasia, screening, management and treatment had been excellent for those training for general practice and specialist positions.[92]

During his cross-examination, Green discussed his approach to teaching. He refuted the suggestion in the *Metro* article that he had controlled teaching on cervical cancer while at the hospital, and his claim is supported by a perusal of the teaching programmes.[93] When it was suggested to him that he had belittled the value of taking smears, he replied, 'I deny that view absolutely. I have always said it is a useful diagnostic aid as well as history, clinical examination, cytology, colposcopy and histology, but by itself, it is not sufficient and any case history by itself is not sufficient. Clinical examination by itself is not sufficient.'[94] Murray Jamieson reinforced this to the Inquiry (he had been a registrar at National Women's from 1976 to 1978), stating:

> Professor Green taught everyone to take cervical smears on every patient they saw, indeed as one of his junior staff, I would be most anxious if I were ever to present a patient to him who did not have documentary evidence in her notes of either the fact that she had a recent normal cervical smear or that one had been taken immediately previously.[95]

Dr Charlotte Paul later commented on Green's widely publicised anti-screening views, adding that, 'Despite this, opportunistic screening [i.e. taking smears when women turned up for any consultation] did become established in many general practices and family planning clinics.'[96] Yet Green had applauded the work done by family planning clinics in promoting screening. This is clear from his comments on the transcript of the 1985 Health Department screening conference. He described the statement by general practitioner representative Dr Ian St George that 'Family Planning Clinics could endanger the success of the (screening) programme' as 'arrogant, selfish and outrageous'. Green claimed that these clinics had emerged out of the failure of GPs (mainly male, he said) and public hospitals dominated by private specialists to provide adequate family planning advice for women. He continued, 'My long experience of referrals from such clinics convinces me that they have done far more in the way of screening women for cervical cancer, especially those in the so-called "high-risk" group, than has the average GP. Dr St George may never have met a GP who would not do a smear, but the clinics at NWH certainly have.'[97]

Opportunistic testing was in fact well established by Green and colleagues at National Women's. In his 1977 recommendations for hospital protocol in relation to CIS, Green advised that, 'Since cytology helps greatly in the detection of cancer in those at risk, National Women's Hospital should continue taking smears of all patients of hospital clinics – including antenatal patients.'[98] Bonham noted that smears were regularly taken at National Women's. Returning from an overseas trip, he reported on a two-day conference he had attended in Britain in 1983, organised by the British Congress, the Royal College and the British Society for Colposcopy and Cervical Pathology: 'Cytology is still not used as much in the UK as in NZ; the GP payments are limited to older women whereas in NZ there is more extensive screening of young women (pregnancy and contraceptive visits).'[99]

In 1988 Dr Ron Jones, pathologist Dr George Hitchcock and others published a paper in the *New Zealand Medical Journal* suggesting that the Auckland hospitals with public antenatal clinics had a rate of cervical screening of just 58 per cent. Noting that he had for many years taught new residents the importance of recording smear results of all clients of the antenatal and gynaecological clinics at National Women's, Middlemore and St Helen's hospitals (the public maternity hospitals in Auckland), Bonham expressed surprise at these results. He asked the National Women's antenatal clinic supervisor to do a spot check of the 500 antenatal bookings in August 1987. The results showed that 98.2 per cent had had smears. Of the nine patients who had not had a smear, one had refused, one was bleeding too much for a smear on that day, and seven were 'found to be missed and already flagged for smear at next visit.'[100]

In her submission to the Inquiry, Sue Neal spoke of the negative influence of National Women's Hospital. She related how her own general practitioner gave her a yearly smear, 'but he always tells me that it's not necessary and that experts have proved that we only need a smear test once every three or so years . . . . If I have to argue and keep asserting my right to yearly smears, where does this leave the quiet woman who goes along for her yearly smear and gets a refusal from the G.P. on the grounds that "experts" at National Women's are saying you only need one every three to five years!'[101] Neal seemed unaware of the fact that Professor Skegg also recommended one smear every three years.[102]

Prior to the Inquiry, the Health Department had funded three pilot schemes for screening Maori and low socio-economic status women, organised by consumer groups (see chapter 10). Yet the Te Ohu Whakatupu submission to the Inquiry still declared that, if it had not been for the influence of Green on a whole generation of doctors, 'every death from cervical cancer could have been prevented with effective screening procedures and follow-up treatment.'[103]

National Women's Hospital had taken smears from all women in its clinics since 1960, and taught medical students the importance of the Pap smear as a diagnostic tool. Population-based screening was a different matter, and its pros and cons prompted lively debates both in New Zealand and internationally. Along with others, Green questioned the economics of population-based screening, and expressed concern that the money would benefit primarily middle-class white women, which to some extent is what happened in the following years (see chapter 10). Yet in the late 1980s, the public increasingly framed the issue of screening in terms of women's rights, and questioning it had become politically unacceptable.

Conclusion

Chapters 5 and 6 have provided a narrative of a profession divided, pathologists and cytologists pitted against clinicians, and clinicians against epidemiologists. Green was at the centre of the disputes. With his outspoken views, he had enemies both within his own hospital and beyond. The Inquiry was dominated by disputes within the hospital, with two factions represented by two different medical protection societies: the Medical Defence Union represented Green and Bonham, and the Medical Protection Society represented the McIndoe estate, McLean and Jones. Cartwright's comments about the arrogance of the National Women's Hospital doctors made it clear that she fully accepted the submissions of Hugh Rennie, counsel for McLean, McIndoe and Jones. Rennie declared, 'The past history of National Women's Hospital is marked by stubborn over-confidence by many doctors, streaks of arrogance that New Zealand could be right and the rest of the world wrong, and a notable absence of the peer group discussion, exchanges and dissent which are the proper mark of an academic research team.'[104]

Rennie's allegation about the absence of academic and peer group discussion at National Women's is not borne out by a review of the Hospital Medical Committee minutes. At the 1966 meeting, there had been a lengthy discussion during which Green answered many questions. The hospital established a Tumour Panel explicitly to discuss controversial cases openly, and a special sub-committee to review the situation in 1975. Dr Bernie Kyle, a senior consultant at the hospital, refused to sit on the committee because he thought it should have comprised medical professionals from outside the hospital.[105] However, when the Health Department had urged the establishment of ethics committees in 1972, the only guideline was that committees should include 'experienced members of the professional staff'.[106] The 1975 sub-committee met this criterion. Dr Alastair Macfarlane, who chaired the committee, was a highly respected gynaecologist with an international reputation and president of the New Zealand Council of the RCOG from 1971 to 1973.[107] Dr Richard Seddon, another member of the committee, had already been appointed to a chair of obstetrics and gynaecology at Wellington Clinical School. Dr Bruce Faris, the third member, was medical superintendent of Auckland's St Helen's Hospital from 1963 to 1983 and a member of the Auckland Hospital Board. Cartwright called the committee's report a whitewash, but there is no evidence that a committee of 'outsiders' would have produced a different result. Following the committee's report, there were further discussions within the Hospital Medical Committee, which now included all senior clinicians at the hospital, as a result of which Green, McIndoe and McLean were each asked to write a report. Following an appraisal of these reports, new protocols were laid down in 1978. Green, along with McIndoe and McLean, was required to be accountable and subject to group discussions and decisions.

A review of the literature also reveals the inaccuracy of the perception that Green was at odds with the rest of the world. As Professor David Cole, dean of the University of Auckland School of Medicine, later pointed out in his unpublished memoirs, there were plenty of people who agreed with Green but they either did not turn up to the Inquiry or else failed to express their convictions effectively.[108] Dr Sadamu Noda from Japan was dismissed by Coney as lightweight and a figure of fun: 'He later wrote to the judge saying he was honoured to have given evidence "in front of such a [sic] attractive and sophisticated person like you. Truly, I would say I have never seen such a very fair and dignify [sic] Judge as you".' Coney reproduced this in her book on the Inquiry.[109] Despite representing one of the world's leading institutions in the research and treatment of cancer, the Gynaecological Cancer Section of the Osaka Cancer Prevention and Detection Center, Noda was given very cursory treatment in the media. Cartwright herself questioned Noda's ethics. In response to his submission, which supported Green and his professional ethics, Cartwright replied among other things, 'You further mentioned in your letter that at National Women's Hospital, you found Professor Green's protocol of CIS management very useful. Could you comment on what you meant by very useful. Was it to yourself or to the patient involved?'[110]

David Collins, who represented Green, spoke of the latter's dedication to his work and his patients: 'He received support not only from his colleagues, but his patients also revered and supported him'. This was shown by the volume of letters he had received since the Inquiry began. Collins declared, 'Dr Green's entire professional life was dedicated toward caring for and treating women with obstetric and gynaecological disorders.'[111] This assessment had very little impact on public perception, however. Instead, the final submissions of Dr Rodney Harrison, who represented Coney, Bunkle and Fertility Action, attracted the most public attention. His address was reported in the local press: 'Women had continued to go to National Women's Hospital for treatment "like lambs to the slaughter" while the medical profession maintained closed ranks and an unbroken silence, the cancer inquiry was told yesterday.' Harrison said that the Hospital Medical Committee had given Green approval in 1966, 'in total disregard of patient wellbeing and in breach of medical ethics'. He further claimed that there had been 'a collective abdication of ethical and professional responsibilities by members of the medical profession outside National Women's, who knew what was going on and who failed to speak out or act to prevent what was at worst a criminal offence and at best a gross breach of the physician's primary duty to the patient.'[112] Harrison called for Professor Bonham to go, claiming that Bonham was 'inextricably linked with past events at the hospital', and that in order to restore confidence at National Women's, a fresh start was needed.[113]

In her submission to the Inquiry on behalf of Fertility Action, Coney said she had been approached by two of the hospital's doctors, Tony Baird and Liam Wright, who assured her there was no evidence that screening reduced the cancer rate or that CIS proceeded to cancer in many cases. But she said she was also approached by other doctors who assured her she was on the right track.[114] The medical profession itself was polarised, and two feminists stepped into the breach, picking up and running with one side, although ironically this was not the side usually favoured by feminists, as will be discussed.

7.

Four Women Take on the Might of the Medical Profession

From the 1950s to the 1980s, disputes around CIS racked the medical profession in New Zealand and elsewhere. The disputes were between clinicians and cytologists, pathologists and public health physicians. They argued about different understandings of the meaning of CIS, its appropriate treatment and the value of population-based screening. As the *British Medical Journal* declared in 1976, these had become the subject of 'fierce controversy'.[1] The debates, at the centre of the 1987 Cartwright Inquiry, were far from new. What changed in 1980s New Zealand was that two feminists picked up on these to advance a core element of the new women's health movement – to attack medical authority and reclaim women's bodies and autonomy over their lives.

In the 1970s the new women's health movement identified National Women's Hospital as a power-house of male patriarchal medicine. With the Postgraduate School of Obstetrics and Gynaecology based at the hospital, it also represented research on women's bodies. Debates over new reproductive medical technologies were pitting these feminists against the predominantly male obstetricians and gynaecologists. As an article in the *New Zealand Listener* put it in 1985, 'Over the last two decades, women's health has become a hot topic and women's bodies have been the focus for some ferocious power struggles, and some of the messiest battles going – abortion, the politics of birth, the status of women's bodies in a teaching hospital, contraceptive crises like the Dalkon shield debacle'.[2] What happened in 1987 can only be understood in terms of the politics of women's health.

By the time of the Inquiry, the women's health movement had moved from being a marginal social movement to commanding considerable political support. The new Labour government which took office in 1984 set up a Ministry of Women's Affairs. The Ministry consulted Sandra Coney as an adviser on health issues. Following Coney and Bunkle's *Metro* article, Health Minister Michael Bassett, in planning the Inquiry, considered it important that it be 'pretty much an all-women inquiry'.[3] He clearly saw it as a gendered issue.

The Women's Health Movement

The women's health movement grew out of the women's liberation campaigns which had reached New Zealand's shores by 1970. In the early 1970s, as Sandra Coney put it, 'Women marched, shouted through megaphones, brandished placards and waved banners emblazoned with the women's liberation symbol – the clenched fist within the biological sign for woman.'[4] Not only did they march, they also organised networks. By the end of 1972, about 20 women's liberation groups had been formed around the country. That year saw the first meeting of the National Organisation of Women, NOW, and in 1973 over 1500 women went to the first United Women's Convention in Auckland. The first issue of the feminist magazine, *Broadsheet*, appeared in 1972. It was founded by Auckland Women's Liberation, one of whose members was 28-year old Sandra Coney, who edited *Broadsheet* for the next fourteen years.

The movement was inspired by events in America and elsewhere, related to the general anti-authoritarian and civil rights movements of the 1960s, which brought about a questioning of the structure of society and powers within society. Feminists did not strive for equality with men but rather sought to break down the whole sexist structure of society. A focus for most groups in the early 1970s was consciousness-raising, a technique borrowed from the Chinese cultural revolution. It aimed to heighten the individual's awareness of her own oppression and engender a feeling of solidarity with other women. Its slogan was 'The personal is political.'[5]

The liberalisation of abortion law was an early platform for the new feminist movement. Abortion was seen as a woman's right to control her own body, and following on from abortion, reproductive health in general became a central focus. A major influence on the women's health movement throughout the Western world was the 1971 publication by the Boston Women's Health Collective in America of a feminist health care book entitled *Our Bodies Ourselves*, which made information on health more accessible.[6] As American historian David Rothman put it, following this publication 'docile obedience was to give way to wary consumerism'.[7] 'Natural childbirth', or a reduction of medical intervention and technology in childbirth, was another goal of the new movement, led by midwives in alliance with feminists.

In 1973 American feminist Lorraine Rothman was invited to the first United Women's Convention in Auckland and to tour the country. Rothman came from the Feminist Women's Health Center in Los Angeles, one of the first female-controlled health and abortion clinics in the United States.[8] She demonstrated self-examination of the cervix and menstrual extraction (a form of self-abortion and period regulation).[9] Two years earlier, Rothman and Carol Downer had toured the United States demonstrating these methods. Downer claimed to have had 'a kind of conversion experience' when she first viewed her own cervix. She called cervical self-examination revolutionary, as it broke two taboos – touching her own genitals and appropriating the tools of the medical professional to reclaim knowledge and control of her own body.[10] In an early study of the women's health movement, Sheryl Ruzek from the University of California, San Francisco, wrote that self-help gynaecology 'more than any other event transformed health and body issues into a separate social movement'. She argued, 'men have had more intimate contact with, and far greater access to, the vagina than women have ever had. The male organ, on the other hand, has always been exposed, and seeing it reinforced its reality. Women now have this possibility.'[11] New Zealand feminist Christine Dann wrote in 1985 that 'the act of cervical self-examination had a symbolic value which far exceeded its actual practical use'.[12] The cervix had symbolic meaning to the new feminist health movement.

The two authors of the 1987 *Metro* article, Sandra Coney and Phillida Bunkle, met at the first United Women's Convention in 1973. Bunkle was a sociologist who taught women's studies at Victoria University of Wellington from 1975. By the 1980s they had campaigned together on abortion and contraception, the rights of midwives, violence in the home, child abuse and sexually transmitted diseases.[13] As Coney told a *New Zealand Woman's Weekly* reporter in 1990, 'Health is where women feel male power the most directly.'[14] She also revealed her own attitude to the medical profession while discussing one of her favourite activities, gardening: '"There's something very satisfying about hacking and slashing gorse bushes. I pretend they're doctors", she laughs.'[15]

In 1991 Bunkle explained that upon becoming pregnant and seeking an abortion in 1970 as a student in Oxford, England, 'the full force of patriarchal medicine was thrown at me, and also the force of patriarchal control'. She had encountered sexual harassment from two doctors, and also experienced unpleasantness from a third – John Stallworthy, professor of obstetrics and gynaecology at Oxford, who was a New Zealander and had been involved in the founding of National Women's Hospital. Her abortion was a very important experience because of her 'acute powerlessness – and because the only people who had shown any sympathy or solidarity were women'. The first political action that she organised on coming to New Zealand in 1973 related to abortion. The New Zealand Obstetrical and Gynaecological Society was having an anniversary celebration, with Stallworthy as an invited speaker. Bunkle organised a protest when he delivered a lecture on the dangers of abortion.[16]

Bunkle recalled in 1988 that her 'first experience of total powerlessness' was at the birth of her first child in 1975. 'As the doctor bullied and yelled and threatened, it became apparent that this was *his* theatre and *his* drama [her emphasis].' She recounted the glee with which he did the episiotomy without anaesthetics and that his face was 'alive with pleasure . . . . My flesh would no longer defy his power . . . . Since then I have listened to hundreds of women abused by the medical system.'[17]

Phillida Bunkle started investigating the dangers of the Dalkon Shield, an intra-uterine contraceptive device, around 1975. This had been withdrawn from the US market in 1974 because of its side effects and world sales were suspended in 1975. By that time, about 2.3 million had been distributed in the US and 2 million overseas, some of which came to New Zealand. It continued to be available in New Zealand. By the early 1980s, American women who had used the contraceptive were successfully suing the manufacturer, A.H. Robin, and Bunkle founded the Coalition for Fertility Action in Wellington to collate information on side effects among women in New Zealand, and to publicise their right to claim compensation. Shortly afterwards, Coney organised a branch in Auckland, and they combined to form Fertility Action. They were joined in their campaign against the Dalkon Shield by The Health Alternatives for Women (THAW), which had been set up in Christchurch in 1980. Fertility Action, according to Sandra Coney, 'soon developed into one of the most visible and active of the modern women's health consumer advocacy groups'.[18]

Feminists and the Doctors of National Women's

The earliest platform for the women's liberation movement was the right to abortion. Under the Crimes Act 1908 and its successor the Crimes Act 1961, which applied until the Contraception, Sterilisation and Abortion Act 1977, abortion was illegal unless the woman's life was in danger. Two prominent medical professionals at National Women's Hospital were core figures in opposing the liberalisation of the abortion laws. Sir William Liley, professor at the Postgraduate School of Obstetrics and Gynaecology and world-famous for having conducted the first intra-uterine blood transfusion, was founder president of the Society for the Protection of the Unborn Child (SPUC) in 1970, and Dr Pat Dunn, a long-standing senior consultant at the hospital, was a co-founder. In 1970 the *Dominion* quoted Dunn as saying that 'it was necessary to quash the idea that a woman had the right to do what she liked with her own body'.[19]

Associate Professor Herb Green was not, as Coney and Bunkle claimed in their 1987 *Metro* article, a member of SPUC. However, he was opposed to abortion and published a three-part article in the *New Zealand Nursing Journal* in 1970 entitled 'The Foetus Began to Cry'. In this series he reviewed the social, ethical and legal issues relating to abortion and concluded that whether to abort the fetus or not was generally a social or ethical decision that should not be left to the medical profession: 'Not by the wildest stretch of the imagination can this be regarded as a medical problem. To put physicians into the position of deciding who shall live and who shall die, as has the UK Act, must and will ultimately be regarded as one of the supreme follies of this age of technology.' He referred to Liley's acclaimed research: 'By permitting such research at National Women's Hospital and publicly acclaiming its chief architect, New Zealand has declared that the unborn foetus has the right to receive treatment in the interests of its life and health.' In his opinion, 'Social problems cannot be cured by surgery and all the enthusiastic abortionists in the world can but scratch vainly at the surface of a problem of which the only solution is better standards of living.'[20] A colleague described Green as a 'conservative socialist', with left-wing political views while at the same time being socially conservative.[21] In the early 1970s, Green fell foul of the feminists such as Sandra Coney who worked as a counsellor at a private abortion clinic, the Auckland Medical Aid Centre (AMAC), when he expressed concern about cases of pelvic infection arriving at National Women's Hospital from this clinic. In 1974 he made a sworn statement about the poor conditions there, which was later used to help close the clinic.[22]

Another target of the new women's health movement was Professor Dennis Bonham, described by feminist and independent midwife Joan Donley as 'the emperor [who] moulds the thinking of obstetrics and gynaecology in New Zealand'.[23] English-born and trained, Dennis Bonham had arrived at National Women's Hospital as the new postgraduate professor of obstetrics and gynae-cology in 1963; a decade later he was awarded the OBE for his services to the community. Immediately upon his arrival he was appointed medical adviser to the Parents' Centre, a consumer organisation that was working for a reduc-tion of medical intervention in childbirth, the presence of fathers at childbirth, rooming-in and breastfeeding.[24] Bonham was responsible for changing the regu-lations at National Women's to allow the presence of fathers at birth.[25] He hosted the first state-sponsored family planning forum at National Women's in 1971, and from 1972 chaired the new New Zealand Family Planning Association (NZFPA) Medical Advisory Council, the formation of which was described by NZFPA historian Helen Smyth as 'another facet of success' for the NZFPA.[26] Bonham included FPA professionals in postgraduate obstetrical and gynaecological courses at National Women's Hospital. He held liberal views on abortion and, unlike his colleague Green, he told the Royal Commission on Contraception, Sterilisation and Abortion in 1975 that he believed the private abortion clinic in Auckland, AMAC, was 'providing an important service and doing it well'.[27] He was largely responsible for the acquittal of the doctor accused of conducting illegal abortions at AMAC in 1978, an event which received much media attention.[28] He was one of the few doctors to get a favourable mention in *Broadsheet*'s 1979 list of doctors and their views on abortion; his entry read: 'liberal, good with patients'.[29]

Bonham claimed in 1980 to be 'delighted to find women questioning the birth service offered'.[30] He thought obstetricians and women should work together to improve women's health services. When Sandra Coney interviewed him about Green's work in 1986, he sought her help as a women's health advocate to increase the time devoted to obstetrics and gynaecology in the medical school curriculum, which he said had been eroded over the years.[31] Bonham also supported women in medicine and had worked to increase the number of women specialising in obstetrics and gynaecology. The first appointed to the hospital had been Cecilia Liggins, the wife of Graham Liggins, in 1967; in 1981 she was asked to join the Hospital Ethics Committee.[32] Bonham recognised that the major obstacle for women specialising in obstetrics and gynaecology was the incompatibility of the hours with the demands of a young family and he organised a job-sharing arrangement between two young female obstetricians. One of the participants, Hilary Liddell, described Bonham as 'very progressive about women in training'. She also commented that during her early years, 'Bonham was very influential in making sure that I got good training and encouraged me'. She explained, 'He felt that women should do advanced training, they should have senior positions – if they were the right kind of people. He didn't tolerate fools gladly. He targeted people, both amongst the nursing staff and amongst the doctors that he thought had the qualities to go ahead . . . '. The other partner in the job-share, Lynda Batcheler, held similar views.[33] Another woman trained in obstetrics and gynae-cology, Lesley McCowan, found him very supportive of women in medicine; when she returned from overseas postgraduate study in 1986 with a young family, he facilitated a part-time academic job for her at the hospital.[34] Bonham's wife, Nancie, worked with Sandra Coney and others to set up the university creche in 1968 to enable young mothers to attend lectures.[35]

Another initiative of Bonham's was to invite Sandra Coney to address medical students on the women's perspective. She received her first invitation to speak at the hospital in 1973; her topic was 'Women in the New Society'. She spoke at Bonham's invitation on further occasions, but eventually was 'dropped after one visit when I had decided to say what I thought instead of being lightweight and humorous. I heard privately that Bonham thought I was "too militant".'[36] Coney told the Cartwright Inquiry of an occasion in the late 1970s when she had been asked to speak on a panel together with someone from the Council for the Single Mother and Her Child: 'Instead of just talking to the students we decided we would do a dramatised presentation of one of my colleagues' experience at National Women's Hospital, and Professor Bonham was present and became hysterical about what we were saying and shouted at us that it was lies.'[37] Nevertheless, she was invited back. In 1987 she said, 'More recently, I have been asked to go to the medical school to talk about women as patients, by Dr Lesley McCowan, as part of an introduction to fourth and fifth year students gynaecol-ogy component and I have been pleased to do that, but I would point out that my input into the whole of medical training is a 15 minute component'. She had also been approached by McCowan to talk to registrars at National Women's.[38]

Despite Bonham's liberal views on abortion and his attempts to work with con-sumer groups like Parents' Centre and the FPA, and to increase the number of women in his specialism, he nevertheless fell foul of the women's movement over the issue of hospitalised childbirth. Opposition to this was a core element of the new women's health movement in New Zealand. Bonham came to New Zealand in 1963 with a reputation of having been involved in a study which, as Coney and Joan Donley declared, had placed 'the final nail in the coffin of home birth in Britain'.[39] Once in New Zealand, they said, he attacked home birth 'with missionary zeal'.[40] The study, which had been designed by Bonham's boss in London, Professor William Nixon, head of the Obstetric Unit at University College Hospital London from 1946 to 1966, aimed to compare home births with hospital births and Bonham's role was to collate the data. British historian Susan Williams has disputed that the study was responsible for promoting hospital births, arguing that the trend to hospital births was occurring anyway.[41] Although the initial goal of comparing home and hospital was abandoned because of the complexity of the data, the study was later identified as a useful early study of perinatal mortality.[42]

Shortly after his arrival in New Zealand, the Board of Health appointed Bonham to its Maternity Services Committee and he persuaded it to sponsor a survey of small maternity homes to regulate and improve the standards of maternity care throughout the country. The survey of 160 maternity hospitals was conducted by Bonham and Dr Joan Mackay (the Health Department's director of maternity services), who published their report in 1976. The report was particularly critical of the conditions of some of the remoter maternity hospitals run by general practitioners.[43] Some of these smaller homes were closed as a result, leading Joan Donley to conclude that the 1976 report 'consolidated O&G power'.[44] Donley dubbed the closing of these smaller homes the 'Bonham squeeze'. She claimed that the 'squeeze' occurred despite a report by Professor Roger Rosenblatt from the Washington School of Medicine who argued that small maternity units were just as safe as larger units. What she failed to add was that Rosenblatt supported small units together with backup services of tertiary hospitals.[45]

Donley argued that the 'Bonham squeeze' had led more and more women to opt for home birth as an act of rebellion against the technological takeover of their bodies. This rebellion, she said, led to the formation in 1978 of the Home Birth Association, which by the mid–1980s was 'a political force to be reckoned with – not only on its own merit, but also because of its proliferation into other feminist/political organisations such as Save the Midwives, and more recently, Maternity Action'. Maternity Action was a coalition of fourteen organisations, set up in 1984 in Auckland to oppose the closing of small maternity homes.[46]

National Women's Hospital obstetricians publicly challenged the home birth movement. In 1976 the *New Zealand Herald* reported Dr Bruce Faris as saying, 'women should be aware that having a baby is not a normal physiological function at all' – 'a bizarre statement', according to Donley.[47] His comment arose out of a dispute within the Auckland Hospital Board, of which Faris was a member, about the future of the city's maternity services. Faris proposed that all future obstetric or maternity beds in the city should be backed up by facilities for adequate specialist anaesthesia, caesarean sections, and specialist paediatric consultation and care, preferably in the environment of a general hospital. This was defeated, he said, amidst emotional and anecdotal arguments that 'our grandmothers had their babies at home' and that 'specialists lived in ivory towers and did not know what the women themselves, the community, and general practitioners and midwives wanted'.[48] Bunkle later quipped in relation to Faris's statement, 'When Mr Bruce Faris . . . said "women should be aware that having a baby was not a normal physiological function at all", he must have meant normal for men, since fertility is so clearly a manifestation of wellness in women.'[49]

Faris was not the only National Women's obstetrician to be embroiled in these disputes. In 1978 Colin Mantell's inaugural lecture as undergraduate professor of obstetrics and gynaecology at the University of Auckland was reported in the local press. Entitled, 'Where should babies be born?', the lecture not surprisingly presented a case for hospital births. A woman reporter described the event: 'Pacing the carpet square round the lectern in his beautiful imported shoes, Dr Mantell looked every inch a successful professional man.' He told his audience that he was an advocate for babies rather than mothers, and that labour and delivery were times of the greatest risk to the baby. Speaking of the unpredictability of labour, he warned that they should 'beware of the rose-tinted nostalgic view of hypothetical societies, young, robust and strong, glorying in the experience of delivering painlessly, simply, safely' which, for most, simply was not true. The reporter noted that he was 'watched approvingly by the doyen of National Women's, Professor Dennis Bonham'.[50]

In recognition of the growing rejection of medicalised and hospitalised childbirth, the Maternity Services Committee published a brochure in 1979, *Obstetrics and the Winds of Change*, suggesting ways of humanising hospital births to make them more attractive.[51] The committee also set up an inquiry into home births, for which it received 36 submissions, although no report emerged. The various branches of the Home Birth Association stressed consumer choice; the medical community saw itself as the baby's advocate in the face of serious health risks.[52] Bonham produced his own blueprint for the future of obstetrics in New Zealand, *Whither Obstetrics*, in 1982. He suggested that all first-time mothers, all mothers aged 30 and over, and anyone with a medical problem in her background should be classified as high risk and channelled into specialist care and delivery at a regional base obstetric hospital. There was no place for home delivery in his plan and no place for independent midwives. *Whither Obstetrics* resulted in the Parents' Centre removing Bonham's name from its list of advisers.[53]

Increasingly, debates over birth pitted obstetricians against feminists. The latter framed this debate as male power versus women's control over their own bodies. American feminist academic Sheryl Ruzek claimed in 1978 that, 'obstetrician-gynecologists fight to keep delivery their exclusive domain, even if it requires transforming what might be otherwise normal births into surgical events. In the hospital – particularly in the delivery room – the obstetrician can become the star of the obstetrical drama. Deferred to and waited on by nurses and other lower-status personnel, physicians enjoy their status and omnipotence.'[54] Sandra Coney argued in 1979 that the transfer of births from home to hospital was the result of men's 'womb envy'. In the hospital, men could take control of the birth, strip women of their traditional support networks – and even pretend they were giving birth; 'symbolically he can birth himself'.[55]

The Board of Health disbanded the Maternity Services Committee in 1983, possibly in response to criticism from the new women's health movement. According to Donley, the committee had become extremely unpopular with women, as it was associated with the closing of small maternity hospitals and in particular the midwife-run St Helen's hospitals. It was no longer politically acceptable to have women's health in the hands of a professional committee, with only one consumer representative among its fourteen members, or as Donley put it, 'one lone consumer voice crying in the wilderness'.[56] Bonham, who had been heavily involved with the Maternity Services Committee since 1964, wrote that its demise occurred without explanation: 'This is something else I protest about. Perhaps I should just continue to grow orchids and stop worrying about the health of New Zealand women and children!'[57]

The feminist movement achieved greater status and power following the election of a Labour government in 1984. The establishment of the Ministry of Women's Affairs in 1985 endorsed state recognition of the women's movement, and Coney was a key contact on health issues. She was the ministry's representative at a symposium on cervical cancer held at National Women's Hospital in November 1986.[58] The Labour government also set up a Board of Health advisory committee in December 1984, the Women's Health Committee, to replace the Maternity Services Committee. Thanking Joan Donley for a letter outlining her work with midwives and the home birth alternative, Labour MP Helen Clark wrote, 'We will have to watch very carefully to see that that committee does not become a vehicle for anti-midwife and anti-home birth advocates.'[59]

Joan Donley announced the formation of the new committee to the New Zealand Women's Health Network. This network, which had been set up following the 1977 United Women's Convention, produced a bi-monthly newsletter, edited by Dr Sarah Calvert, a psychologist in Child, Youth and Family services in Tauranga.[60] Donley warned readers of the newsletter to beware of the new committee:

> Before you start to cheer, have a look at its composition. There are 12 members. Two of these are MALE O&G SPECIALISTS: Professor Colin Mantell and Professor Richard Seddon. Neither of them have been supportive of women's issues in the past; on occasion they have fought and opposed the ideas and actions of women health workers. Prof Seddon was a member of the now defunct Maternity Services Committee . . ., and consultant in Obstetrics and Gynaecology to the Department of Health – and a lot of good that has ever done us.[61]

Donley described the Women's Health Committee as 'controlled tokenism . . . a way of being seen to give lip service to women's demands and needs without doing anything that will bring about change'.[62] In the next issue, she queried how Labour would achieve its goal of ensuring that the interests of women and children were paramount 'when the power to make the changes has been placed back into the hands of those who developed the maternity services in their own interest in the first place?'[63] Donley and Calvert insisted that they did not want 'any more little cliques of doctors telling us what we need'. They maintained that women were exploited by drug companies, which controlled almost all the funds for research and were only interested in financial returns. The Health Department, they believed, was responsive to the pressures of the medical profession, who were only 'interested in lining their pockets with the profits of our exploitation'.[64] The network sent a letter to the Women's Health Committee declaring that the Board of Health's Maternity Services Committee had been opposed by women through-out the country because it ignored general women's health needs and promoted obstetrical interests. However, the new committee, the letter continued, proved to be no better, and the writer concluded, 'I would request that you not insult us further by suggesting that you represent the interests of women.'[65]

Christine Bird from THAW was equally scathing of the Board of Health's Women's Health Committee. She noted that only three of its members repre-sented consumers. Seven of the sixteen worked directly or closely with the Health Department and a further five were health professionals. Three had been part of the Maternity Services Committee that had recommended 'centralisation of maternity services in machine-dominated and drug-happy maternity hospitals . . . . It is when women are in most need, with drug companies, health officials, vested interests and moralistic woman-haters ranged against them, that our health system totally fails them – and this is no coincidence. Only women's health groups put women and their experience first, not that of "experts", hospital effi-ciency, or profit.'[66] These feminists totally distrusted experts and professionals and questioned their motives for involvement in women's health.

Childbirth was not the only contested area. Phillida Bunkle's research into the contraceptive, the Dalkon Shield, brought her into contact with National Women's Hospital gynaecologist Ron Jones. She found two articles that he had written in 1974 and 1975 supporting the Dalkon Shield, despite the fact that it had by then been withdrawn from the American market. She phoned him and found him hostile: '[He] appeared affronted that I should ask him questions about the implications of his work. He said "I abhor your approach" . . . . My effort he said was "misdirected" because had I been thoroughly informed I would know that evidence against Dalkon was refuted because other IUDs [intrauterine devices] were as bad.' As an indication of his attitude to medical accountability, Bunkle added, 'Furthermore he believed that if women could sue their doctor it would be bad for medicine in New Zealand.'[67] (With the no-fault Accident Compensation legislation introduced in the 1970s, the New Zealand public was in the unusual position of not being able to sue doctors for malpractice.)

Bunkle also claimed that doctors did not take women's complaints seriously. She referred to a seminar by Bonham at National Women's in which he suggested that pelvic pain from the Dalkon Shield was 'likely to be psychophysologic'.[68] Coney noted that it was on Bonham's advice that the Health Department was prepared to allow the Dalkon Shield to be kept on the New Zealand market when it was being withdrawn overseas.[69] Christine Dann wrote in 1985 that Coney and Bunkle's article on the Fertility Action battles with the Health Department over the Dalkon Shield 'showed how little credibility the best-informed of feminist health groups [were] given by the system'.[70]

Another issue that mobilised feminists in the early 1980s related to the contraceptive Depo-Provera, and again the target of the campaign was research undertaken at National Women's, regarded with particular suspicion because it was funded by a drug company. Depo-Provera was a hormonal injection that inhibited ovulation for three months. It had been developed by an American multinational pharmaceutical company, Upjohn, in the 1960s but was not approved for use in the USA by the Food and Drug Administration and was mainly used in developing countries, where it was also tested.[71] Depo-Provera was available in New Zealand from 1968, and by 1980 there were about 35,000 users. Some health professionals apparently favoured it because, unlike the contraceptive pill, it was easy for women to use. As Coney pointed out, the principal researcher for the study was Professor Graham Liggins.[72]

Graham Liggins began researching the effects of Depo-Provera, along with other contraceptives, in 1980. The New Zealand Contraception and Health Study with funding from Upjohn was an observational study following three groups of 7500 women using Depo-Provera, IUDs or the contraceptive pill for five years. When questioned by Bunkle about his link with Upjohn, Liggins 'vehemently maintain[ed] that the study is independent and was initiated by him'. Liggins had been studying the contraceptive pill since the 1960s, and saw this as an extension of that work.[73] Bunkle commented that Upjohn seemed less concerned to maintain the appearance of independence for the study; their 'media package' stated: 'Upjohn is currently conducting long range studies in New Zealand.' The 'primary objective' of the study, the media package explained, was 'to examine the relative association between contraceptive practices and the development of dysplasia, carcinoma in situ, or invasive carcinoma of the cervix'.[74] A Pap smear was taken at each annual examination and a questionnaire completed by the doctor. While the cervical smears were initially processed at National Women's Hospital, all abnormal smears and a selection of normal ones were sent to a laboratory in Chicago for further analysis.

Broadsheet had featured a critique of Depo-Provera in 1976, reprinted from the British feminist magazine *Spare Rib*.[75] In December 1979 Sue Neal wrote about the dangers of Depo-Provera in *Broadsheet*,[76] and in 1980 Coney and Neal launched the Campaign Against Depo-Provera, claiming that New Zealand women were being used as guinea pigs. 'From our observations it is used particularly on Maori and Pacific Island women (who are most at risk from some of its side effects), women in lower socio-economic groups and mental patients. Health activists overseas have described it as "the contraceptive for second-class citizens".'[77] The *New Zealand Listener* printed a feature about Depo-Provera in 1980. It noted that *Broadsheet* originally 'broke the story' of the survey, which was being run by specialists including Liggins and Seddon. It revealed a 'suspicion that this country's women were providing an American multi-national with a test ground for the product – to be emotive, the Bikini atoll of hormonal contraceptives'.[78]

Others were also suspicious of links with drug companies. In a scathing *Broadsheet* article dated January 1980, Dr Ruth Bonita accused New Zealand's Medical Research Council of 'encouraging the production of pseudo-scientific facts for financial gain by a multi-national corporation'.[79] While Liggins had been researching contraception before he got funding from Upjohn, the extent of public criticism of the links with Upjohn led the Medical Research Council to suggest that Liggins ask the MRC Standing Committee on Therapeutic Trials to assess the protocols of the 1980 study. Headed by professor of medicine Sir John Scott, the committee commented in its report on the 'lack of objectivity' in public discussion of the study, which it described as 'trial by media'. It saw no problem with drug company sponsorship, provided the protocol was sound, which it concluded it was, as far as possible in the 'non-ideal world of real-life communities'.[80] What had really changed, according to the 1980 *Listener* article on Depo-Provera, was the attitude of 'a section of educated, liberated women . . . to pharmacologically based, largely male-dominated medicine'. The article ended with the comment by one 'activist woman' that 'if men had to take a hormonal contraceptive the Pill would still be under test in a laboratory'.[81] The Depo-Provera story reveals the concern among feminists that doctors experimented on women's bodies for profit, and that such experimentation was possible because their subjects were women.

Hysterectomy was another area of concern to feminists in New Zealand as elsewhere. In 1981 Lyn Potter had a hysterectomy at National Women's Hospital, an experience that led her to become interested in the subject from a feminist perspective. She later explained how she was shocked at the information available, or lack thereof. She told *Broadsheet*, 'Most gynaecologists neglect to tell women that there is a relatively high rate of complications after hysterectomy . . . . Doctors can be very manipulative.' She commented on the high rate of hysterectomies in New Zealand compared to Britain (212 per 100,000 in New Zealand compared with 122 per 100,000 in the UK in 1982). She reasoned that the profit motive could be ruled out at National Women's because it was a public hospital, but wondered why hysterectomies were so readily done there. 'Is it because it's the easy option, less time-consuming, or is it an interesting procedure for teaching purposes perhaps? The underlying reason is – so what! Who cares! It's only a uterus. As they say in the official support pamphlet at National Women's, "We're only taking out the carry-cot and leaving the play-pen"!'[82]

The pamphlet referred to had been reprinted from an article in *Thursday,* an alternative magazine for 'progressive' young women which dealt openly with issues such as sexuality, sex education and fertility, and was edited by feminist Marcia Russell.[83] The article, based on an interview with Bonham, was entitled 'Hysterectomy Hangups' and was undoubtedly a poor choice for a hospital information sheet. Bonham told the reporter that when he started practising gynaecology people occasionally died after a hysterectomy. Since then there had been a huge drop in the mortality rate and it was now about as risky as having a baby and was done for a better quality of life rather than merely to save lives. While it was carried out to enhance quality of life, it was not, he explained, used for contraceptive purposes as in the United States. New Zealand followed Britain in preferring to do tubal ligations for sterilisation rather than hysterectomies. He made the carry-cot statement in relation to the 'common fear among women considering hysterectomy that they won't be able to have sex after the operation'. Taken out of context it could be construed as referring to a 'playpen' for men's enjoyment, though Bonham, addressing a female audience, was presumably referring to women's sexual satisfaction. Asked about psychological after effects, Bonham told the reporter, 'Any ordinary person who isn't a psychiatric cripple will be perfectly all right. In all the years I've been doing hysterectomies I have never come across a woman who has regretted having one.'[84]

Potter disagreed. Following her own experiences, she organised a Hysterectomy Support Group and set up a phone line from which she collated information on women's experiences of hysterectomy. Over several years, she 'accumulated quite a lot of feelings women had about lack of information about their emotional side not being well cared for'.[85] Potter wrote a new pamphlet which the hospital agreed to use in place of the one reprinted from the *Thursday* article.

Early in 1987, Sandra Coney wrote an article on hysterectomy for the *New Zealand Woman's Weekly* in which she claimed:

> One obstacle to support and information-sharing for women who have had children and are undergoing hysterectomy has been the male medical profession's attitude. They see it merely as the removal of a redundant organ. . . . An extreme example of this view was the doctor who wrote that after the last planned pregnancy 'the uterus becomes a useless, bleeding, symptom-producing, potentially cancer-bearing organ and therefore should be removed'.

She pointed out that in 1984, 6811 New Zealand women had undergone hysterectomies. This was twice the British rate but half the American. She said that it had been estimated that up to 40 per cent of Australian women and 50 per cent of American women would undergo this operation at some point, and that it was one of the operations most frequently performed in the Western world.[86]

Like Potter and Coney, Herb Green was also sensitive to the woman's perspective on hysterectomy. He explained to the Inquiry that the physical effect of hysterectomy could be variable: 'One woman takes it quite easy and never worries about it, and the other woman feels that she has been savagely mutilated.' He explained that it depended on the individual patients, and that there could also be complications such as anaesthetic problems and bleeding afterwards. He added that there was also 'considerable psychological morbidity for hysterectomy in some women' which should be taken into consideration when recommending hysterectomies.[87]

Feminism and Carcinoma in Situ

In 1984 Sarah Calvert provided a feminist interpretation of CIS and cervical cancer when she complained in a *Broadsheet* article entitled 'Cervices at Risk' that women's 'promiscuity' was often cited as the 'cause of soaring rates of cervical cancer'. She also noted side effects to treatment of CIS, pointing out that cone biopsy was suspected of causing cervical incompetence and premature labour in future pregnancies. In an effort to provide women with more information, Calvert outlined the five classes of cervical smears.[88] One reader, Marion Kleist, responded that Calvert's listing was misleading in some respects. For instance, she said, both Class 3 and 4 smear categories could mean 'mild dysplasia', and a Class 3 result did not necessarily mean 'carcinoma in situ'. She had received a Class 4 diagnosis, and had obviously studied current overseas literature on the subject as a result. She noted that many dysplasias and carcinomas in situ never developed any further and that many regressed. Researching feminist writings, she found that not much had been written on alternative treatments, 'although I have heard there are women trying various methods'. She wrote, 'The medical profession urges treatment. Because of the possibility of it developing into a malignant cancer it is difficult to leave it for long and feel confident that it will be all right. It is a very vulnerable time for many women and the medics' sense of urgency does not help matters.' She asked *Broadsheet* for a more in-depth article to be written, 'which could help women understand the result and possible options they have more fully'.[89] She was seeking more information and control, and less medical intervention.

Further feminist interpretations of the kind requested by Kleist were not hard to find. The British feminist magazine *Spare Rib* had featured an article 'Cervical Cancer: The Politics of Prevention' the previous year. The author, Lisa Saffron, commented on the medical profession's tendency to link cervical cancer with promiscuity, and the effect this had on women's self-image when they had a positive smear. But she said there was also a 'deeper dilemma about cervical smears', and this related to the uncertainty about the natural history of the disease and the tendency of doctors to 'over-treat'. She pointed out that 'the claim that vast numbers of women are being cured of a fatal illness is entirely misleading'. She listed the physical and psychological side effects of treatments, adding,

> It is the nature of Western medicine to start with the most drastic measure, to use a sledgehammer to crack a nut. Coupled with contempt for women's bodies, our reproductive organs don't stand a chance. The unspoken rule of thumb seems to be that there is no cervix so healthy that it isn't better treated and no testis so diseased it isn't best left intact. Men's reproductive organs simply do not get that kind of attention that women's do. Their importance is recognised by a predominantly male medical profession.

She wondered whether there were gentler ways to make a positive cervical smear go back to normal and referred to an American study of the 'sheath treatment'.[90] She concluded, 'Only in a male dominated society could doctors have such disregard for a woman's physical integrity.'[91]

The 'sheath treatment' study was also referred to in a Women's Health Information Centre leaflet from London which the Christchurch women's group THAW sent to Silvia Cartwright during the Inquiry. It described an American study in which 136 out of 139 women with positive smears got better without treatment when their partners started using a condom. This same article referred to the 'potential problem of overtreatment', claiming that, 'As more and more women are found to have abnormal cells, doubts have been raised about whether medical intervention is always the best approach. Even with the new methods, this can be a distressing experience with both physical and psychological consequences.'[92]

Long-term British patient advocate Jean Robinson published an article on cervical cancer in *Spare Rib* in 1985, in which she regretted that, 'Neither the ethics, the efficacy, nor the adverse effects of screening have been adequately discussed by women's organizations.' She too commented on the medical tendency to blame women for bringing about their own cervical cancer through promiscuous behaviour.[93]

A feminist construction was also provided by Tina Posner of the University of Oxford Department of Community Medicine:

> Women have been blamed for not 'coming forward' for screening, in order to be 'saved' by medical intervention. However, once screened and found to have 'pre-cancerous cells', they may then be blamed for bringing the condition upon themselves by their own behaviour. This campaign, so replete with moral meanings, is based on the deception that it is fundamentally a matter of life and death – a construction which involves the medical profession's utilisation of cultural meanings and lay fears for its own purposes.[94]

She claimed that screening could be 'seen as an attempt to assert the dominance of the medical definition of reality based on objective signs rather than subjective accounts'. She also pointed out that in this area the ability of modern medicine to diagnose had outstripped its ability to offer a prognosis, with a tendency to over-react, although the uncertainties were not conveyed to patients. Patients still commonly believed that a positive smear result meant cancer, and saw this in terms of black and white: 'The greyness of the area between perfectly normal cells and frankly cancerous cells, and the normality – the commonness of some degree of possibly passing abnormality – is not generally known.'[95] Some feminists were indeed criticising the medical interpretation of, and an invasive medical approach to, cervical smears.

By the time Kleist had urged *Broadsheet* readers to adopt a critical feminist review of the treatment of CIS and to be cautious of 'doctors' sense of urgency' (or over-reaction) in treating the lesions, Dr Ruth Bonita had sent Coney a copy of McIndoe's 1984 paper.[96] This paper was based on what Tina Posner called the medical (or pathological) view of reality, in other words, assessing women's health according to whether they had a positive or negative smear, with 'the success of treatment . . . judged almost entirely in terms of the disappearance of [patho-logical] signs of disease – the obliteration of the abnormal cells; the fulfilment of the medical aim of intervention', regardless of the costs to the woman's general health.[97]

The feminist literature in this area was not picked up by Coney and Bunkle when they came to write their *Metro* article. Fundamental to the general women's health movement was the claim that women's knowledge of their own bodies was as legitimate as the expertise of the medical profession.[98] Coney inadvertently revealed how Green's daughter had put her finger on the inconsistency at a talk on women's health which Coney delivered to the YWCA immediately prior to the Inquiry. During question time, Green's daughter said, 'You say women know what's the matter with them and that doctors should listen to them more. Why then do you put such store on pathology, rather than symptoms like bleeding?' Coney recalled, 'Fortunately I did not have to deal with this [question] myself. Some young nurses present took up the question . . .'.[99]

Dr Erich Geiringer, described as 'a Wellington GP and frequent commentator on public issues', similarly drew attention to what he called a 'feminist flip-flop'. He wrote:

> For years they had built their reputation on warning women against the horrors of male medical technology and against misogynist womb snatchers. Now they would champion the cause of women who had been deprived of the benefits of male medical technology by a man whose main motivation was to put an end to womb snatching.
>
> Indeed had his first name been Hermione he would at this point have been a shoo-in for the pantheon of fearless feminist fighters against male medical mutilators. Her

opposition to wholesale womb snatching, her questioning of the medical establishment's determination to 'upend all women' for an annual smear, would have earned her a rave in *Broadsheet*.[100]

Coney and Bunkle and subsequently other women's groups in New Zealand and the Ministry of Women's Affairs did not embrace the alternative feminist interpretation of CIS and its management. On the contrary, they aligned themselves with those members of the medical profession who favoured an aggressive interventionist medical approach. Having feminists as allies in the mid–1980s was to prove a bonus for these professionals, as the feminists' political and public support was on the ascendancy. Fertility Action later claimed the credit for initiating the Inquiry into the Treatment of Cervical Cancer and pointed out that Fertility Action members also 'acted as consultants to both the Ministry of Women's Affairs and Health Department on topics such as cervical cancer screening [and] hospital services'.[101] The wide acceptance of the views and pronouncements of these feminists silenced the medical debates and allowed the views of those doctors who favoured an aggressive interventionist approach to CIS to predominate.

Conclusion

The politics of women's health had indeed become a 'hot topic', as noted in the 1985 *Listener* article on maternity.[102] The women's health movement attracted considerable political and media support. National Women's was a particular target of the movement. The feminist magazine *Broadsheet* was to hail the Cartwright Report as a feminist victory.[103] Celebrating the anniversary of the report's release one year later, the Auckland Women's Health Council gave Silvia Cartwright a presentation. The picture of Sandra Coney, Phillida Bunkle, Clare Matheson and Silvia Cartwright was captioned, 'Four Women Mark Medical Milestone'. The article began, 'Four women who took on the might of the medical profession . . .'.[104] The next chapter will investigate how they took on the profession.

8.

The Cervical Cancer Inquiry
and the 'full story'

The *Metro* article by Sandra Coney and Phillida Bunkle focused specifically on Associate Professor Herb Green's research at National Women's Hospital and sparked the setting up of a committee of inquiry. The interrogation of Green at the Inquiry demonstrated that the Inquiry was not just an exercise in 'fact finding'. As the Inquiry proceeded, Green's work was increasingly sidelined as wider issues were canvassed. Sandra Coney's own book, which appeared at the same time as the Cartwright Report, turned the spotlight back on Green and his colleagues at National Women's in order to reinforce the more general issues aired during the Inquiry. Coney promised her book would reveal 'the full story behind the Inquiry'. This chapter focuses on the Inquiry itself, public reactions and submissions, and concludes by revisiting Coney's book.

Setting up the Inquiry

Labour's Health Minister Michael Bassett later recounted that he had no prior knowledge of the 'unfortunate experiment' before the *Metro* article hit the streets at the end of May 1987. His wife, Judith Bassett, who was then deputy chair of the Auckland Hospital Board, spotted the article while standing in a supermarket queue. She suggested to her husband that 'you may well need to have some sort of inquiry' and after reading the article Bassett agreed. He phoned Prime Minister David Lange who also agreed. Bassett commented, 'Remember we are only two months out from an election. Naturally we don't want a huge kafuffle. Everybody wants to be seen to be doing the right thing.' Bassett also stated that it was 'pretty obvious to me' that they needed to have a woman to chair it. He told Auckland Hospital Board chairman Sir Frank Rutter that it would be 'wise to have a pretty well all-women inquiry into the whole thing'. Silvia Cartwright, at that time a district and family court judge, was approached, as was another female lawyer, Lowell Goddard, who provided the legal back-up.[1]

The *Metro* article was about Green's treatment of women referred to National Women's Hospital with a positive cervical smear, and was based largely on Bill

McIndoe and his colleagues' 1984 article and Clare Matheson's story. In setting up the Inquiry, Bassett said he did not want it to be a witch hunt. The Inquiry had nine terms of reference, relating to issues such as research approval, patient consent, teaching and lines of responsibility. Bassett's idea was that the Inquiry was not to be about one (or several) doctors and their practices, but was to be concerned with wider issues relating to medical practice and accountability.

The Labour government itself was not averse to an inquiry into medical practice and accountability. Its past relationship with the medical profession had been fraught, dating back to the 1938 Social Security Act passed by the first Labour government.[2] The third Labour government, 1972–75, had issued a White Paper (*A Health Service for New Zealand*). This proposal for an administrative reorganisation of health services was broadly construed as an attempt to undermine the independence of the medical profession. Widespread opposition to the White Paper had contributed to Labour's defeat in the 1975 general election.[3] Immediately following its return to office in 1984, the fourth Labour government tried to alter the general practitioner subsidy system. While the New Zealand Medical Association endorsed the government's proposal to increase the benefit for children, 'it engaged in "open warfare" with Bassett to overcome the difficulty associated with State intervention in the realm of professional fees'.[4] In 1985 Joan Donley congratulated Bassett 'on your determined stand against the Medical Association'.[5]

The Inquiry could be seen as an attempt to court women's vote, particularly in light of the failure of the Health Department's Women's Health Committee to be seen as representing women's interests. Bassett received a series of letters from women and women's groups in the early days of June 1987 urging that an inquiry be set up; one representative of the Auckland Women's Health Collective wrote to him: 'We would like to offer our opinion on the current NWH issue regarding the dubious conduct of Dr Green. We have heard repeated testimonies to his arrogant treatment of women and would like a Royal Commission. We consider him to be a danger to all women and his retirement seems to be long overdue.'[6]

Bassett received several other requests for an inquiry, including one from a woman who stated that it was 'high time the medical profession were taken down out of their ivory towers and made to account for their all too frequent errors in diagnosis of the most basic medical problems often through lack of interest or care'. She said she was shocked that a specialist could be so arrogant, and hoped an inquiry would ensure such errors would never happen again. She concluded, 'There are many very caring doctors in the health field who work very hard for their patients and they should not have to have their profession tarnished in this way.'[7] Another woman wrote even more passionately, 'The gravity of the experimental nature of the treatment at this clinic can be described as barbaric, inhuman, living guinea pigs. I trust that the force of public opinion ensures that a Royal Commission is called.'[8] A further plea for an inquiry argued that, 'For too long, doctors have been able to protect each other and hide their arrogant decisions over women's health issues.'[9] These letters reflected the general anti-doctor/anti-expert feeling of the time, but with little knowledge of the facts of the case.

Bassett also received a request from the Cancer Society of New Zealand for an inquiry to be set up.[10] Professor David Skegg wrote to Director-General of Health George Salmond enclosing the letter to the *New Zealand Medical Journal* in which he had referred to Green's research as 'the unfortunate experiment'. He added, 'No-one replied to the question about whether the patients had been warned! I have had one or two thoughts about people who might be suitable for a Commission of Inquiry.'[11] One of the members of Skegg's Community Health Department at Otago, Dr Charlotte Paul, was appointed medical adviser to Cartwright.

The broadening of the terms of reference of the Inquiry to encompass more than Green's management of CIS cases suited the feminist lobby who saw it as a unique opportunity to canvass those issues relating to women's health for which they had been campaigning and about which they felt so passionate. When Cartwright called for submissions from any women who felt they had something to say about their treatment at National Women's, Fertility Action publicised this in *Broadsheet* and offered to help write their submissions.[12] Sandra Coney took the credit for suggesting to Bassett that he appoint Silva Cartwright to head the Inquiry. According to Coney, Bassett had asked for her views and when she suggested Cartwright, he said that her name had been put forward by several others. He also asked Coney for suggestions for other personnel on the Inquiry, for the legal counsel and for medical advisers.[13] Bassett himself later gave a different version of events to journalist Jan Corbett. He claimed that Coney had been trying to badger him, and only once he had made up his mind about the judge did he agree to talk to her. 'I thought it was prudent to discuss with "her ladyship" [Coney] who the judge would be. The inquiry would have been flawed if she, who had made the accusations, thought she had been given a hanging judge.'[14] Bassett clearly saw Coney as politically or publicly influential.

At the time of the Inquiry, Cartwright was 43 years of age and had recently completed an inquiry into social science research. She was an Otago University law graduate, and had been a district court judge since 1981. Her judicial work had been mainly in the family court in Auckland and Hamilton. She was 'interested in women's issues, particularly those relating to the law'.[15] In a 1985 report to the United Nations, she stated that, 'the rights of women and children are firmly entrenched in legislation in New Zealand, but full social acceptance of these rights is probably still some time away while those in the most powerful positions in the country continue to be predominantly middle aged, middle-class and male'.[16]

Yet the feminist lobby undoubtedly had support from those in positions of power in this instance. The Minister of Women's Affairs and most senior woman in government, Ann Hercus, was the niece of pathologist Jock McLean, one of the authors of the 1984 paper. He had recounted the story to Hercus over the years and she knew about the *Metro* article in advance. Coney noted that her support and that of the ministry was important to them at the Inquiry.[17] The Health Department granted Coney and Bunkle $750 per day up to a maximum of $15,000 for legal assistance. This was initially declined by Bassett, but he eventually agreed to it on the advice of Cartwright, the Chief Judge and the Deputy Solicitor-General.[18] The Inquiry lasted a lot longer than anticipated and in 1988 the new Minister of Health David Caygill gave Fertility Action an additional $52,000 for legal fees at the Inquiry.[19] Caygill's support went further; on 5 August 1988, the day the Cartwright Report was released, Phillida Bunkle went to Caygill's office where they drank champagne in celebration.[20]

'A failure adequately to treat CIS': The Inquiry's First Term of Reference

The first term of reference of the Inquiry was: 'Whether, as alleged in the Metro article, there was a failure adequately to treat cervical carcinoma in situ (CIS) at the National Women's Hospital, and if so, the reasons for that failure and the period for which that failure existed'.[21] While there was apparently no intention to apportion blame, Green was clearly considered central to an exploration of this question. He was cross-examined from 25 August to 7 September, non-stop apart from a two-day interlude; the record filled 1000 pages of the Inquiry's transcript. For most of this time, he was cross-examined by Rodney Harrison, acting for Coney, Bunkle and Fertility Action. When Cartwright suggested to Harrison that he was attempting to 'imply the question of fault or blame or the possibility of fault or blame on one particular person and that is not what this Inquiry is about', the latter rightly answered that 'the term of reference squarely asks whether patients were adequately treated and in my submission, it is quite proper for questioning directed at the person who was involved in the treatment to ask whether the particular treatment in a selected number of particular cases was adequate'.[22] He was allowed to continue.

Harrison's cross-examination of Green could only be described as aggressive, a stance lapped up by the press. When Harrison put it to Green that at least six of the patients in interviews had described themselves as 'guinea pigs', Green replied that he was not surprised, adding, 'And some of that came from the television stunt of a picture of National Women's Hospital with 100 guinea pigs in front of it, and if you stress the word "research" the word guinea pigs automatically arises in the patient's mind.' During cross-examination, Green explained that he preferred to use the term 'medical cartography' to describe research on patient data as less emotive, a term which had been suggested by Superintendent-in-Chief of Auckland Hospital Board Dr Wilton Henley. Harrison retorted, 'Dr Green, you may wish to have yourself seen as the Marco Polo of the cervix.'[23]

Harrison questioned Green closely on the case files of the 72 patients whom Cartwright interviewed in private, as discussed in chapter 3. He appeared to regard Green with the utmost suspicion. For example, when Green said that he was not involved with a particular case in 1971 because he was on sabbatical leave that year, Harrison asked him to produce his passport as evidence to the Inquiry.[24]

Discussing one patient who developed cancer, Green referred to certain variables associated with the advent of cervical cancer, and his belief that the oestrogenic hormone present in the contraceptive pill was a factor, although he was unsure if it was a factor in this particular case. He continued,

> Green: I don't know about the herpes virus, which was just being talked about I think by then in 68, 70, no possibly a little earlier, and certainly we knew nothing about the human papilloma virus, or Chlamydia. I am saying those could have been the causes.
> Harrison: None of these conditions appear to have been present according to the notes, do they?
> Green: None of them were looked for, because we didn't know about them.
> Harrison: That's convenient timing.[25]

Harrison interrogated Green about his move to less conservative management from around 1974, or, as Harrison put it, the end of the 'trial'. Green explained that the reason that he began to do more cone biopsies rather than punch biopsies was that he no longer had the services of the colposcopist McIndoe as a support and therefore 'could not afford to follow those criteria such as were laid down by Dr Richart in excluding invasive cancer'. Harrison suggested, 'But you didn't decide to stop the study because you were concerned at the danger to patients. You stopped it because they took your colposcopist away.' Green responded, 'No, because I realised that it would be dangerous to continue in the same way without colposcopy.'[26]

Discussing the emotionalism associated with cancer, Green stated that if death from cancer were abolished, the average life expectancy for a person of 50 years would be increased only by about 18 months. Drawing on his interest in epidemiology he continued, 'I've always thought that cancer in the medical services is given too much priority and if it can't improve expectation of life by more than a year or so, then shouldn't we be concentrating on diseases, whose correction will improve life expectancy by more than a year or 18 months?' This was a philosophical reflection and did not represent his treatment of individual patients. However, it allowed Harrison to suggest that Green had no regard for the welfare of his female patients: 'It is a question of the value you put on a woman's life, isn't it, Dr Green?' Green replied, 'It's a question of the value you put on anybody's life, Dr Harrison.'[27]

Recalling events many years later, Auckland University professor and head of the Department of Obstetrics and Gynaecology in the 1990s, John France, reversed the popular simile of women coming to the hospital like 'lambs to the slaughter': 'When the Commission started it seemed to me, as an outsider, that Dennis [Bonham] and Herb [Green] thought all they had to do was go down and talk to them and explain their motives and everything and it would all be fine. They had a junior legal counsel, this was his first big experience . . . . and these guys were lambs to the slaughter – Herb and Dennis – lambs to the slaughter.'[28] The dean of Auckland's School of Medicine, Professor David Cole, also commented on Green's 'tough cross-examination' by Harrison who seemed to treat Green 'like a naughty schoolboy and contrived to make Green appear totally out of step with world opinion'.[29] Despite the intensity of the interrogation, a review of the transcripts suggests that Green appeared to hold his own against Harrison's attacks on his integrity and professionalism.[30]

The Widening of the Net: 'Other related matters' and Public Reactions

The short title of the final report, 'The Report of the Cervical Cancer Inquiry 1988', was not a true reflection of the scope or subject matter of the Inquiry. It began as an inquiry into the conservative treatment of CIS at National Women's Hospital, a matter that had arisen directly from the *Metro* article; Bassett had added to the terms of reference 'other related matters', which were indeed wide-ranging.

One related issue was vaginal examinations on anaesthetised women by medical students. In September 1987 an *Auckland Star* editorial predicted that however much damage had been caused by the claims of inadequate treatment of women with precancerous and cancerous systems, it was unlikely to match the loss of confidence in National Women's, and other hospitals that treated women, provoked by the 'appalling revelation that medical students practise vaginal examinations on anaesthetised women without seeking their consent, and without even their knowledge'. The editor believed that Auckland Hospital Board chairman Sir Frank Rutter reinforced the image of the insensitive male doctor with his 'unforgettable and unforgivable dismissal of the whole matter as a simple case of women being "naïve" if they enter National Women's unaware its teaching hospital status means they give "implied consent" to such procedures'.[31]

This procedure was not unique to National Women's. It had become common practice around the Western world for medical students to undertake vaginal examinations of anaesthetised women. It was believed to be a sensitive way for students to learn the technique, and was implicit in patient attendance at a teaching hospital. Correspondents to *The Lancet* in 1988 from the University of Birmingham, England, noted that it was not common at that time to seek consent for vaginal examinations by medical students when the patient was under anaesthesia.[32]

Doctors training at National Women's in the 1970s remembered IUD insertion demonstrations in theatre on anaesthetised women, with six or eight doctors present. A nurse said it was 'common knowledge' that vaginal examinations were performed under general anaesthesia at National Women's Hospital: 'It used to be on Monday mornings', she explained.[33] In 1980 a patient at the hospital complained to the Hospital Board about the embarrassment caused to her by the presence of medical students in the operating theatre when she was wheeled in for minor surgery. She had not been informed or asked if she objected to their presence. Medical superintendent Ian Hutchison explained that students were normally asked to leave the theatre between cases and return after the patient was anaesthetised to avoid embarrassment and that unfortunately on this occasion the students had remained in the theatre. He apologised to the patient on behalf of the hospital. Board chairman Dr Frank Rutter also apologised to the patient on behalf of the Hospital Board and the hospital for the oversight in not informing her, pointing to Section 7 of the Board's Code of Rights and Obligations for Patients and Staff: 'Patients have a right to be advised of the reason for the presence of any persons not directly involved in their care. If they are not so advised why someone is present, they are entitled to ask and be informed.' The discussion of the complaint was also minuted by the National Women's Hospital Medical Committee:

> . . . it was generally felt that, although not compelled to do so in the code of rights, patients should be afforded the courtesy of being informed of the presence of students and the reasons why they were present. It was felt that very few patients would object to the presence of students if the situation was explained to them. It was agreed that there was a difficulty in the theatre situation and it was felt that the Surgeon-in-charge of the theatre should be responsible for seeing that observers do not stand around in the theatre between cases.[34]

While examination under anaesthesia for teaching purposes was not a new issue, during the Inquiry it received attention like never before. Rae Julian, who had just been appointed Human Rights Commissioner with Special Responsibility for Women, called it a 'form of rape',[35] and *Broadsheet* editor Pat Rosier, who described herself as 'a radical lesbian feminist with both socialist and separatist tendencies', commented that it was 'after all symptomatic of what many women perceive as the patronising and de-humanising attitude of many doctors to them'.[36] Lyn Potter believed that, 'Their attitude stems from not having much regard for that part of a woman's body.' Phillida Bunkle asked, 'when could you be more objectified, more powerless?', and thought it demonstrated the failure of internal processes to regulate medicine: 'The autonomy of the medical profession to regulate itself has to end . . . . These abuses go on in the dark. The moment you can get openness through the system you'll begin to get public accountability.'[37] To these feminists the practice epitomised male medical insensitivity to women.

Judy Larkin, spokesperson for Maternity Action, pointed out, 'We knew in 1984 that they were doing vaginal examinations under anaesthesia, but there was very little that we could do.'[38] Coney also said that she and Bunkle had known of it for some time and, 'This was a chance to raise it publicly.'[39] According to Coney,

'The public went wild when news of this broke.' She recounted how the National Council of Women and 30 nurse tutors protested to the Minister of Health, who went on television saying he was instructing hospital boards that the practice must stop.[40] Coney added, 'If the public had the idea that the events being examined by the inquiry were in the past and couldn't affect them, it was shattered by these revelations.' 'God, I'm glad I'm not a woman', one of the reporters told Coney.[41]

Not all women believed they were being victimised, however. Following the burst of adverse publicity relating to examinations under anaesthesia, National Women's medical superintendent Dr Gabrielle Collison received considerable support from ex-patients. One was a secondary school teacher in her forties who claimed to have had 'quite a number of vaginal examinations' and who believed that she was 'imparting the attitude of the silent majority of women patients'. She wrote, 'I would hate to be a patient with problems in the reproductive system, having to put my trust in a doctor who because of the inhibitions of the society in which he was trained had little or no experience of investigating my condition.'[42] Another informed Collison that she had received excellent treatment when she had a hysterectomy in the 1970s: 'Also I was asked and gave my permission for medical students to be present when I was examined and I was told they would be present whilst my operation was in progress. I was not embarrassed. I was not humiliated. I did not feel degraded.' Referring to Rutter's statement in the *Sunday Star* about the teaching hospital, she added, 'I heartily agree, for Doctors have to learn.' She concluded, 'Keep up the good work!'[43] M. W. also wrote in support, stating that her permission was asked for a medical student to examine her internally and that she was treated with the 'utmost respect and courtesy'.[44] M. K. wrote that she had been in Ward 9 (the cancer ward) in January of that year to have caesium treatment and six weeks later underwent a hysterectomy. 'I met wonderful Nurses and Drs, who treated me with respect as a person not just a patient. Everything that was going to happen to me in both operations was made crystal clear and at no time did I feel uncomfortable or exploited.'[45] However, the publicity did frighten those with no experience of the hospital, such as V. S. who wrote to Collison that as a 'potential patient' she would now view admission to National Women's with great anxiety.[46]

Another story that hit the press contained further revelations about Green's research. 'Secret smears for new babies', ran a press headline, which reported that between 1963 and 1966 more than 2200 baby girls had smears taken with small cotton swabs inserted into their vaginas.[47] This was a research programme initiated by Green into possible cellular forerunners of CIS.[48] When counsel assisting the Inquiry looked into the matter, they reported 'for public distribution' that:

The swabs were collected during the day when the babies were with their mothers. Mothers were told of the tests and the Special Duties Sister can remember no protests. She is adamant there was no selective process and that no distinction was made between the babies of public and private patients. No racial or other bias applied. The babies were not harmed by the procedure.[49]

However, the press described it as 'sick, the vilest practice yet to be revealed at an inquiry which has uncovered some gross practices'.[50] Human Rights Commissioner Rae Julian was 'appalled'.[51] Medical superintendent Gabrielle Collison explained to distressed callers to the hospital that not a smear but rather a vaginal swab specimen had been taken.[52] While Bonham explained publicly that there was no risk of damage to the hymen and no instrument like the speculum was used,[53] the *New Zealand Herald* cited a request from the 'Regional Maori Health Board' for the suspension of 'those responsible for the deaths of women and female infants'.[54] Despite the statement from the special duties sister that mothers were told, Cartwright concluded in her report that 'there was no provision made to comply with the fundamental requirement that children are not included in research without the consent of their guardians'.[55]

In presenting the Fertility Action submission, Coney referred to 'Dr B' whose patient had died under anaesthesia, but who was still practising, and noted, 'Peer review ultimately failed to deal with this doctor.'[56] Coney need not have referred to him anonymously as 'Dr B'; the 1982 case involving Dr Ian Barraclough had been reported in the *Herald* at the time, which mentioned him by name.[57] He was charged with 'professional misconduct', though the Medical Council had also received a petition with 131 signatures attesting to his professionalism, dedication and sensitivity. In referring to other such doctors, Coney made it clear that Green was not the only doctor under scrutiny.[58]

One 29-year old woman who said she had no children as yet, wrote to Collison to express her horror at what had been revealed by the Inquiry: 'I am absolutely disgusted at the way the medical profession treats women and defenseless female babies. In my view these women and babies have been raped. This man [Green] is sick.' She found it 'almost beyond belief' and concluded, 'I will do my utmost to avoid National Women's Hospital and any male doctor if I decide to have a child. Thank God for Sandra Coney who exposed this criminal situation.'[59]

Coney described the level of public support during the Inquiry as 'phenomenal'. She explained that she could not even go shopping without being stopped by complete strangers and encouraged to keep it up. 'We had said in our opening submission that we saw ourselves as representing women, and it was clear this was the perception of our role out in the community . . . . Some days when I came home from the hearing there would be flowers, cooked meals and bottles of wine waiting.'[60]

Day 57 was Coney's 'favourite day of the whole inquiry, because for the first time, and when I had most despaired it would ever happen, we were hearing from women themselves'.[61] These women were not patients, who had given evidence at the beginning of the Inquiry, but were part of the new women's health movement. They included representatives from THAW and the Glenfield Women's Community Health Co-operative. Sue Neal, who had worked with Coney on Depo-Provera, represented the Auckland Young Women's Christian Association. The day's proceedings were reported in the press. Neal said National Women's had stood out as a hospital which was 'extremely resistant' to any consumer participation. 'The vast majority of women I know have not particularly felt any great respect for National Women's for quite some time', she said. She thought the hospital was 'notorious for insensitivity'. She said that her comments were mainly directed at the medical staff, and one or two well-known members in particular, though she did not mention any names. Regarding examinations under anaesthesia without consent, she said, 'We feel affronted by the medical profession's indifference to our dignity as women and our vulnerability at such a time.' A Glenfield representative, Phillipa Thompson, said that when National Women's came up in discussion among women, 'inevitably there would be a series of horror stories'. Another speaker, Therese Weir, said there had been anger, frustration and also celebration that women were, at long last, talking about some of the experiences which had affected them for years.[62]

The consumer representatives did admit, when asked by Cartwright, that they had heard of some positive experiences with the hospital. The Judge said she had received many letters during the Inquiry and most of them, largely from the middle-aged and elderly, had spoken very highly of their experience there.[63] Mrs C. M. Purdue sent a submission to the Inquiry in which she discussed Green's role as patient advocate, including helping a group of women to draw up a Maternity Patient's Bill of Rights in the 1970s. Mrs Purdue stated in her submission to the Inquiry that, as chairwoman of the Auckland Hospital Board's Investigation Committee and a member of the Appeals Committee, she had heard nothing but praise from Green's patients. Further, 'In conversations with him, I felt he had the best of intentions towards his work with women and was considerate beyond the measure of ordinary professional care for their future.' She was convinced he did his best 'and perhaps more than that'.[64]

Some ex-patients wrote to the press: Mrs J. M. Snow declared, 'From the gentle and caring surgeon right down to the cheerful tea lady, everyone was wonderful to me, as they were to all the ladies who had only praise for them.'[65] The New Zealand Herald also published a tribute from Mrs R. Truscott of Grey Lynn: 'I, for one, take strong exception to Sue Neal's assertion that National Women's Hospital was notorious for insensitivity. I have been going there for many years, and have found the doctors and staff not insensitive, but kindness itself. I look forward to going there knowing I am being well looked after. It does not matter to me who examines me. How else are young doctors to learn. To me, doctors are unsung heroes.'[66] Others wrote directly to the hospital, just as they had done following the revelations about internal examinations by students. S. A. had given birth to five children at National Women's and had suffered one or two gynaecological disorders over fifteen years. She wrote to Collison, 'It has been said that "Delivery suite" is like a baby factory. It may well look like that, but only because of the phenomenal workload the staff seem to get through. I have always felt very much an individual [there].' She ended her letter, 'So chins up National Women's, you are not without your supporters, despite the recent bad publicity. Good Luck to each and every one of you, and many thanks for super care.'[67] P. W. of Palmerston North whose wife had died the previous month of cervical cancer wrote to the hospital telling them that there was 'a big silent majority of the public that admire you all for the dedicated work you do so don't let this present publicity get you down. I know you guys are truly unsung heroes.'[68]

However, far from seeing the doctors at National Women's Hospital as 'unsung heroes', Cartwright wrote in her report, 'With some regret I have concluded that I cannot leave the encouragement of new habits and practices to the medical profession alone.' [69] She found National Women's staff 'extraordinarily insensitive', with 'a pervading atmosphere of defensiveness and even arrogance, [which] does not bode well for the future care of patients at National Women's Hospital'.[70] Cartwright told the press, 'I don't pretend that my report is going to please some of the doctors who read it, and I don't think for a moment communication will be enhanced in that small group, but I am aware, as no doubt everyone in this room is that there are very many doctors – male doctors – who are very caring in their attitudes towards women.' The 'small group' were National Women's doctors – clearly differentiated by their 'lack of caring' from the majority.[71]

Cartwright referred to one woman who contrasted her treatment at National Women's with the 'helpful' treatment at Auckland Hospital, where, Cartwright said, there had been no attempt to gloss over problems or to patronise.[72] Cartwright chose to highlight a negative perception of National Women's when there was considerable evidence presented that supported an opposing view. Correspondence received by medical superintendent Gabrielle Collison, letters to the press, submissions to the Inquiry itself and letters received by Health Minister Michael Bassett suggest that there were many ex-patients of the hospital who were prepared to speak up in its defence.[73]

Submissions from Women's Groups

Some feminists saw the Inquiry as an opportunity publicly to air broader concerns about the behaviour and attitudes of male doctors, and the tone of their submissions was one of anger. The Auckland Women's Health Collective sent a submission to the Inquiry: 'we submit that there is every evidence of arrogance and . . . inadequate training and guidance for Doctors in virtually all women's health care, and most particularly to gynaecological health. There would also appear to be inadequate research being done into the CAUSES of the incredibly high rate of cervical cancer, and we would like to know why.'[74]

The Wellington Women's Health Collective alleged that the doctors at National Women's, with the exception of McIndoe and McLean, had closed ranks and did not challenge Dr Green, even when they did not approve of his 'experiment', because 'mateship' was more important to them than patient welfare. They also maintained that, 'Medical practitioners, especially obstetricians and gynaecologists, disclaim women's experiences and the knowledge they have of their own bodies', explaining that women were frequently labelled as emotional or neurotic without the doctor exploring the cause of the reaction. On a more practical level, they were 'concerned that Fertility Action should be incurring such high personal costs for bringing the "Unfortunate Experiment" to our notice'.[75]

Writing on behalf of the New Zealand Women's Health Network, Sarah Calvert who had completed a Waikato University PhD on women and mental health in 1980, referred to the evidence given by 'Professor Dennis Bolham' [sic], and maintained that the Network believed that this information further supported the contention of the writers of the *Metro* article and other women's health groups that the attitudes towards women held by the medical profession in general, and by doctors at National Women's in particular, were 'horrendous'. She did not provide details on what Bonham had said that had offended so much. The Network opened its submission with the following statement: 'It is the view of the New Zealand Women's Health Network that in general the New Zealand Medical Profession displays sexist attitudes and racist attitudes towards its patients.' It claimed that doctors believed that women were not sufficiently intelligent and capable human beings to be given full information and 'secondly the medical profession has a general contempt for its patients and in particular, women patients and treats them inadequately and badly'. The submission explained that it was their general experience that the views of women consumers or professional women who were critical of the medical profession were treated with derision and amusement, an attitude which they believed pervaded National Women's. In the opinion of the network, the medical profession was an elitist group and reflected the views and opinions of white middle-class males. Noting that women found it hard to obtain information on alternative health practitioners, the submission also argued that the concept of informed consent meant that clients should receive all available information, even if it was regarded as 'frivolous or non-scientific'.[76]

The submission of the Ministry of Women's Affairs was on similar lines to that of the Women's Health Network. The submission opened with a statement that they spoke on behalf of the women of New Zealand, and that the events at National Women's Hospital did not stand alone but were part of the way in which the attitudes of medical practitioners to 'women as case studies' influenced treatment practices to the detriment of 'women as human beings'. They believed that women all over the country had been outraged by what they had heard, 'especially as they have identified with the powerlessness of the women who entrusted the medical experts with their bodies and their lives'. Pointing out that women were the major consumers of health care, the ministry regretted that women's reproductive health was dependent on the attitude and skills of men who dominated medicine, in policy-making and practice.[77]

Te Ohu Whakatupu produced a separate submission. Prior to the submission, Miriama Evans, its director, had held hui around the country to solicit Maori women's views. She reported on 'a conservatism amongst many Maori women about getting involved in an issue such as this because they see it as the feminist type issue and they don't want to get involved with the political implications of that'. Evans expressed her concern that Maori women with whom she had spoken said, 'I'm alive today so I must be okay.' She was concerned for these women because of their absolute trust in the medical profession, and because they had no access to their medical records. She explained that in trying to encourage Maori women to come to the Committee of Inquiry, she would discuss what had happened to the babies (smears of newborn baby girls) and the women were 'horrified and shocked' about this, 'and I say to them "well that happened to you" and they said, "but that is different." I said, "how is it different, you weren't given the choice to know whether you are well or not well?"' Evans believed that it was particularly important for Maori women to have advocates 'to help them in that empowering process' as 'they won't insist on their rights'.[78] When Ruth Norman subsequently followed up Maori women who had been to National Women's following a positive cervical smear, she told Cartwright, 'I am bound to say that a number of the women said they felt sorry for Professor Green after all he had been doing the best for them. Others also noted that they felt the only reason they were still alive was because of his care. Some enquired after his health. Others felt he had provided them with adequate explanations about the care they received and were satisfied.'[79] Many Maori women did not appear to subscribe to the dominant feminist political perspective.

Te Ohu Whakatupu's submission was, nevertheless, highly ideological. It stated that the inequalities of power uncovered in the 'unfortunate experiment' were reinforced by the hierarchy of race: 'From a perspective which acknowledged the sacredness of the female genital tract as te whare o te whenua the enormity of the systematic violation of the dignity of Maori women as women and as people was clear.'[80] Cartwright included this viewpoint in her final report: 'The implications of the sacredness of the genital area for Maori women cannot be underestimated [sic]. They will have repercussions not only for population-based cervical screening but also for the treatment and monitoring of CIS as well as invasive cancer. There seems to have been little cultural understanding of these mores on the part of the profession.'[81]

The Cartwright Report was responding to the greater sensitivity to Maori culture and identity that had emerged from the 1970s. Again, it addressed a much broader issue than the specific approach of Green. The report's castigation of doctors as insensitive to Maori women patients could even be seen as ironic in an investigation that centred on Green. Te Ohu Whakatupu's submission included a reference to the prevailing epidemiological belief that promiscuity was a causal factor in cervical cancer which, it said, had led one doctor to ask his Maori patient with a positive smear, 'and how many pricks have you had?'[82] This is not an approach Green would have condoned.

Broadsheet considered it 'very appropriate' that the Fertility Action submission was the final one at the Inquiry and that it was presented by Sandra Coney who had fought constantly to keep the perspective of women patients before the commission.[83] Coney played an important part in the Inquiry, including cross-examining some of the witnesses such as Ministry of Women's Affairs representatives.[84] She appeared as a witness for five days in total, first in her own right, when she read the text of the *Metro* article, and later as the Fertility Action representative. A student who analysed the media representation of the Inquiry later commented that, 'Throughout the Inquiry Coney, who sits very still, constantly wore an expression of careful unsmiling attention. Any stills shown of her at the time capture this expression which gives the impression that she regarded the proceedings with the utmost seriousness and wanted to see justice done.' The student noted that in calling for evidence from overseas doctors, Coney was 'seen challenging both the professional competence of the doctoral team at National Women's and the ability of the institution itself, the seat of their power, to cope with the situation'. She concluded, 'Coney's stand against the power of doctors called for women to be empowered by changing the structures and practices that caused their subjection to the whims of the medical profession in the first place.'[85]

Coney was a formidable opponent, something she attributed to a childhood training in which she was constantly challenged by her father. In her book on the Inquiry she noted, 'On one school report, my headmistress wrote: "Sandra takes unkindly to any form of criticism and is quickly on the defensive to prove she could not have been in the wrong. She would be wise to conquer this tendency before it spoils an otherwise pleasant nature".' She added, 'I never had conquered the tendency, and as a consequence I was better placed in the witness box than I might have been.'[86]

In her own submission to the Inquiry, Coney stressed that the issues were not historical, a distinction she felt was important in order to initiate change. She declared that it would be erroneous to give the impression that Green was solely responsible for either the 'trials', or the general management of patients with cervical carcinoma in situ and/or invasive carcinoma. In her view, it was important for the Inquiry to recognise that many other doctors were involved in what happened so that they too could be held accountable. She gave the example of a woman seen by eight doctors apart from Green, including Mont Liggins who first saw her in 1965, before she died of invasive cervical cancer ten years later. Another patient, originally under the charge of Bill McIndoe in 1971, died of invasive carcinoma in 1976, and Green was at no stage involved in her management. Nine doctors apart from Green saw 'Mrs M.' over the years, and twelve other doctors saw 'Gladdy' before Green's retirement.[87] Dr Graeme Duncan in Wellington was also implicated, as was Bonham.

Coney noted that Bonham was asked in the interview with herself and Bunkle, 'When you are talking about CIN3, do you see that as cancer?' The reply, 'I don't think you should', was given as evidence that Bonham, like Green, belittled the diagnosis of CIN3.[88] National Women's Hospital medical superintendent Gabrielle Collison was also accused of downplaying its importance. Coney cited Collison's view, expressed in a 1987 *Herald* article, that there were two separate conditions of the cervix: 'One is the malignant progressive disease which is managed in a standard fashion throughout the world by surgery and/or radiotherapy. The second is carcinoma in situ – a pre-malignant disease which may progress over a period of many years to malignant invasive cancer.'[89] This was disputed by pathologist Dr George Hitchcock, who claimed, 'The statement is quite incorrect as cancer in-situ is cancer whether it occurs in tongue, skin, cervix or any other site.'[90] Yet this statement about CIS of 'any other site' was not without its detractors. Historian Barron Lerner wrote in relation to ductal carcinoma in situ (DCIS) and lobular carcinoma in situ (LCIS) of the breast:

> Although physicians characterized the removal of LCIS as the preventive elimination of a precancerous lesion, much of the impetus to be aggressive resulted from the conflation of LCIS with actual cancer . . . . Yet as critics began to argue in the middle 1970s, discovering that a procedure was 100 per cent curative called into question what was being treated in the first place.

According to Lerner, this led some eventually to adopt 'watchful waiting'. He further wrote that in correlating DCIS and LCIS with cancer, 'physicians pathologised the healthy breast, going so far as to term it "precancerous" or a "premalignant target organ"'.[91] Yet Coney quoted Hitchcock and concluded that, 'Ultimately, perhaps, one of the very most important issues is that of current practice. Certainly the fact that Professor Bonham and Dr Collison express thinking at odds with the rest of the world should make us suspicious of the way in which carcinoma in situ of the cervix is being treated currently at National Women's Hospital.'[92] As discussed earlier, they were not 'at odds with the rest of the world' in questioning the meaning of a diagnosis of CIS of the cervix.

Sandra Coney wrote the Fertility Action submission along with Phillida Bunkle, their counsel Dr Rodney Harrison and Dr Forbes Williams, a young medical graduate who was at that time taking a women's studies course in Bunkle's department at Victoria University of Wellington.[93] The submission was 77 pages long, plus appendices, and ranged widely through medical education, medical research and informed consent, to medical accountability. It was viewed as an opportunity to canvass all issues relating to women and male doctors generally and Green's research itself was rarely mentioned. Echoing the Ministry of Women's Affairs, Fertility Action's submission declared, 'It is our view that the inquiry has been about power: the power of the medical profession and patients' lack of it.'[94]

The submission included a section on teaching medical students, and particularly 'hidden agendas'. The authors stated that, 'Anaesthesia is an extreme example of the relations of power and powerlessness which characterise relationships between doctors and patients. This practice [of examining women under anaesthesia] "teaches" that exploiting this power for convenience is acceptable.' They further noted that Professor Mantell's evidence showed that students at Auckland Medical School were currently assessed on 'Passing a Speculum' three times and 'Examining a Pregnant Abdomen' ten times. But, they continued, 'abdomens are not pregnant, women are. Women are not ambulant or non-ambulant uteri, which may or may not grow significant lesions.' They warned, 'While doctors are trained to see only collections of cells and not whole, individual people nothing will change.'[95]

The submission also included a statement on 'Drug Company Funded Research – Upjohn's New Zealand Contraception and Health Study', in yet another attempt to extend the Inquiry well beyond the research of Green, who never engaged in this sphere of research. They pointed out that Associate Professor Robert Beaglehole of the Auckland University School of Medicine and Professor David Skegg of Otago University had declined to be consultants to the study because of the lack of independence from the Upjohn Corporation. They reported that there had been a great deal of criticism of the study, including a television documentary in 1983, to which Liggins had responded in the *New Zealand Times* (6 March 1983): 'I don't give a damn whether Depo Provera is the safest drug or the worst drug . . . . What I am interested in is finding out about it. Were it not for the Depo Provera controversy, we would be able to do that without all the problems we are having.'[96] The Fertility Action submission was not impressed with the conclusions of the MRC Scott Committee, which had assessed the protocol.[97]

The Fertility Action submission concluded with a section on 'Cervical Cancer Screening' in which they advocated annual screening, particularly for younger women. They recommended that 'Professor Bonham and Dr Jamieson be professionally confronted. If necessary be subjected to disciplinary proceedings; [and] that doctors holding anti-cervical screening views be removed from teaching courses where the subject matter is cervical cancer and screening.'[98] Thus, no dissension from the dominant or 'correct' view was to be tolerated and those holding opposing views were to be removed and even disciplined.

These submissions – from draconian suggestions like disciplining wrong-thinking individuals to more generalised proposals to 'humanise' medical training – were part of a broader agenda to change society and specifically gender relations within medicine. As one member of a women's group told Bassett in July

1987, 'Although men no doubt also experience a problem having their medical needs and rights met, women are more at risk.'[99] The reason was, she believed, the patriarchal nature of society.

Nurses' Submissions: 'less than brave'

Cartwright complained during the Inquiry that nurses had been 'less than brave' in coming forward to give evidence. She was concerned that if nurses were to be patients' advocates they would have to be more courageous than they had demonstrated to her so far.[100] Giving evidence to the Inquiry, one nurse claimed that medical superintendent Gabrielle Collison had issued a memorandum stating that 'nurses currently employed by National Women's Hospital are advised not to go forward to the hearing and it is in their hospital's best interest and their own interest that they do not take part in the hearing'.[101] Collison denied issuing any such memorandum. The only related document she could think of was one issued on 26 June 1987 on the direction of superintendent-in-chief Dr Leslie Honeyman, reminding staff of their obligations under the Hospitals Act concerning the confidentiality of patient information.[102] However, one nurse told the Inquiry that her colleagues had felt they would be victimised and their career prospects hampered if they spoke out.[103] The Nurses' Society of New Zealand national director David Wills responded that no nurses had approached that society to say they were scared to make submissions to the Inquiry.[104] Pam Hayward, who had first worked at National Women's as a staff midwife in 1971, and was charge nurse from 1981 to 1996, later commented that nurses at the hospital were 'quite incensed' at being called 'less than brave'. They told Silvia Cartwright at a meeting she convened that most worked on the obstetric side in any case.[105] Another nurse, Barbara Smith, interviewed in 2004, did not believe National Women's Hospital nurses were 'less than brave'. She said, 'I actually don't believe that, because believe you me, some of the nurses at National Women's, they would speak out all right and they would speak out loud and clear if they thought anything was going wrong.'[106] Nevertheless, Cartwright expressed reservations in her report about the ability of nurses to stand up for their patients: 'Nurses who most appropriately should be advocates for the patient, feel sufficiently intimidated by the medical staff (who do not hire or fire them) that even today they fail or refuse to confront openly the issues arising from the 1966 trial.'[107]

While some nurses working at the hospital clearly believed that Green and other doctors at National Women's did their best for patients, the view put forward by Cartwright and repeated by others, that 'doctors had failed in their duty to patients', was generally accepted by the nursing profession along with other members of the public. This alleged failure on the part of doctors was of more than academic interest to some nurses, however; they construed it as a specific opportunity for them as a professional group. A letter to the *New Zealand*

Herald in November 1987 declared, 'In this inquiry, nurses can only improve their professional integrity by discarding the acquiescent handmaiden role they have accepted for far too long, and standing up to be counted as an equal and vital part of the health team.'[108] Another letter to the *New Zealand Nursing Journal* stated that Cartwright's remarks regarding nurses' silence were a timely reminder to nurses that they were meant to be patients' advocates, and advised that, 'If nurses wish to be thought of as independent professional practitioners then surely this situation must change.'[109] Nurses were currently in the process of professionalising and discarding the 'handmaiden' image which extended back to the influence of Florence Nightingale in the nineteenth century. This process can be seen in the transfer of nursing education from the hospital to tertiary institutions that had started in the 1970s; nurses were no longer trained on the wards.

The nurses' professional organisation, the New Zealand Nurses' Association (NZNA), presented two submissions to the Inquiry, by the northern regional officer Carol Mitchell and the association's professional officer Joy Bickley.[110] The submissions emphasised the need for changes in structure and practices at National Women's Hospital. Bickley declared that the events at the hospital could only be understood in terms of power relationships.[111] Responding to Cartwright's claim that nurses had been 'less than brave', the submission stated that the NZNA was committed to the concept of a more assertive nursing professional but believed that no amount of assertiveness training would suffice if structural power imbalances were not addressed. As women, she said (and 94 per cent of their members were women), they identified with the patients at National Women's and their powerlessness. Recommendations included 'a form of education that corrects the imbalance of power between doctors, other health professionals and patients'.[112] As Bickley later wrote, both nurses and patients at National Women's had felt relatively powerless, though they supported each other. What they had in common, she said, were their experiences as women: 'Nurses and consumers share the same struggle to have their voices heard.'[113]

The New Zealand Nurses' Industrial Union of Workers, whose 10,000 members were drawn from the private sector, also made a submission. Like the NZNA, it believed the central issue, which it regretted had not been included in the Terms of Reference, was the power structure at the hospital and the fact that women were denied equality in participation in decision-making in the health care system.[114]

Nurses thus aligned themselves with female patients who experienced male domination. This intensified during the public discussion of internal examinations and IUD insertions performed on anaesthetised women by medical students. The NZNA annual general meeting in September 1987 condemned 'the violation of women's rights perpetrated by medical personnel carrying out vaginal examinations under anaesthetic without the informed consent or knowledge of the woman, and expects all nurses to ensure that the NZNA Patients'

Code of Rights and Responsibilities is upheld'.[115] Thirty nurse tutors sent a letter to Health Minister David Caygill criticising the practice. Nurse tutor Nicola Hill said the letter was to 'express our abhorrence and to make clear that nurses do not condone this practice'.[116]

Despite this attempt to distance themselves from the practice, some commentators cited this as another example of nurses failing to act as patients' advocates. One nurse told *Nursing Journal* readers that the exposé was 'a salutary lesson' for nurses everywhere and that nurses should take responsibility to ensure that patients' rights were not subordinated to the teaching needs of doctors.[117]

S. Pemberton, Charge Nurse of the Operating Theatre at National Women's Hospital, responded to the adverse publicity on behalf of her staff in a letter to *The Listener*. She wrote firmly that they wanted to clarify one area of concern in the Cartwright Report as reported in Coney's recent article in that magazine. She wished to assure women that the nursing staff were very aware of patients' rights and were not intimidated by medical staff when assuming the role of patient advocates.[118]

There appeared to be a division between nurses who worked at National Women's Hospital and those nurses and their organisations who took the opportunity to reflect on general relations between nurses and doctors in health care services. Similar divisions appeared between women who reflected on gender relations in medicine in general and those who had personal experience of National Women's and wished to defend its doctors. Maori women too appeared divided – not all supported the Ministry of Women's Affairs perspective and indeed some did not wish to become involved in what they perceived as a 'feminist' matter. Women were far from united on this issue.

The Unfortunate Experiment: The Feminist Perspective

In 1988 Sandra Coney published *The Unfortunate Experiment: The Full Story Behind the Inquiry into Cervical Cancer Treatment,* for which she received a 1989 Goodman Fielder Wattie Book Award. At that time she explained that she had written the book because the Cartwright Report was 'not readily accessible to ordinary people'.[119] She also felt 'the need to record everything from a feminist perspective . . . [and] to demonstrate that the whole process had been a feminist effort'.[120] The book was advertised in the same issue of *Broadsheet* which announced the Cartwright Inquiry as a feminist victory.[121] Featured in the 'Women's Book Festival', it was reviewed by Pat Rosier who had reported monthly in *Broadsheet* on the Inquiry. At the same time, Rosier reviewed Phillida Bunkle's book, *Second Opinion*, which, she said, provided a context for the Cervical Cancer Inquiry and included the unedited version of the 1987 article in *Metro*.[122]

Reviewing Coney's book in the *New Zealand Listener*, Joy Bickley described it as 'a tale about those who hold power and how they retain it. It is also, however, about the courage of those who are victims of that power and of those who challenge it. It is the full story so far. Its most important lesson is that the power structures still exist and that changing them is going to require even more struggle.'[123] Judith Medlicott, a Dunedin lawyer and later chancellor of the University of Otago (1993–98), reviewed Coney's book in the *Dominion Sunday Times*. She declared, 'Coney has delivered a tour de force. It records a feminist achievement not far short of those women a century ago who worked for women's suffrage.' On reading it, she claimed, 'it takes a little time to register each new horror'; and she concluded, 'All women should be grateful to Sandra Coney and Phillida Bunkle.'[124]

Others reviews were equally complimentary. Michael King declared in *Metro* that, 'With its persuasive indictment of the high-handed way in which the medical profession has conventionally dealt with patients and banded together to protect itself from outside attack, this book ought to have far-reaching consequences. By presenting a patient's-eye view and review of medical procedures, it will shame doctors (or at least those of them sufficiently open-minded to be influenced by the revelations of the Cartwright inquiry).'[125] Margot Roth, the editor of the *Women's Studies Journal*, commented in the *Sunday Star*, 'the narrative's main theme is, in the words of Linda Kaye [the lawyer appearing for the Ministry of Women's Affairs at the Inquiry]: "To those who hold power, its retention in their hands is an absolute priority."'[126] A visiting epidemiologist from Oxford, Professor Klim McPherson, described Coney's book as 'amazingly moderate', and advised that, 'The medical profession must explicitly admit abusing women.'[127]

Coney herself had concluded *The Unfortunate Experiment* by stating that 'the real problem was medical power and its exercise. It could easily have been another doctor, another hospital and another city altogether. Without radical change, in the future it could be.'[128] She explained that the feminist organisation Fertility Action was seeking 'sweeping changes in the health scene: legislation to improve patients' right and control over research; patients' advocates in institutions; independent avenues of redress; consumer representatives to be involved in decision making about health'. It also wished for a 'fresh start' with new personnel at National Women's Hospital, and a nationwide cervical cancer screening programme.[129]

While Coney argued that the Inquiry was about wider issues, her book focused closely on Green and his research. To understand the thrust of the book, it is necessary to consider two areas: one is the ways in which the main protagonists, the medical profession, were portrayed, and the second is the manner in which the experiences of Green's patients were recounted.

The Unfortunate Experiment: The Doctors and 'overwhelming misogyny'

Sandra Coney made her views of the medical profession very clear when she commented on the 'overwhelming misogyny of everything we had heard over the sixty-five days of the inquiry'.[130] Coney and Bunkle had already profiled Green in the *Metro* article, and, according to the Medical School dean David Cole, the Judge 'seemed to accept some of the accusations by Coney and Bunkle about Green, which painted him as the archetypical male chauvinist doctor accusing him of being anti-abortion and a founder member of SPUC . . . which he was not'.[131] While Coney's book omitted the SPUC reference which had appeared in the *Metro* article, she nevertheless sketched him as chauvinistic. As Cole wrote, 'It was . . . said maliciously that his first reaction to the news of Silvia Cartwright was to say he would not testify in front of a female judge; not a good start'.[132] This allegation was repeated in Coney's book.[133]

Describing her interview with Green, Coney commented that, 'He did not have the quickness, the verbal sprightliness, that accompanies an agile mind'.[134] She also found him to be physically slow at the Inquiry, which showed she said a 'slow mind' and an 'inability to assimilate new information'; she did not mention at that stage that he was 71 years of age and that he ended up in hospital with pneumonia immediately after the cross-examination.[135] At the end of the chapter, however, she did note that he was 'old and a heavy smoker'.[136]

Regarding Green's medical standing, Coney wrote disparagingly that he belonged to 'a one-man medical flat earth society'.[137] Yet in her evidence to the Inquiry she cited Green's 1972 claim that conservative treatment was gaining support, and used this to show that he had influenced other doctors. There she wrote, 'Although Professor Green was basically the source of the thinking and clinical practice regarding carcinoma in situ of the cervix, these ideas were adopted by others as their own, and then not only in National Women's Hospital'.[138] The Fertility Action submission had also noted, 'Among the medical profession and hospital administrators, Professor Green's views still continue to have their supporters as evidenced by the letters supporting his views among the volume of correspondence produced to this inquiry by his counsel. In the written submissions, a Christchurch doctor calls the belief that cancer is preventable by screening "false" and "alarmist propaganda"'.[139] It was hardly a 'one-man' society.

Coney also wrote that during Green's cross-examination, 'words simply ceased to have their normal meanings', giving as one example that diagnosis also meant treatment.[140] Yet in this instance it was common knowledge that cone biopsy was both diagnostic and a treatment.[141] The 1987 edition of *Jeffcoate's Principles of Gynaecology* unequivocally stated, 'The cone biopsy essential for the diagnosis of the disease is usually curative'.[142]

Nor did Bonham fare any better in Coney's account. She wrote, 'There was hardly an expression of concern for the patients in the whole of Bonham's evidence'.[143] She called for Bonham's retirement. She complained that he should not be allowed to exercise power in academic and clinical matters when the Inquiry had shown how he had tried to shift blame to one protagonist who was conveniently dead (McIndoe). Even more importantly, she claimed that he 'value[d] academic and intellectual honesty so little that he is prepared to put his name to an editorial in a medical journal expressing views he claims not to agree with'.[144] The 'editorial' was a reference to a joint article by Bonham, Green and Liggins, questioning the value of population-based screening. Bonham told the Inquiry that 'he supported screening, he just had a few reservations about whether it would work'.[145]

Coney was equally disparaging of Dr Gabrielle Collison, describing her as someone 'who had a talent for turning awkward situations to her advantage'.[146] Collison, who had taken over as medical superintendent at National Women's Hospital in 1985 and was the first female incumbent, was not regarded as part of the sisterhood. In a selection of her writings published in 1990, Coney opened the section on the start of her feminist career with a 1973 article which began: 'On 29 January in the *Auckland Star* Dr Gabrielle Collison (29), formerly a school medical officer in Timaru, and now newly appointed to the post of Deputy Medical Officer of Health for Auckland said she was not "an advocate of women's lib"'. The reason, according to Coney, was that, 'For most women, being liked by men is what being a woman is all about'.[147]

Sandra Coney reiterated that it was not unusual for eight or more doctors to see a given patient; all doctors at National Women's were therefore implicated.[148] She wrote that there was little evidence of challenges to Green's management, except for Bill McIndoe who sometimes took the opportunity when Green was overseas 'to inaugurate proper management'.[149] In Coney's book, McIndoe emerges as a selfless campaigner to save women. Some of her assumptions regarding him are, however, open to question. She wrote, 'Even McIndoe, as a relatively new doctor, who was not a member of the elite HMC [Hospital Medical Committee] and who was not an academic, felt he could not make himself heard, nor muster support for his cause. It was, said Hugh Rennie, like the "office boy" challenging "the managing director" . . . . McIndoe was a cytologist and a colposcopist, rather than a clinician'.[150]

Her description of McIndoe 'as a relatively new doctor' is puzzling. McIndoe arrived at the hospital just five years after Green. He was, like Green, a fellow of the RCOG. As Coney noted elsewhere in her text, he was a gynaecologist who did 'perform extensive surgery' on patients.[151] As an obstetrics and gynaecology specialist, McIndoe was also on the emergency duty roster for the B and C teams at National Women's, though he did not usually perform caesareans or hysterectomies.[152] The analogy of the 'office boy versus the boss' does not work. In this case, the 'office boy' had the same professional qualifications as the 'boss' and it would be an unusual office boy who claimed more expertise in a professional matter (colposcopy) in which the boss was vitally interested, and who described the boss as 'lacking in maturity' and 'over-enthusiastic' in his advocacy of colposcopy.[153] At the Inquiry, Green pointed out that McIndoe was a recognised specialist, and that Green expected McIndoe would carry out the treatment he thought correct and 'not just what somebody else wanted him to do and in fact, he did so'.[154]

Coney reported a claim by Jock McLean that, 'A member of the professorial unit . . . came down to my room one time and [said] quite bluntly everything was under his control including me.'[155] If this referred to Green, the summation could not be applied to McIndoe; he exercised choice in refusing to perform colposcopies for Green, as Coney noted.[156]

McIndoe's version of events was accepted without question in Coney's account. A phone call with McIndoe 'confirmed' to Coney and Bunkle that they 'were on the right track'. McIndoe told Bunkle that a professor called Green had been 'treating people in all sorts of different ways and observing the results'. McIndoe's claims that Green stopped publishing in the mid–1960s when the results were 'discouraging', and that others were still following Green's conservative management in the 1980s, have already been discussed (see chapter 5).[157]

Coney also portrayed McLean in glowing terms, stating that he was the only doctor who really appeared to put the women's interests first and could see the human suffering, not just the statistics. She argued that he was the only New Zealand doctor at the Inquiry to talk about the accountability of the profession.[158] In addition, 'he clearly saw patients as people', and 'came through it [the Inquiry] with flying colours, and through it all managed to inject some humour. It is the only place in the transcripts where the typists noted laughter.'[159]

While Coney used this incident to show McLean's personable side, a hint of levity by another witness was frowned upon at the Inquiry. Richard Seddon later recounted –

> . . . minutes of committee meetings were produced, some of them I'd never seen before, and I was asked to comment on them. I can remember the judge actually asking me, did I recall a particular meeting when I think something about the study had been discussed, and I couldn't recall that. But I noted that at the meeting they had discussed the idea of husbands being present at the birth which produced a ripple of amusement around the court, and upset the judge, who thought there was no room for levity in this inquiry.[160]

Apart from McLean and McIndoe, National Women's doctors fared no better than Bonham and Green in Coney's account. She wrote, 'Other clinicians were prepared to let Green try out his theories on patients, because such variations were acceptable within the doctrine of clinical freedom.'[161] She maintained that there was 'something larger than human, almost saintly, about the image doctors at the inquiry had of their profession', and described it as 'a medical mutual admiration society'. In her view, concepts like 'integrity', 'sincerity' and 'good faith' were the doctors' ultimate defence, and were 'the most overworked words at the hearing'; in her view they did not amount to much.[162]

While Coney was adamant that Green was not the only doctor under investigation, she still criticised other doctors at the hospital for not distancing themselves from him. Coney wrote that the hospital 'adopted a siege mentality, pulling up the drawbridge against what seemed to be perceived as a malicious attack from outside which threatened all within'. She noted that far from distancing themselves from Green, as some commentators had suggested would occur, the hospital staff formed 'an almost solid phalanx of support'.[163] Coney claimed that, 'The collective silence continued up to 1987, even though the legacy of Green's experiment was still coming into the cancer wards', and asked 'How anyone could be reassured it couldn't happen now?'[164] She described the 'fatal scenario':

> We have the institution, as hierarchical as the Vatican, ruled over by the pope professor with his priestly clinicians, and serviced by the acolytes/nurses. We have the eminent senior clinician, protected by his shroud of clinical privilege and the doctrine of brotherhood; the resident 'expert' theologian on the subject of CIS, world traveller, resplendent in his professorial title, trailing his entourage of lesser doctors. Brought to the altar of clinical freedom are the women, working women, mothers, ordinary folk, without degrees, gender status, titles, white coats, or entourages (unless we count children), but possessing bus timetables, the patience of Job and trust; above all, trust.[165]

Overseas doctors were given a completely different press by Coney. She explained that while she and Bunkle had initially been regarded as the 'lunatic fringe', the tide turned when their analysis was substantiated by overseas experts. She compared the two groups in terms which were disparaging of the New Zealand gynaecologists, few of whom, she believed, 'had a clear understanding of the disease and how to diagnose and treat it'. By contrast, 'the overseas doctors came across as straightforward likeable human beings; there was none of the arrogance, evasiveness, or the self-importance of most of the New Zealand gynaecologists'.[166] It is hard to believe that New Zealand gynaecologists were somehow a different breed from their overseas counterparts, and there is no evidence to support this. Indeed, most New Zealand gynaecologists had trained overseas. New Zealand was affiliated to the RCOG until 1982, and most doctors sat their fellowship examinations in London following overseas postgraduate experience.

Coney was particularly impressed by one overseas doctor who appeared as an expert witness at the Inquiry. She first made contact with American pathologist Ralph Richart at the suggestion of an American gynaecologist who attended the 1986 symposium on CIS at National Women's. Coney explained, 'We had been in touch with Richart while writing the *Metro* story; he was one of the few doctors who agreed to be quoted and who had actually encouraged us . . . . Richart was the perfect expert witness for the cancer inquiry.'[167]

Coney also referred in her book to a conversation in 1988 with an unnamed doctor who had not been involved in the Inquiry, a meeting which she described as 'another depressing occurrence'. This doctor apparently told her that some of the overseas experts 'had a personal animosity' towards Green, and that at a meeting in the USA in the early sixties, Richart had introduced Green as 'that bastard from New Zealand'.[168] Some of Green's colleagues were later to speak about this to a journalist who was writing an article for the *New Zealand Woman's Weekly* following the Inquiry. One said that when Green had visited Richart in the 1970s he was introduced as, 'the bastard from New Zealand who has been trying to undermine our cytology programme'. The journalist was told, 'Herb wrote that in his diary but wouldn't allow it to be used at the inquiry.'[169]

The Unfortunate Experiment: The Patients and 'chilling reading'

As the Inquiry was getting under way, pathologist George Hitchcock alerted the press to another medical paper on CIS patients at National Women's Hospital, published in 1986. Entitled, 'Carcinoma in Situ of the Vulva: A Review of 31 Treated and Five Untreated Cases', this was authored by Jock McLean and Ron Jones. Of the five 'untreated' cases, all developed invasive vulval cancer. Four died, and the fifth died of another illness. From the notes provided in the article, it is hard to see why they were described as not treated. Two had had hysterectomies, four had been given 'local irradiation', three had what the authors called 'mutilating vulvectomy' and one was subjected to 'vulvovaginectomy'. The authors of the article in fact recommended 'conservative local excision', and wrote, 'Although a small proportion of recurrences will occur, mutilating vulvectomy [should be] avoided.'[170] It was in regard to a case where the CIS had spread to the vulva that Dr Joe Jordan sympathised with Green's 'dilemma' in having to decide how invasive to be in the absence of symptoms.[171] Green told the Inquiry, 'Unfortunately, all these five patients were reported in the media as having died untreated'; he maintained they had had recurrent or intercurrent disease for which they had been treated and gave the details.[172] However, this paper by Jones and McLean made 'chilling reading', according to Coney.[173] Green was not mentioned in the paper and he himself had no knowledge of who 'managed' the cases,[174] but it was assumed that this was a continuation of his 'conservative' treatment.

Chapters 6 and 7 of Coney's book, about 'the women', also make chilling reading – if one accepts as an underlying premise that CIS is invariably an early form of cancer, and that the failure of Green and his colleagues to treat the lesion at the earliest possible moment, preferably by hysterectomy, made them responsible for any ultimately unfortunate outcome. Coney attributed every instance of CIS advancing to cervical cancer at National Women's Hospital from 1964 to 1987 to the management of Green and his colleagues, as if medical intervention could inevitably have prevented this.[175] She believed unequivocally that untreated CIS led to invasive cancer and that CIS was 'eminently treatable'.[176] While reading through case files, Coney wrote that they had to 'remind [themselves] that a single ignored positive smear was dreadful, that leaving diagnosed CIS for six weeks or six months was bad, let alone years. In any other hospital in the world these would have resulted in prompt curative treatment.'[177] Yet, as discussed in chapter 3, there were many 'thinking gynaecologists' who questioned radical treatment for CIS and dysplasia. Professor Kolstad, for example, stated in 1976 that his work in Norway did not support the frequent claim that there was no risk to CIS patients 'if treated properly'.[178] Many hospitals left women with positive smears alone provided there were no other symptoms, such as the 145 cases followed for ten years in Scotland and reported in *The Lancet* in 1978.[179]

In chapter 3, Coney gave the example of 'Mrs J.' who was first diagnosed with CIS in 1968, and on whom Green performed a hysterectomy in 1972 following two histology reports, even though there were no clinical signs of invasion. McLean advised radiotherapy as well as hysterectomy; in the absence of clinical signs, Green opted for hysterectomy only. Coney wrote that fifteen years later this patient 'probably' had invasive cancer of the vaginal vault and vulva, and clearly placed the blame for this fact at Green's door.[180] In the narrative, McLean's diagnosis was presented as inviolate. Yet, as noted earlier, five experienced pathologists could come to five different diagnoses. Moreover, as Jordan explained in discussing Green's cases, radiotherapy carried a high morbidity.[181] Radiotherapy was a highly invasive treatment and in the absence of clinical signs it is debatable if it would have been beneficial for this patient. Green spoke at the Inquiry of the reluctance of radiotherapists to give such treatment on the basis of a positive smear alone because of the known side effects.[182] In his Norwegian trials, Kolstad noted a case where radiotherapy had been given, but the patient developed cancer fifteen years later.[183] There were no certainties, as implied by Coney.

Coney recounted the story of Joy as a typical example of Green's mismanagement. Joy explained, 'His manner with me was excellent. I felt good. It was really I suppose the father image. He listened to all the things I'd done in my life . . . . He always said to me, why remove everything when there's no need to . . . . So I completely believed [that] if it didn't have to be removed, why dispose of something in your body which is still OK?' Others too, such as Lyn Potter, argued that hysterectomy had major physical and psychological effects on women and should be approached cautiously, as did Coney herself.[184] After Green retired, Joy had a hysterectomy and wished that she had had one years earlier – 'What are the bits useful to me for? What are you achieving by keeping them in?' In fact, Green's approach was more in accord with the feminists' attitude to hysterectomy, and Joy's subsequent change of heart more in line with the 'medical model'. Moreover, a hysterectomy would not necessarily 'solve' the problem, as was clear from McIndoe's article, which showed that some women who had had hysterectomy still developed cancer.[185] Coney narrated Joy's story as an example of a patient who was 'unsuspecting and powerless'.[186]

Conclusion

The Inquiry was designed to be a general inquiry into patient welfare and medical accountability, as indicated in its terms of reference. It extended its remit well beyond the research conducted by Green, and the National Women's doctors as a group came under fire in the report produced by the Inquiry. Coney made this indictment much more individualised in her book when she discussed the personalities of key figures at the hospital. While she discussed patients' experiences and their 'abuse' under Green's management, she also sought to show that the 'problem' was current and not historical, a campaign she would continue, following the publication of the Cervical Cancer Report. She wished to see the stables swept clean, and called for a fresh start with new personnel at the hospital.[187]

9.

Media Wars
The Report's Reception

In 1985, discussing the campaign against Depo-Provera, Christine Dann had written that, 'Women's health activists were made painfully aware of what a formidable system they were challenging when doctors, the manufacturing company, the Health Department and ultimately the media closed ranks against them.' In particular she referred to '"the fourth estate", the media, which claim to be independent of the other powers in society, [but] turn out to share the general reverence for members of the medical profession. A man with M.B.Ch.B. Dip Obsts. [*sic*] after his name will be given far greater credence than an independent feminist researcher.'[1] In the months immediately after the publication of the Cartwright Report, Sandra Coney put this theory to the test. This chapter traces the immediate public responses to the report, as articulated primarily through the popular, medical and nursing press. Coney's views were given considerable exposure but she also had to defend her ground.

Keeping Up the Pressure – Sandra Coney and the Press, 1988–1990

Three days after the Cartwright Report was made public, Sandra Coney and Phillida Bunkle were interviewed by the *New Zealand Woman's Weekly*. The headline read: 'We had to speak out about what we knew'.[2] A week later, Coney declared that she was 'not impressed so far at the feedback she [had] received regarding attitudes from the medical profession', describing the profession as 'still incredibly defensive'.[3] A month later she asserted, 'Cancer death No 28 demands an "answer"'. She stated, 'We, the consumers, want to know they acknowledge they made mistakes, that they were wrong. The consumer group Fertility Action is considering taking action against doctors and professors mentioned in the Inquiry's report. This includes Colin Mantell, Murray Jamieson and Professor Bonham from the medical school and National Women's medical superintendent, Gabrielle Collison.'[4]

Just one month after the report's release, Coney wrote a three-page article for the *New Zealand Listener*, in which she reminded readers of Cartwright's conclusion that 'The medical profession failed in its basic duty to its patients' and that

26 lives had been 'wasted'. What, she asked, was the medical profession doing about it?

> With mind-boggling arrogance they have rejected the evidence, the facts, the science and an independent arbitrator's assessment and are prepared to place more weight on gossip from their colleagues. I am forced to conclude that many New Zealand doctors are intellectual and scientific hacks. The fact that so many of these reports come from a major teaching hospital and the university is a cause of despair. . . . Never good at listening, they have failed to detect the mood of anger and distrust in the community. People will no longer rely on faith, for it has been so thoroughly betrayed.

She stressed that the problems were not 'historical', noting that some doctors 'continued Green's mismanagement after his retirement in 1982', and that of the 140 women Cartwright demanded be recalled, 54 had first attended the hospital between 1977 and 1986, 'that is, relatively recently'. In Coney's view, 'The current reaction of the medical profession does not bode well. The mode is denial. Most doctors have simply gone to ground.'[5]

Coney said that Colin Mantell had been criticised in the report for giving 'incorrect information to the public outside the inquiry'[6] but had since been elevated (without the job being advertised) to the position of head of the Department of Obstetrics and Gynaecology. She was presumably unaware that headships of university departments, as opposed to chairs, are not normally advertised but rotated within departments.[7] Mantell himself appealed to the university Vice-Chancellor to issue a statement to correct all the misinformation being circulated, asking that this include the reassurance that 'I was appointed in the usual University way and did not become Head by some unsavoury manipulation'.[8] One woman told the Vice-Chancellor that it was inappropriate for Mantell to be head as he was a 'supporter of Green'. She explained that the patients at National Women's Hospital and students 'must not be subjected to anyone who has been, or is in sympathy with Dr Green's practices'.[9] Another letter of complaint about Mantell's appointment came from Lynda Williams of Maternity Action, written on paper with a letterhead that included the Auckland Childbirth Education Association, Auckland Home Birth Association, Auckland Central Parents' Centre, Auckland Women's Health Collective, Caesarean Support Group, Helensville Hospital Community Committee, Manukau Parents' Centre, New Mothers' Support Group Inc., Obstetric Watch, Papakura Parents' Centre, Save the Midwives Association, West Auckland Parents' Centre and West Auckland Women's Centre.[10]

Coney further lamented that Bonham was still professor in charge of the Postgraduate School: 'It seems barely credible that, despite the serious criticism of him, Bonham can just sit it out.' She noted that David Cole, dean of the Auckland Medical School since 1975, was 'very cool about the Cartwright Report and articulated the dominant medical position'.[11]

Two weeks later, the *Sunday Star* reported Coney's belief that 'doctors at Auckland's National Women's Hospital were still studying Dr Herbert Green's discredited cancer theory as late as last year'. Perusing the 1987 directory of ongoing research in cancer epidemiology published by the International Agency for Research in Cancer, Coney had found an entry pertaining to National Women's which read: 'The aim is to ascertain whether cervical cancer commences in the pre-invasive form and whether the discovery of carcinoma in situ by cytology will decrease the mortality rate of invasive cervical cancer in the population.' Coney observed that this exactly matched the aims of Green's study, which had been discredited in the Cartwright Report. The directory said the study, started in 1962, was headed by Dr Murray Jamieson, with contributions by Drs Liam Wright and Herbert Green. It was in 1962 that Green started compiling statistics on the outcome for patients with CIS.[12] When contacted by the press, Wright denied any involvement. Jamieson and Green declined to speak to the *Sunday Star*. Mantell's explanation that it was a database of women registered with CIS at the hospital did not appease the journalists. The editor of the *Sunday Star* declared that the Inquiry had 'uncovered a horrifying disregard for the patient rights which led directly to the death of at least 28 women', and complained that the ongoing research programme was 'further evidence of the utter insensitivity of the medical dinosaurs'.[13]

When Jamieson refused to comment, Coney wrote to the editor: 'Part of coming clean would involve the principal researcher, Dr Murray Jamieson, accepting that he owes some accountability to the public. Declining to comment is not on when such serious questions are being raised.'[14] Jamieson was not prepared to engage in debate with Coney, but he did respond to an inquiry from Nicolas Holford, chair of the Auckland Hospital Research Ethics Committee. Jamieson assured him that there was no contact with patients in this study, which utilised a database on patients with CIS and invasive cervical cancer at the hospital to construct various reports.[15]

In an oration to the 1990 General Scientific Meeting of the Royal Australasian College of Surgeons in Wellington, Sir Graham Liggins commented on the irony of the fact that 'the famous 1984 article which emanated from the National Women's Hospital and on which the Metro article which stimulated the cervical cancer enquiry was based, was misinterpreted by the authors of the Metro article and by the judge. Once rolling, such minor matters became irrelevant to the course of the juridical inquiry which allowed its brief to expand to encompass every possible area of medical practice (fees excepted) about which there was public concern.'[16] No-one took issue with Liggins's questioning of the Inquiry, but other National Women's staff did not fare so well.

In October 1988 Coney had reminded her readers that it was 'important the public can see the medical profession is trying to clean up its own act'. She said that doctors other than Green and Bonham had to be investigated as there were a number of doctors who continued Green's 'mismanagement'.[17] In March 1989 Coney stated that Collison had reaffirmed her staff's intention to implement the recommendations of the Cartwright Report, but this commitment was undermined by one staff member's claim that the report's findings were based on a faulty interpretation of the figures.[18] She labelled this latest resistance, 'Part Two: The Empire Strikes Back'.[19]

The 'resistance' referred to an article in the *Dominion Sunday Times*, entitled 'Cartwright report based on a scam'.[20] This reported the views of Dr Graeme Overton, who later explained, 'It would have been probably nine or ten months afterwards and I hadn't shown a great interest in the inquiry to be quite honest.'[21] Having belatedly read the Cartwright Report, 'my only interest was that the charges against Herb Green that he had divided the patients into two groups and treated them differently, was then, and still is, entirely false.' Two days after the appearance of the *Dominion Sunday Times* article, the *Auckland Star* editor proclaimed:

> The intransigence and arrogance of some sections of the medical profession is sadly evident in the bid by Dr Graeme Overton, a senior obstetrician and gynaecologist at National Women's Hospital, to revive examination of major findings of the cervical cancer inquiry . . . . The rearguard action by senior doctors to discredit the findings of the Cartwright report indicates a dangerous inability to accept the serious deficiencies in patient care and medical controls exposed. The medical profession generally has had difficulty accepting any creed other than 'doctor knows best', and indifference [*sic*] to the notion of patients' rights.[22]

The *Dominion Sunday Times* issued a statement explaining its belief that the Cartwright Inquiry had properly examined the statistics submitted to it, and had analysed an extremely wide range of other evidence, including patients' files and many witness statements. It averred that, in printing the previous article, 'There was no intention to infer that the Cartwright report was based on anything other than a thorough and totally objective analysis of the facts presented to the inquiry and if such inference was taken, we withdraw it unreservedly and apologise to Judge Cartwright.'[23] The recently appointed Health Minister, Helen Clark, advised that arguments about cervical cancer treatment at National Women's should be 'put behind us', and that attention should focus on implementing the recommendations of the report, 'and not on academic arguments'.[24]

Yet the question about two groups was far from 'academic'; it went to the heart of the public perception of Green's so-called indifference to patient welfare. Shortly after the publication of the report, the *New Zealand Woman's Weekly* reiterated that, 'women with CIS had been separated into two groups. Some received normal treatment, of which 1.5 per cent developed invasive cancer. The rest were allowed to continue to produce abnormal smears. 22 per cent progressed to invasive cancer of the cervix or vaginal vault. Seven died.'[25]

Drs Charlotte Paul and Linda Holloway critiqued Overton's claims in a *New Zealand Medical Journal* article headed, 'No New Evidence on Cervical Cancer Study'. They continued to advance the hypothesis that there were two groups based on differential treatment, as discussed in chapter 3.[26] Petr Skrabanek from the University of Dublin sought to correct some 'errors' in Paul and Holloway's article: 'To present Group 2 as without treatment, is a travesty of fact', he wrote. With 25 per cent and 78 per cent of women in Group 2 having had hysterectomies and cone biopsies respectively, 'Do Drs Paul and Holloway now agree that even hysterectomy or complete excision by cone biopsy . . . is not sufficient to eradicate the disease?'[27]

Referring to Overton's critiques, Coney concluded that the resistance of National Women's Hospital staff to the findings of the Inquiry meant that women would not be treated safely at the hospital.[28] She told the *Dominion Sunday Times* in March 1989 that there were persistent reports of patient rights being violated, rude and uninformative doctors, and sub-standard medical care. Coney stated, 'I get reports from women all the time. They (some doctors) don't understand what informed medical consent or patients rights are. There is still arrogance in attitudes.'[29] A week later, again in the *Times,* she admonished, 'Women will only trust again when there is an honest acknowledgement of past wrong and a public commitment to change. An apology from National Women's Hospital is long overdue. Firm action is needed by the Auckland Area Health Board, the Medical Council and the Minister of Health.' The situation, she continued, had implications for doctors and their clients everywhere.[30] Patients, or 'clients', were demanding their rights.

When interviewed in early 1989 for the *New Zealand Nursing Journal*, Coney was even more vehement: 'For years National Women's has been notorious among Auckland women for violation of women's rights – not just physically but in their attitudes to women.' She informed her readers that the hospital had been known for its 'horrible comments about sex – it was a real meat market' and that women for years had avoided the place at all costs.[31] Hospital statistics, however, do not support that claim, with the hospital working at full capacity.

In May 1989 Coney wrote another lengthy piece for the *New Zealand Listener* on the aftermath of the Inquiry. Entitled, 'Doctors in Charge', this reiterated many of the concerns she had flagged in the daily press.[32] In 1990 she entered into debates with two National Women's Hospital doctors regarding the number of deaths that could be attributed to the 'unfortunate experiment' – Dr Tony Baird in the *Sunday Star*[33] and Dr Bruce Faris in the *New Zealand Medical Journal*. Faris took issue with Coney's use of the following quotation from the Inquiry: 'For a minority of women their management resulted in persisting disease, the development of invasive cancer and in some cases death.' He argued that this generalisation was 'unfortunately true for any study of any treatment protocol for potential lesions in other parts of the body such as breast, larynx, bowel, skin'.

He added that Cartwright had studiously avoided giving any figure, thus leaving the assertion concerning deaths virtually unchallengeable. Coney responded by reaffirming her belief that CIS was 'an eminently treatable disease and the chance of a woman with adequately treated CIS subsequently dying of the disease is very small indeed'. The view of one authority, Professor Per Kolstad, on that point has already been discussed.[34] In his 1991 letter to the *New Zealand Medical Journal*, Skrabanek noted that even McIndoe's 1984 paper did not support this statement. The study had shown that five of twelve patients with normal cytology after initial management later developed invasive carcinoma despite complete removal of the original lesion. He quoted statements by McIndoe and colleagues that invasive cancer could and still did occur following 'apparently successful' treatment.[35]

While Cartwright had not mentioned any figure, Coney claimed that 29 women had died because of the so-called experiment.[36] In her 1988 *Listener* article, she referred to '26 lives wasted'.[37] Tony Baird wrote to the Director-General of Health, George Salmond, to inquire where this figure had come from, to be told that it was a 'commonly used figure' rather than an 'official figure'.[38] There was no 'official' figure. Sixteen years later the feminist activist and director of the Auckland Women's Health Council, Lynda Williams, reminded readers of the *New Zealand Medical Journal* that, 'We must not forget that over 30 women died as a result of being part of the "unfortunate experiment at National Women's Hospital" and their untimely deaths were entirely avoidable'.[39] Again Baird responded claiming the figure was 'a contrivance of AWHC and it cannot be substantiated'. He referred to appendix 12 of the Cartwright Report listing 24 women who died between 1973 and 1987, pointing out that in only eight of them was cancer of the cervix recorded as the cause of death and 'there is no way of knowing whether or not those women were part of the study of Associate Professor Green'.[40]

Coney concluded her response to Faris's 1990 letter in the *New Zealand Medical Journal* by bringing the discussion back to the patients: 'Of course the people who have become invisible in this "debate" [about numbers] are the women and families of women who died. The current attempts to justify what happened at National Women's Hospital deny what for them was a very tragic reality'.[41] The doctors were thus portrayed as callous and uncaring in responding to the accusations and questioning the statistics.

During the Inquiry, Green's family wrote letters to the daily and weekly papers, based on their own research. They questioned the independence of the American witness, Ralph Richart, after discovering an article in the *Wall Street Journal* which showed that his private cytopathology laboratory in New York processed three times the recommended number of cervical smears per year.[42] Some two months after publication of the Cartwright Report, they drew attention to a recent article in the *British Medical Journal* by Joe Jordan, who had 'helped to discredit Professor Green's work at the inquiry' and who now argued for 'a more conservative approach' which he believed was 'medically valid and socially less damaging'. They pointed out that this was what Green had been saying for many years, and added, 'We find it ironic that a caring man who worked hard to save women from unnecessary surgery (surely a feminist goal) has had his reputation destroyed by an unjust campaign.'[43]

Valerie Smith, a retired secondary school teacher who was Green's neighbour and friend, also raised questions about the allegations against him. She sent a submission to the Inquiry complaining of inaccuracies in the media portrayal of its proceedings. Smith maintained that Cartwright had 'failed the public of New Zealand by allowing collusion of reporters, and inadequate or distorted reporting on the Inquiry by the media to pass unchecked'.[44] Early in 1990 she wrote a critique of the Inquiry entitled 'Cartwright Inquiry or Inquisition'; this was prefaced by the picture of Cartwright celebrating the anniversary of the report with Coney, Matheson and Bunkle. She wrote, 'The Caption "A Medical Milestone" could just as readily have been "A Legal Milestone" as never before has a presiding judge been photographed celebrating with the victors.'[45]

In March 1989 Coney had issued a challenge to those who questioned Cartwright's findings: 'If National Women's Hospital doctors have disagreements with the report, why haven't they taken the appropriate step, and initiated a judicial review? Presumably because their criticisms wouldn't stand up in a court, and they know it.'[46] At the time, doctors had good reason not to become involved. As journalist Jan Corbett later put it, 'For doctors to speak out in that climate would have made them look foolish, defensive and chauvinistic.'[47] The atmosphere was highly charged. Valerie Smith did, however, attempt to challenge the findings of the Inquiry in the High Court and Dr Bruce Faris joined her at the last moment. In the event, the case was withdrawn because of her lack of status in relation to the Committee of Inquiry and because of Faris's delay in initiating proceedings. Solicitor-General John McGrath, while agreeing that Coney, Bunkle and Cartwright had misinterpreted the McIndoe paper, explained that he was disallowing it because of the time lapse since the Inquiry findings in August 1988.[48] He also commented that too much money and effort had been spent on implementing the findings of the report for these to be challenged as invalid. That expenditure included plans and discussions for government legislation to establish a health commissioner, patient advocates and ethics committees, and $14 million to launch a nationwide cervical screening programme – 'we have to face the fact that the world has moved on', he said.[49] At least, Smith observed, the Solicitor-General did acknowledge that the McIndoe paper, which was central to the Inquiry, did not describe a prospective division of patients into two differently treated groups, as the original *Metro* article had claimed.[50]

One *Auckland Star* correspondent responded angrily to the letters from Smith and Green's daughter and son-in-law: 'After World War II Nazi doctors were condemned for performing medical experiments on people in concentration camps without their consent. Now Professor Green's friends and relations think he should be praised for doing the same thing at the National Women's Hospital.'[51]

Smith's was a lone voice, until Auckland journalist Jan Corbett published a lengthy article for *Metro* magazine in July 1990 entitled 'Second Thoughts on the Unfortunate Experiment at National Women's'. This again included the photograph of Coney, Bunkle, Matheson and Cartwright, with the comment, 'It was the first time a presiding judge has been photographed celebrating with the victors. The photo got us thinking.'[52]

The article questioned whether Green 'got a fair go' and whether the Inquiry was a 'witch hunt'. Corbett interviewed Overton, and expressed concerns about the independence of Richart as a medical expert. While it was a well-researched article, *Metro* editor Warwick Roger was heavily criticised for allowing it to be published. Roger's motives were questioned, including an alleged personal animosity between him and Coney. According to Coney, '*Metro* had attacked feminists from its first issues and anti-feminism has characterised the magazine's editorial policy. With the establishment of the inquiry, and the major impact it had on the medical profession, Roger may have felt he had unwittingly played a role in advancing a feminist campaign. About this he may well have not been pleased.'[53]

Mary Varnham's article in the *Sunday Star*, which followed this second *Metro* article, referred to the challenges to the Inquiry's findings by some of Green's colleagues at National Women's and, 'understandably enough', by some of Green's family and friends. She also claimed, 'It is no secret that *Metro* editor Warwick Roger is not enamoured of Sandra Coney, despite (or perhaps because of) winning the top journalism award of 1987 in part for publishing her National Women's story.' She described the recent *Metro* article as 'a disturbing male backlash. The Government owes it to Judge Cartwright and the women of New Zealand to see that *Metro*'s slurs and misleading information are thoroughly and publicly corrected.'[54]

Warwick Roger was not prepared to let Varnham's article stand unchallenged, and wrote to the editor of the *Sunday Star* the following week. He believed that the Cartwright Inquiry had been 'taken over by people much exercised by what they see as the evils of male power'. He claimed that Varnham, like Coney and Bunkle, was a propagandist for the newly emergent women's health movement, 'the speciality of which is making mountains out of medical molehills (117 deaths from cervical cancer per year, compared to 7614 a year from heart disease)'. He also noted that Varnham was a prominent member of the media women's pressure group and former press secretary to the Minister of Women's Affairs. He referred to Health Minister Helen Clark's statement that *Metro* had been incorrect in saying that no-one at National Women's was left with untreated invasive cancer. But, he added, 'as she acknowledged, in cases where medical records showed that Green continued to watch cases where microinvasion was suspected it was because he didn't agree with the pathologist's diagnosis. Again, two overseas experts also disagreed with the pathologist's diagnosis; where he had seen microinvasion they saw carcinoma in situ.' Roger concluded, 'It's all very well now for Cartwright supporters to back away and say that the inquiry didn't rely on either the McIndoe paper or the Coney Bunkle *Metro* article. What they conveniently forget is that the judge enthusiastically endorsed the validity of both documents.'[55]

Joan Donley wrote to the Sydney publishers of *Metro* and *North and South* magazines, expressing her 'outrage' at the recent publications, telling them that, on principle, she would never again purchase either *Metro* or *North and South,* and alerting them to the fact that she had written to 24 of their advertisers to say so. She told them, 'I and many others have been concerned for some time about the glib and scurrilous articles aimed at damaging the reputations of well known and respected community members',[56] by which she meant Sandra Coney, or as the author of the *North and South* article had called her, 'the saint of the New Zealand feminist movement'.[57]

In October 1990 Jan Corbett revisited the issue in another *Metro* article entitled 'Have You Been Burned at the Stake Yet?' The synopsis began, 'Jan Corbett tells what happens when you question militant feminists and reveal some further strange goings-on.' She reminded readers of the Solicitor-General's acknowledgement that the original *Metro* article was wrong in its interpretation of the two groups of National Women's patients, adding that this was the first time this had been officially recognised. She pointed out that as Smith and Faris were not given legal standing, none of this could be challenged in court, and explained how Helen Clark had filed an affidavit to the court seeking to prevent any judicial review on the grounds that implementing the recommendations of the Cartwright Report was too far down the track to be stopped. Corbett also cited the Solicitor-General: 'The disruptive effect of reopening what was decided two years ago in relation to events which began taking place over 20 years earlier is such that it is an abuse of court processes to seek to do so.' Corbett observed, 'Investigating events which took place 20 years ago didn't stop anyone establishing the Cartwright inquiry in the first place.' She related Justice Barker's conclusion that the court was no place to review what was only a legal opinion and not a judgement, adding that this meant that if you had been publicly maligned by a 'legal opinion' you had no place to go to seek redress.[58]

Shortly thereafter the Auckland District Law Society's public issues committee produced a report in which it concluded that public concern over the validity of the Cartwright Inquiry was 'unfounded': 'The only evidence to support the witch-hunt theory is that the judge, a number of professional advisers assisting the judge, and the victims of the conduct under investigation were of the same sex', said the group. Its findings appeared in the *New Zealand Herald* in October 1990.[59] Valerie Smith subsequently pointed out that this committee was convened by Dr Rodney Harrison, the lawyer for Coney, Bunkle and Fertility Action during the Inquiry.[60]

The Responses of the Medical Profession

Health Minister David Caygill announced at the press conference to mark the release of the Cartwright Report that, 'future confidence in doctors would be largely dependent on the way they responded to the report. If they can respond in the same forward looking response that I can discern in the report, we can avoid a repetition of what has happened.'[61] Before the report was issued, Caygill had told the New Zealand Congress of Obstetricians and Gynaecologists, 'I would not be surprised if you are feeling somewhat bruised and battered as a profession . . . . As one of my staff recently summarised the inquiry hearings, "The profession has had a cold speculum shoved rather abruptly into a very sensitive place."'[62]

While Coney claimed that National Women's medical staff closed ranks,[63] the evidence suggests otherwise. Coney could not understand why Overton came out in support of Green, since he was the one who finally advised that Matheson have a hysterectomy and commented that it should have been done years before.[64] The most plausible explanation is that Overton was not motivated as a 'Green supporter', but simply saw an injustice in the interpretation of events. Other doctors, however, did not agree with Overton. Responding to his claim that there were not two groups, National Women's gynaecologist Dr Ron Jones wrote,

> Those who have difficulty with the concept of the division of patients into two groups based on adequate and inadequate treatment are not familiar with Green's published papers. In 1970 he states 'the only way to settle the question as to what happens to carcinoma in situ is to follow adequately diagnosed but untreated lesions indefinitely . . . it is being attempted at NWH by two series of cases'.[65]

In this paper Green explained that these 'two series' were: first, 27 cases who had no symptoms other than a positive smear and, second, five women who had a hysterectomy and then developed vaginal CIS.[66] Both groups were carefully monitored for any change and treated accordingly. They were not two groups differentiated by conservative versus conventional treatment.

Within days of the report's release, the Royal New Zealand College of Obstetricians and Gynaecologists (RNZCOG) sent a letter to the Minister of Health supporting implementation of the recommendations and the following month it called a special general meeting to discuss the report. Fifty fellows attended the four-hour meeting. Ron Jones later described it as a 'tense meeting'.[67] There were sharp divisions between the participants, and the president, Professor Richard Seddon, recalled that some even suggested removing the word 'National' from the title of the hospital. At the meeting, Dr Ian Ronayne 'made a personal statement as an obstetrician and gynaecologist working at National Women's Hospital during the whole period under review', and his motion that McIndoe, McLean and Jones were 'morally and scientifically justified in publishing their

1984 and 1986 papers' was passed unanimously.[68] A second motion by Dr Gerald Duff from the Christchurch Women's Hospital, that the RNZCOG express its sympathy with the women involved and abhorrence at the events as reported in the Cervical Cancer Inquiry Report, was lost.[69]

Following the meeting Dr Tony Baird, one of the National Women's gynaecologists who had been unable to attend, informed Seddon that he took exception to the use of 'morally' in Ronayne's motion; he wished to dissociate himself from it, claiming, 'To me it is humbug.' He believed that McIndoe and colleagues had a 'moral duty' to inform the hospital and possibly the RNZCOG, and to ensure that the women were looked after. 'Instead they publish a paper in a Journal overseas that very few people in New Zealand read regularly and then do nothing more. To wait for journalists to show concern for the subjects of the so-called trial is anything but moral in my view and it could be a dereliction of care, which, of course, is most unethical.'[70]

The RNZCOG itself was concerned with the profession's image. One of the Special AGM's conclusions was that they needed to project a more positive image and that they might need professional advice on this. Members approved a motion to appoint a public relations consultant. Seddon commented in the following newsletter that he had sought advice from several active media people, but found there was currently 'little interest from the media in any statement concerning the actions and attitude of the College'. Colin Mantell also discovered that the media showed little or no interest in the Postgraduate School's responses to the Inquiry.[71]

The RNZCOG meeting indicated that doctors within National Women's and beyond were polarised. This was reinforced by statements in the medical and popular press. Wellington general practitioner Dr Ian St George was outspoken in the *New Zealand Family Physician*. St George had already come into conflict with Green two years earlier when the former opposed smears being taken at family planning clinics rather than by general practitioners. Green had called St George's views 'arrogant, selfish and outrageous'.[72] St George now wrote that 'every doctor should read Coney and Bunkle'. He explained that what they were seeking was a little humility from the profession, some relinquishing of power and greater accountability. He averred, 'We need to remember that health and wellness and wholeness and healing come from empathy and not from arrogance', maintaining that 'it took hard-headed women to demand change, but if we can feel a thorn in the side as a prick in the conscience rather than just a pain in the neck, perhaps we can make progress. Perhaps even together.'[73]

Eleven doctors from Wellington, mostly young and female but with no obstetrical and gynaecological specialists among them, wrote to *The Listener* about the Cartwright Inquiry. They declared that they wished 'to acknowledge the debt we owe Coney and Bunkle for bringing to public attention the disastrous experiment at National Women's Hospital'. They also said they were indebted to Cartwright 'for the wisdom, integrity and compassion with which she conducted the inquiry', and concluded their letter with the statement that they felt specially for the women and their families who had suffered as a consequence.[74]

Dr John Neutze, a cardiologist at Auckland's Green Lane Hospital, also wrote to *The Listener*, admitting that it had generally been agreed there had been 'some errors of judgment, failures of communication, and other practices requiring change – some of them historical'. However, he was concerned that the 'incessant publicity' had given the impression that inadequate medical practice was common at National Women's, which was totally misleading. He was concerned that the public would lose sight of two critical facts, that the overall standard of practice at National Women's Hospital had always been high, and that clinical research at National Women's had resulted in medical advances of inestimable value to many mothers and babies. In his view, the publicity was 'beginning to poison the relationship between the great majority of patients, who are understanding and sensible and the great majority of doctors, who are caring and conscientious'.[75]

Dr Peter Cairney, who had qualified in medicine in 1973 and completed his Diploma in Obstetrics at National Women's in 1977, cited recent FIGO (International Federation of Gynecology and Obstetrics) statistics which showed that National Women's Hospital ranked third in the world in the treatment of gynaecological cancer. He believed that Green's name deserved to be alongside those of Liley and Liggins as instigators of world-class research occurring within the hospital.[76]

Dr William Pryor of Christchurch (a student contemporary of Green) regarded the different responses to the Cartwright Inquiry by medical professionals as a generational divide. He described the Inquiry as 'a well orchestrated campaign by two militant feminists, aided and abetted by the media', and continued:

> It has saddened me to see the subsequent orgy of public breast-beating by some of my younger colleagues jumping on the trendy bandwagon and saying how terrible we have been . . . . My generation entered medicine with a sense of vocation and have served the patients well for 40 years to the best of our abilities. I see no reason at all to apologise for doing our best over this period, nor do any of my contemporaries, just because doctor bashing seems to be the flavour of the month at present.

He thought the view that patient advocates were necessary was 'a lot of codswallop' as it presupposed an adversarial situation between doctors and patients. He believed it would take years to recover from the fallout from Cartwright and that medical research had been set back by decades, screening programmes costing millions and of doubtful value had been embarked on, unworkable consent forms had been produced and majority lay committees rather than doctors were now to decide on treatment protocols. He told his colleagues, 'It is time we stood tall again and told the bureaucrats and feminists that many of these things are not right and need to be corrected. The first thing we need to do is to support a judicial review of the Cartwright findings.'[77]

Following her 1990 *Metro* article, Jan Corbett related, 'By day two after the magazine went on sale I was continually getting calls of support, but my supporters didn't want to go public because most of them were doctors, and doctors just don't go public.' She cited a registrar from Auckland's Middlemore Hospital: 'A lot of us here feel very strongly about it', he said, 'but we can't speak out because of the politics of the situation.'[78] An atmosphere had been created in which it was extremely difficult for doctors to criticise the report. Dr Michael Mackay from Dunedin wrote that the Inquiry had 'put the whole medical profession on trial and found it guilty. I do not believe I have had a fair trial.'[79]

The Central Ethical Committee of the New Zealand Medical Association was unequivocal in its response to the Inquiry. It declared at its August 1988 meeting that 'the clinical trial of treatment modes of cervical cancer undertaken at National Women's from 1966 [was] patently unethical', again overlooking the fact that it was treatment of CIS and not cervical cancer which was under investigation. Further:

> The Committee finds that there was total failure to observe the ethical guidelines of the Nuremberg Code of 1947 and the Declaration of Helsinki 1964. Whilst the Committee acknowledges that Research Ethical Committees were not established in 1966 in New Zealand, the guidelines for biomedical research were well in place by at least 1964. Consequently, the Committee set up in 1977 (3 years following the Auckland Hospital Board's resolution to establish such Committees) continually failed to observe the Declaration of Helsinki with regard to the ongoing 1966 experiment.[80]

In claiming that the hospital's Ethics Committee from 1977 'continually failed to observe the Declaration of Helsinki with regard to the ongoing 1966 experiment', the Central Ethical Committee appeared to have been unaware of the 1978 protocols which the Medical Council later acknowledged as signalling the end of the 'trial'.[81] The suggestion that National Women's was tardy in setting up an Ethics Committee was also misleading – such a committee had been set up in 1973; the change in 1977 was constitutional, as noted earlier, when the entire Hospital Medical Committee was reconfigured.[82] The only guideline in the Department of Health's 1972 directive on such committees had been that they should include 'experienced members of the professional staff'.[83]

Dr Roger Hilliker, president of the New Zealand Medical Association, requested an investigation into the conduct of doctors involved in the cervical cancer research, which he hoped would 'help restore public confidence in the profession'.[84] As Coney had declared, 'It is important the public can see the medical profession is trying to clean up its own act.'[85] Responding to Hilliker's suggestion, Dr Lyall Varlow of Gore wrote that it reminded him of 'let's sacrifice a few to appease the natives . . . . doesn't really matter who . . . just get them off the top of the pile . . . it will make us feel good and clean.'[86]

The Medical Association set up a working party chaired by Dr Glenys Arthur, a Wellington neurologist, to study the recommendations of the Cartwright Report.[87] A summary of its findings was released at the beginning of 1990. Dr Arthur noted that there was 'considerable anger amongst the profession that the 1966 trial was allowed to occur' and said that the working party found it regrettable that the trial deteriorated scientifically and ethically and did not change as scientific knowledge advanced or as adverse results were observed. She claimed that it was 'inexcusable and deplorable that all patients involved did not know they were part of a trial, and that it took a magazine article to bring about an investigation.'[88]

In June 1990 Dr Doug Baird from the Freeman's Bay Medical Centre in Auckland informed readers of the *New Zealand General Practice* that the Medical Council had 'reassur[ed] the public that more than cosmetic changes had been made' and that 'Once again women can be referred to Auckland's gynaecological services safe in the knowledge that competent, "mainstream" medicine will be practised on them.' He thought morale in medicine and in obstetrics and gynaecology in particular had been boosted as new hands took over the reins.[89] This was not the view of the incoming president of the NZMA, Dr Jeremy Hopkins, a Wellington surgeon, who declared in his first presidential address, that, 'since the Cartwright Report, the medical profession has faced hostility and loss of trust from the public which is impinging on daily medical practice.'[90] After attending the 1991 Royal Australian and New Zealand College of Obstetricians and Gynaecologists Congress, the president of the RCOG, Sir Stanley Clifford Simmons, commented on the low morale of the New Zealand profession following the Inquiry. He added that Green 'undoubtedly had a profound effect on reducing the degree and number of unnecessary interventions' and that 'many thousands of women would be grateful for that.'[91] However, the medical profession in New Zealand was divided on the issue, and many regarded the restoration of public confidence as the top priority.

A 'Visible Fresh Start' – Judging Bonham and his Staff

Five years before the Cartwright Report, Keith Sinclair's history of the University of Auckland described the head of the Obstetrical and Gynaecological Unit, Professor Dennis Bonham, as 'a human dynamo and a great success. The O & G Unit became an excellent centre of research.'[92] Cartwright disagreed with this assessment. Part of her indictment of Bonham rested on a 1976 letter from former medical superintendent Dr Algar Warren who commented on 'the overbearing attitude of the University staff, particularly the outspoken hectoring manner of the Head of the University department'.[93]

Cartwright accepted Warren's views as a fair assessment of the relations between the academic and part-time specialist sides of the hospital. Yet Warren was hardly an objective observer in this dispute. Tensions between university and hospital board staff dated back to the 1950s. When Harvey Carey was appointed professor in 1954 he also became head of the hospital until the 1957 Hospitals Act enabled the Hospital Board to appoint a separate medical superintendent alongside the professor. An agreement was reached with the university that the professor was still responsible for teaching, research and the clinical running of the hospital. Conflicts arose immediately upon Warren's appointment as medical superintendent in May 1960. Warren cancelled a Hospital Medical Committee meeting called by Carey in November, and the following April Carey contested the legality of a meeting called by Warren. Warren referred to the 'right of the medical superintendent [to] discuss with his staff all matters pertaining to his hospital'.[94] Carey found the situation impossible and resigned in mid–1962; Bonham walked into an already tense situation. In his 1976 letter, cited by Cartwright, Warren again stated, 'Surely it is axiomatic that the Head of this hospital must chair all meetings of his senior medical staff.' He was sure the Hospital Board would not 'capitulate to the university' by allowing the professor to chair meetings, implying that this was something which Bonham was attempting afresh. According to Warren, the part-time consultant staff 'need protection from the academic side and this can only come from a strong medical superintendent who has not had his power stripped from him'.[95] In the same year, Warren accused Herb Green of unethical conduct for reporting the names of septic abortion cases to the medical officer of health, an accusation which was not upheld by the Medical Association's Central Ethical Committee.[96] The friction involved much more than 'overbearing' academic staff and a 'hectoring' professor.

Nevertheless, on the day the report became public, Coney and Bunkle stated publicly that Cartwright, even though she had not been asked to apportion blame, had concluded that the 'hospital leadership' (by which they meant Bonham) had been 'severely deficient'. Fertility Action called for Bonham's immediate retirement, saying public confidence in the school and the training hospital would not be restored without a 'visible fresh start'.[97]

On 31 January 1989 Bonham retired as chair and head of department following pressure from the Medical Association. The Auckland Women's Health Council wrote to the Chancellor of the University of Auckland, telling him that many members of the public, particularly women who had passed through National Women's Hospital, were dismayed at the university's failure to examine Bonham's conduct following the revelations. They claimed that it appeared to many people that Bonham had not been called to account for his role in the 'unfortunate experiment' at National Women's and for his responsibility for the 'patronising and insensitive treatment which many women had received at National Women's Hospital from both Dr Bonham himself and clinicians under his leadership'.

The Auckland Women's Health Council warned the Chancellor that bestowing emeritus status on Bonham would be an endorsement of his conduct and that it would distress many women and would be 'an insensitive step on the part of the University'.[98] The university bowed to this pressure and Bonham was not granted emeritus status.

In October 1990 the *Sunday Star* proclaimed: 'Bonham faces $150,000 bill'. The accompanying article claimed that Bonham 'looks certain to be struck off the medical register and could face costs of up to $150,000 after being found guilty of disgraceful conduct by the Medical Council'.[99] In a decision announced on 7 December, Bonham was fined $1000 – the maximum allowable under the 1968 legislation governing the Medical Council. He was found guilty of four counts of disgraceful conduct and two of professional misconduct, relating to his role as head of the Postgraduate School of Obstetrics and Gynaecology at the time of Green's research.[100] Bunkle claimed that Bonham had merely been given 'a slap on the hand . . . . The women hurt by the cervical cancer experiment were the only ones paying the price'.[101]

The evidence for the Medical Council charges read as a continuum to the Inquiry, though the council's committee, which was set up to investigate the case, claimed that it did not rely on the Cartwright Report but 'went back to the source of the complaint and conducted its own investigation'. The committee also stated that it did not impose 1990s standards on the 1960s.[102] Presided over by Dr Stewart Alexander, a pathologist and chair of the Medical Council, the committee included Drs Bob Gudex, Judith Treadwell, Murdoch Herbert, Ian St George and John Broadfoot. Three of its members (Treadwell, Herbert and St George) possessed diplomas in obstetrics but Gudex was the only one on the New Zealand Register of Specialists for Obstetrics and Gynaecology. The overseas experts included British specialist Dr Joe Jordan who had given evidence at the Inquiry. Dr Jim Hodge, director of the MRCNZ, and Dr Charlotte Paul, a medical adviser to the Cartwright Inquiry, also gave evidence.

The committee concluded that Green's 1966 proposal 'was dangerous in the light of prevailing world opinion, that is, that CIS is a precursor to invasive cancer. We repeat that we have considered the evidence carefully to ensure that this opinion is that of the 1960s and not only that of more recent times.' They appeared to be unaware of the major international controversies that had raged in the 1960s. The committee also concluded that colposcopy in New Zealand at the time was in its infancy, and should not have been relied on as a safeguard. This ignored the facts that Dr Malcolm Coppleson's 1968 book in which he advocated the use of colposcopy in the diagnosis of CIS had been highly commended by Oxford University's professor of obstetrics and gynaecology, John Stallworthy, that Coppleson had been awarded a RCOG prize for his work with colposcopy and that McIndoe had trained under Coppleson. In addition, the committee stated, 'The belief that patients with positive smears could be followed without risk was unjustified.' Again, international authorities such as Drs Per Kolstad and Leopold Koss had argued that this was not so. Koss had placed patients with positive smears in the non-urgent category, and Kolstad posited in 1970 that with the use of cytology and colposcopy, cases of invasive cancer would be unlikely to be missed and that conservative treatment of CIS was justified.[103]

The committee's report stimulated further debate. A July 1990 editorial in the *New Zealand General Practice* stated, 'No matter what is written now, there were obviously goings on at National Women's which were not right. Would the NZMA have asked the Medical Council to investigate and would the Medical Council have spent two years doing so, before pressing charges, if there was not a case to answer? I think not.'[104] Others disagreed. During the Medical Council's deliberations, Bonham's counsel produced letters of support from 33 patients and 51 doctors, including Professors Sir John Scott, Sir Graham Liggins and Colin Mantell, Associate Professor Ross Howie and dean of the Auckland Medical School Derek North.[105] One woman wrote to the *Sunday Star* that the council's decision was 'utterly inappropriate'.[106] In November 1990 four of Auckland's female gynaecologists wrote to the editor of the *New Zealand Medical Journal* declaring their support for Bonham. They were concerned that the 'negative publicity overshadows his major contributions to obstetric practice in New Zealand'. They pointed out that he always insisted on the highest standard of care for all his patients and in this regard was an excellent role model to his junior staff, and that he had actively encouraged women who wanted to train as obstetricians.[107]

When the judgement on Bonham came out, Dr John Malloch of Hamilton (an Edinburgh medical graduate of 1934) wrote to the *Medical Journal* expressing his surprise at the verdict, declaring, 'No one but the most stony hearted could be unmoved by the harsh verdict recorded against Professor Bonham by the savants of the Medical Council. To say that it is other than monstrous is, I consider, to lack a sense of proportion.' He believed that when the General Medical Council (UK) introduced the term 'disgraceful conduct', it did not contemplate applying it to circumstances such as these. He revealed that a legal friend had told him that in law such a charge had a quasi-criminal connotation, and concluded, 'The very last thing Professor Bonham could be accused of is criminal intent. He has broken no law. The most he can be accused of is persistence in an error of belief.' Malloch personally did not even believe that to be the case. Citing a 1990 *British Medical Journal* article describing a situation in which women with positive smears were being kept under surveillance, he added, 'Many other sources could be quoted to show there is still uncertainty.'[108]

By 1989 Professor Dennis Bonham had had a long and successful career in obstetrics and gynaecology. He had promoted women in medicine, had done his best to improve the safety of maternity services in New Zealand, had been perceived by his patients as a conscientious and caring practitioner, and had played a major role in setting up the Medical School at the University of Auckland. He had been awarded the OBE for services to medicine in 1973. He founded the New Zealand Perinatal Society in 1980, was largely responsible for the 1968 Maternal Mortality Research Act and chaired its assessment committee from 1969 to 1988. Sir John Scott wrote that Bonham 'instituted a remarkably innovative and modern national register of maternal mortality. This achievement led to his appointment to the WHO Advisory Panel on Maternal and Child Health.' Scott also commented that, 'Undergraduate and graduate students, nurses and staff at National Women's regarded him with awe and elements of fear, but equally with recognition that here was a man who got things done, a man on the University payroll who walked the wards and attended the clinics day and night.'[109] He was an excellent surgeon, had very high standards for his staff and a very caring attitude towards his patients. He was married to his job. Yet as a result of the Cartwright Inquiry and the publicity following it, he was forced to retire from his post at the university, was denied emeritus status and was charged with professional misconduct by the Medical Council. When he died in 2005 his part in the 'unfortunate experiment' was highlighted in the press.[110]

Green himself had been regarded as too unwell to come before the Medical Council; he had collapsed and been taken to hospital during the Inquiry, and remained unwell. Obstetricians Bruce Faris and Richard Seddon, the two remaining members of the 1975 committee, were found guilty of 'conduct unbecoming a medical practitioner' and fined in July 1995, 20 years after the event.[111] The charge was laid ten years after Faris had retired. The former dean of the Auckland Medical school Professor David Cole wrote in his unpublished memoirs. 'The sad thing was that one colleague, Murray Jamieson, a Rhodes scholar, got embroiled, wrote an inappropriate letter [the one on screening, which he co-authored with Skrabanek] and damaged his medical credibility and finally decided to leave medicine to become a lawyer (and now a coroner).'[112]

In 1991 Dr Christopher Harison, from the Department of Obstetrics and Gynaecology at Thames Hospital and a fellow of the RCOG, wrote to the *New Zealand Medical Journal* that he had followed 'the saga of the Cartwright Inquiry with dismay'. He believed that the media had played a vital part, 'revelling in the tide of criticism and adverse comment on the profession'. In his view the Cartwright Inquiry had become 'a trial by media, especially television, swaying the public to the sensational'. He stated, 'Recently I wrote to the chief executive of a body concerned with matters medical expressing my distress at the way Professors Green and Bonham were treated in terms of truth and justice, to be told that "public opinion would accept no other outcome". Political expediency? May be we do in fact live in a house of cards after all.'[113]

Media Reporting of Compensation

In August 1988 Clare Matheson announced that she was going to sue for damages of $1.5 million, and was soon joined by others.[114] They took their case to the High Court, the defendants being Bonham, Green, Algar Warren, the Auckland Hospital Board and the University of Auckland. The right to sue for medical misadventure had been overturned by the passage of the 1974 Accident Compensation Act which provided compensation for personal injury by accident, including medical misadventure, occurring after 1 April 1974. However, in a landmark decision the Court of Appeal confirmed in 1989 that the Accident Compensation Act did not bar the women's claims for exemplary damages, only for compensation. The court stated that the purpose of exemplary damages was to punish the defendant rather than to compensate the claimant. In 1992 an out-of-court group settlement of just over $1 million was shared among the nineteen claimants; 29 also applied for accident compensation for loss or impairment of bodily function and received sums ranging from $4900 to $1700 by 1993.[115]

In a letter to the *New Zealand Medical Journal,* Dr Desmond Purdie of Auckland, another of the older generation of doctors (he had qualified in 1947), complained that the journal's report on the out-of-court settlement for the nineteen women was 'neither accurate nor full'. The report had stated that 'an out of court settlement totalling $1.2 million to all women had been reached'. Purdie explained that the women received $1.02 million between them, from which they had to pay costs, not the $25 million which they had sought. But, Purdie continued, 'More importantly, like the popular press, you withheld the information that actual settlement states that no fault or liability was admitted by the doctors or the institutions involved.'[116] He was right about the popular press. The *Herald* did not mention 'fault or liability' in its report of the settlement.[117] The *Sunday Star* downplayed the liability; its report of the payment titled 'Women Take $1m Cancer Restitution' did not mention that the agreement had recorded no admission of fault or liability by the doctors, health board or university. This was only revealed at the foot of a separate article in the same issue on another compensation case.[118]

Given that the agreement did not admit fault or liability, Purdie asked:

Why then the shameful scramble by the dean of the Auckland Medical School, Professor Derek North via radio news, talkback, television and the press to apologise to the women concerned . . . ? Is the dean in a state of bewildered cultural shock and gender cringe, or is it that, having encouraged his school to jump enthusiastically on the Cartwright screening bandwagon, he is trying to delay recognition of the fact that we are spending an indefensible amount on the detection and overtreatment of a relatively rare disease, about which we still know little that in our zeal, we are doing grievous harm to women's minds and bodies?

To support his views, Purdie referenced three recent articles in the *British Medical Journal* questioning the status and management of CIS.[119]

The Responses of the Nursing Profession

Not surprisingly given its submissions to the Inquiry, the New Zealand Nurses' Association fully endorsed the Cartwright Report and its recommendations. The association's press release indicated that it would be seeking guidance from its Auckland branch on the possible censure of medical and administrative personnel at the hospital. Letters of congratulations and gratitude were forwarded from the NZNA to Cartwright, Coney and Bunkle.[120] Carol Mitchell explained to fellow nurses that the 'cervical cancer fiasco' at National Women's Hospital was all about the power of 'a very strong macho clique' that could not be stopped. 'We were powerless in that situation . . . [but] nurses have got to be prepared to take a stand and start fulfilling their role as a patient advocate. And when they do . . . the NZNA will be there to ensure those nurses are not victimised.' Mitchell declared that nurses had to shake off a hundred years of believing that doctors reigned supreme.[121]

The following month Joy Bickley addressed the significance of the cervical cancer inquiry report for nurses in a *New Zealand Nursing Journal* article. She regretted that the report had almost side-lined nurses. While it stated that patients with complaints should have access to a patient advocate, a board committee, the medical superintendent, the Medical Council, a complaints committee or a health commissioner, there was no mention of nurses. She reminded nurses that Cartwright herself had expressed reservations about their ability to stand up for their patients. However, Bickley asserted, the report did provide nurses with an opportunity to establish a client-oriented service freed from medical dominance. She concluded, 'The NZNA will be conducting a campaign to support the implementation of the Cartwright report.'[122]

'Nurses must act', declared one Wellington nurse in September 1988. Referring to Bonham and Green's research, she thought it was intolerable that Bonham was still employed at the hospital and paid a salary by the taxpayer. She urged nurses to join together and, as care givers, show their disapproval by writing to the medical superintendent at National Women's. She considered it inappropriate for Bonham to 'retire on a fat pension' and called for his resignation.[123]

Bickley told the *Auckland Star* that the early retirement of Bonham could 'only be good for the health of women in New Zealand'. The same article, published on 6 September 1988, declared, 'Nurses happy at Bonham's leaving: Christmas cannot come quickly enough for nurses wanting to see the back of Dennis Bonham.'[124] This evoked an indignant response from nurses employed at National Women's, who wrote to the *Auckland Star*: 'We the undersigned [31 nurses], are just a few of the nurses at National Women's Hospital who deeply resent the statement

published in your paper pertaining to the resignation of Professor Bonham . . . .
Opinions expressed by Joy Bickley are not representative of the majority of the
nursing staff.'[125] This was followed by a separate letter from M. W. Bullock, a nurse
employed at the hospital for several years, who declared she could 'no longer
remain silent' and was 'disgusted' by Bickley's remarks. She added, 'Nobody
approached me before the article went to press.' Another cheered, '"Well done
and courageous" is what I have to say to those nurses from NWH who had the
"motzie" to stand up and be counted on behalf of the much-maligned Professor
Bonham.'[126]

This correspondence did not go unchallenged. Later that month the *Star*
printed a heated rebuttal by two nurses who claimed the 31 signatories did not
comprise a majority of National Women's nursing staff. The authors regretted that
these nurses were 'supporting the hitherto unquestioned and unchallenged power
of the medical profession over women's health' and warned that with nursing
support other experiments at National Women's could occur. They claimed that
Coney and Bunkle had 'earned the support of thinking nurses by academic argument, reasoned judgement and compassion for women patients'. They continued,
'As Professor Bonham, still unrepentant, despite Cartwright's report, is to retire/
resign, perhaps his supporters will follow him'; they saw this as a step to ensure
National Women's patients were 'afforded basic human rights, so suffering and
death for the sake of an experiment never happens again'.[127]

Valerie Fleming, the Auckland Hospital Board's quality assurance coordinator for obstetric hospitals, was 'dismayed' at the 6 September *Star* article
which claimed that nurses were 'happy at Bonham's leaving'. She had spoken to
over a hundred National Women's Hospital nurses and none agreed with Bickley's
comments. She added that to 60 per cent he was only a name in any case, not a
doctor with whom they had any dealings. She stated that the NZNA evidence at
the Inquiry mistakenly referred to doctors interfering in the placement of nursing
staff, and that it now seemed the NZNA wished nurses to be instrumental in the
removal of medical staff.[128]

A group of Christchurch midwives including Karen Guilliland, who became
the inaugural president of the New Zealand College of Midwives in 1989, asked in
the December 1988/January 1989 issue of the *New Zealand Nursing Journal*, 'Why
is it that as soon as someone makes any criticism of the medical profession, there
always seems to be a nurse who jumps up to defend them?' She believed that,
as one of New Zealand's most powerful political lobby groups, they were quite
capable of defending themselves and, in fact, she said, they were 'flat out doing so
at present over the Cartwright Report'. Guilliland concluded that she would have
been more impressed to have seen letters from nurses supporting the 'women
involved in the unfortunate experiment at National Women's rather than to its
perpetrators'.[129] When Jan Corbett's 'Second Thoughts' *Metro* article appeared in
July 1990, Bickley called it 'scurrilous' and declared that the NZNA stood by its

view that the Cartwright Inquiry was an independent inquiry conducted in the most objective and thorough manner.[130]

Conclusion

Health Minister David Caygill declared on the publication of the Cartwright Report that future confidence in doctors would depend on the way they responded to its contents. Any criticism was interpreted as chauvinistic, defensive and reactionary. The media played a large part in stifling debate with its strong words about the 'horrifying disregard for patients rights' and its description of the doctors at National Women's Hospital as 'medical dinosaurs'. Sandra Coney too played an important role in keeping up the pressure on the doctors following the report's publication, and her views were given high exposure in the press. Those who attempted to question her interpretation, such as consultants Graeme Overton, Tony Baird and Bruce Faris, 'Green supporter' Valerie Smith, *Metro* journalist Jan Corbett and *Metro* editor Warwick Roger, were vigorously condemned as anti-feminist and insensitive. When Overton pointed out that Green had not divided his patients into two groups, this was described as 'a dangerous inability of doctors to accept the serious deficiencies in patient care and medical control'.

Doctors (and male doctors in particular) had come to be perceived as a threat to the public; patients needed to be protected from them by other people or by nurses.[131] In particular, a generation of mainly male obstetricians and gynae-cologists who had entered medicine with a sense of vocation in the immediate post-Second World War period was now branded as uncaring and patronising. As one commentator put it, 'Dr Kildare had become Dr Death.'[132] Coney suggested that staff resistance might mean women were not treated safely at the hospital. She noted that, 'Some of the doctors who publicly displayed disbelief and defi-ance are still stalking the corridors at National Women's', and asked, 'What should patients do?'[133]

Changing the doctor-patient relationship had been a goal of the feminist movement as they sought to attack what they saw as the intertwined issues of men dominating women and doctors dominating patients. The other professional groups involved in this area, nurses and midwives, firmly aligned themselves with the feminist movement, with some of them seeing the foment as an opportu-nity for their own professional development. The Cartwright Inquiry did indeed appear to be a feminist victory, as *Broadsheet* proclaimed.[134] It remained to be seen what changes to the medical system would emerge as a result.

10.

New World, Better World?
Implementing Cartwright

The Labour government embraced the report wholeheartedly. The *New Zealand Medical Journal* reported within two weeks of its publication that the government intended to implement the recommendations that directly concerned the agencies under its control and follow up those directed at other parties.[1] This included restructuring ethical committees, setting up a patient advocate and health commissioner system to protect patients' rights, developing consent procedures, improving medical teaching and initiating a national screening programme.[2]

Sandra Coney and Fertility Action had fought to enhance the power of consumers in the medical encounter and in health services, and they remained vigilant following the Inquiry to ensure that they would not lose any gains achieved. Immediately after publication of the report, the Auckland Women's Health Council was established to maintain a watching brief to ensure that women's needs were met. On 5 August 1989, the council organised 'A Day of Reckoning' in Auckland to celebrate the first birthday of the Cartwright Report and to discuss its implementation. Over 300 women, including Silvia Cartwright, attended. Despite the government's initial endorsement, Phillida Bunkle declared at the meeting that the government's response had been one of 'determined resistance' and she reiterated the requirement for the government to involve consumers in its plans. Those present also regretted that the attitudes of doctors had not changed.[3] A national network of women's health groups was formed following this event.[4] By 1993 there were 23 councils or affiliated regional groups, all belonging to the Federation of Women's Health Councils of Aotearoa/New Zealand, launched in 1990.[5] These consumer groups were concerned with monitoring the recommendations of the Cartwright Inquiry through involvement on committees, maintaining a watching brief over the development of medical ethics and patient rights, and screening, each of which will be discussed in turn.

The Auckland Hospital Board's Monitoring Committee

Two weeks after the release of the Cartwright Report, the Auckland Hospital Board set up a committee to oversee the implementation of the recommendations.

This ad hoc Monitoring Committee comprised three board members and a representative from each of the Ministry of Women's Affairs, the National Council of Women, Pacifica and the Maori Women's Welfare League. The board's representatives were Dr Peter Davis, Judith Bassett and Sue Greenstreet. Board chairman Sir Frank Rutter explained that it would be inappropriate for a medical doctor to be on the committee (Davis had an arts doctorate) and that the majority of members should be women.[6] The choice of non-medical people to oversee the implementation of recommendations concerning patient care was indicative of a lack of public trust in the medical profession following the Inquiry.

At a Hospital Board meeting, Lyn Potter suggested that the committee should include a representative of the Auckland Women's Health Council instead of one from the National Council of Women, but her suggestion was defeated. Nevertheless, following a 500-signature petition organised by the Auckland Women's Health Council, the board created an additional position for a council nominee and Lynda Williams was appointed.[7] The chairperson of the committee was Mrs Ruth Norman, and Coney was the Ministry of Women's Affairs representative. The committee coordinator was Sallyann Thompson, who had been the Inquiry secretary and who later declared her belief that those who questioned the Inquiry Report suffered from 'persecutory delusions'.[8]

One of the tasks of the committee was to oversee the recall of 139 patients about whom there was still a question mark over their treatment, or 'Special Duty' patients as Cartwright had called them.[9] This recall process had started just two weeks into the Inquiry. Following Ralph Richart's evidence, Rodney Harrison suggested that Cartwright issue a preliminary report, to enable those who had been 'mismanaged' at National Women's Hospital to be contacted immediately and provided with independent examination. While some, such as Dr Bruce Faris, claimed this was a slur on the professionalism of the 20 or so consultants at National Women's, and others felt it pre-empted Green's evidence (he was only half-way through his period of cross-examination), Harrison retorted that those who opposed it showed no concern for the patients who were 'in imminent danger' and the proposal was adopted. Cartwright called for the immediate recall of patients who had 'been referred to or treated for CIS at National Women's Hospital'.[10]

Initially, the patients were recalled to National Women's Hospital and seen by the doctors there. The monitoring committee, however, did not believe that was in the best interests of the women, and took the process away from the hospital. When a colposcopist from the South Island withdrew his offer of services, the committee drew up a list of private colposcopists in Auckland, along with counsellors. According to Coney, when this happened, 'some of the doctors were walking around "with steam issuing from their heads". They objected to other people looking at "their" files. They objected to the absence of doctors on the monitoring committee and the recall. They objected to outside doctors being brought into their territory. They particularly disliked the involvement of counsellors, whom they had decided were going to be making medical judgments about cases.'[11]

In her *Listener* article, 'Doctors in Charge', Coney reported on a meeting between Ruth Norman, Judith Bassett and some senior National Women's Hospital doctors to discuss the recall. Coney suggested that Norman and Bassett 'had a torrid time . . . Bassett later described them as "like naughty boys . . . . They cannot hear what you say."' Coney pointed out that not all doctors were hostile, and that at the end of one meeting, about a third of the 40 doctors present had applauded her. She maintained, 'Everyone agreed that the NWH doctors were difficult to deal with. Nevertheless, it was suggested that they be involved in providing the medical treatment as a way of "building bridges" and "making them feel involved".' She said that she had argued strongly against this. Coney reported that her own presence was a particular irritation to some National Women's Hospital doctors: 'My own situation was becoming untenable. For me, the key features of the cervical cancer inquiry were the powerlessness of the patients, the unchallenged power of senior doctors, and the excessive self-protectiveness of the board. I was seeing it all again.'[12]

In March 1989 Health Minister Helen Clark sacked the Auckland Hospital Board. While this was unrelated to the Inquiry, it meant that the various committees, including the monitoring committee, were disbanded. Harold Titter, appointed as commissioner to replace the board, announced a 'new look' Cartwright taskforce to replace the monitoring committee. Coney believed that the monitoring committee, based solely on consumer representatives, had been doomed to fail. The 'bold experiment', as she said Judith Bassett had called it, was too bold for some. Coney commented that doctors were not willing to share power, let alone be subject to supervision by a lay group.[13]

The members of the new task force included three doctors, two nurses and three lay advisers. The Auckland Women's Health Council was 'very unsettled' over the commissioner's choice of task force members.[14] The committee reported in August 1989 that the recall of the 139 patients was almost complete and that none of the women was found to have a malignancy.[15] Two months later, all but eleven had been located. Four had died of other causes and 24 had 'cervical abnormalities'; four of this subgroup had hysterectomies and the others were 'given treatment appropriate to their condition'.[16] No dramatic consequences emerged from this exercise.

Safeguarding Patients' Rights

Patients' rights were an integral part of the discussion during and after the Cartwright Inquiry. Cartwright had concluded that the medical and nursing staff could not be trusted to protect patients' rights. She recommended that 'full information on patients' rights and responsibilities should be freely available to all patients at National Women's Hospital', and that a patient advocate be appointed, supported by a health commissioner to oversee patients' complaints.[17]

The Auckland Hospital Board had had a 'Code of Rights and Obligations of Patients and Staff' since the 1950s which stressed respect and courtesy by all concerned.[18] National Women's medical superintendent Ian Hutchison believed, however, that it was only since about 1980 that a patients' 'Bill of Rights' had been displayed around the hospital, pinned to wardrobe doors in the wards.[19] Cartwright did not think staff took this document seriously; she explained that a patient she spoke to in the obstetrics ward at National Women's in early 1988 had never heard of it.[20] Judith Bassett remembered that one of the major roles of the Board of Health committee set up to implement the Cartwright Inquiry was to establish a patients' code of rights, which she thought was novel.[21]

Like other hospitals, National Women's Hospital had a consent form for major operations, which patients signed. The hospital also gave maternity patients a consent form for routine tests for their babies.[22] In the mid–1960s, there was an awareness that patients needed to consent to tests and medical procedures, but it was still believed that the ultimate decision lay with the doctor, acting in the best interest of the patients. New Zealand was not out of step in this belief. Medical committees were greatly exercised about the danger of undermining the doctor/patient relationship in the 1960s. For instance, when the Medical Committee of Glasgow's Royal Maternity Hospital in Scotland was asked in 1969 to consider a new system for managing complaints, the committee resolved that they 'would not wish to formalise machinery for complaints as it is felt that this would be detrimental to the doctor/patient relationship'.[23]

By the 1970s patients were demanding more information and more say in medical decisions. Maternity was particularly politicised in this respect, as noted earlier. In 1977 a feminist group sent a 'Maternity Patient's Bill of Rights' to National Women's Hospital; this stressed informed consent with layman's language provided for medical explanations, and the right to make decisions about childbirth.[24] One hospital doctor, Pat Dunn, objected that this document framed doctors and patients in opposition: 'When the patient engages a doctor to care for her she must give him freedom of action and decision, but always within the limits set by justice and prudence.'[25] Coney dismissed the words 'integrity', 'sincerity' and 'good faith', used by doctors in their defence during the Inquiry, as empty terms.[26]

By the 1980s the public called for greater accountability by the medical profession, and the profession was responding to these calls. In 1986 the Medical Council set up a working party to look into Part III (Discipline) of the 1968 Medical Practitioners Act, for Conduct and Discipline in the Medical Profession. Non-medical organisations involved in the discussions included the Ministry of Women's Affairs, the Consumers Institute, Zonta, the Citizen's Advice Bureau,

Fertility Action, the Royal New Zealand Plunket Society, the National Council of Women and the Ministry of Consumer Affairs. Proposals called for greater lay participation in procedures.[27] Tony Baird also pointed out that medical ethics and patients' rights were a major item of discussion at the 1987 centenary celebrations of the New Zealand Medical Association.[28]

In its submission to the Inquiry, the New Zealand Women's Health Network highlighted the importance of bringing a third 'independent' party into the medical encounter. The network believed that 'independent counselling' was required for women to make choices. It seemed to them 'undesirable for the medical practitioner who has clearly a vested interest in the choice the client will make to be the person undertaking counselling and it is particularly inappropriate where the medical practitioner is a male and the client is a female or where the medical practitioner is a pakeha and the client is a Maori/Polynesian person'.[29]

Cartwright argued that in the light of the arrogance and defensiveness uncovered within the medical profession during the Inquiry, 'the encouragement of new habits and practices' could not be left to the profession alone. She recommended the appointment of a patient advocate as 'an independent and powerful advocate for the patient'.[30] A year after the report's release, the Health Department appointed a patient advocate to National Women's and Green Lane Hospitals. The Department planned this as the first in a series of such posts throughout the country, which would eventually come under the jurisdiction of a health commissioner.

The *New Zealand Herald* described the first patient advocate, Lynda Williams, as a 'long-time campaigner on health issues'.[31] She was 39 years of age, had four children and had worked in private practice as a childbirth educator for ten years. She told the journalist that one of her strengths was that she knew what it was like to be a victim of the health-care system, as a result of experiences in the birth of two of her four children and attendances at Green Lane Hospital. She knew what it was like to feel helpless and powerless and not have the support needed to make choices.[32] She had also been involved in the new women's health movement, having worked part-time for Fertility Action and as a member of the Auckland Women's Health Council. Members of the council 'delivered' her to her new post on 13 September 1989, assuring her of their ongoing support.[33]

The following month Williams told the *New Zealand Nursing Journal* that some doctors were still ignoring patients' rights, despite the Cartwright Inquiry.[34] Williams was patient advocate at National Women's for the next two and a half years, until December 1991, when the Auckland Area Health Board called for competitive tenders from private companies to provide advocacy services. Patient Advocacy Services Auckland Ltd won the tender, and signed the first two-year contract with the board. The service employed eight advocates and a manager, and liaised closely with the board which, according to Bunkle in 1993, cast doubt on its independence and hence authority.[35]

Williams, who continued her career as a feminist health activist as coordinator of Women's Health Action, later wrote about her experiences at National Women's. She complained that she had been 'constantly undermined. Sometimes it was obvious and sometimes subtle. The Health Department was simply unable to provide the guidance, support and backup promised.' The government failed to legislate for a health commissioner during her tenure there.[36] Williams was already known to staff at National Women's as a childbirth educator who had often been very outspoken about her 'opposition to the high rate of interference in the birth process, and the lack of informed consent that existed around many routine high tech interventions'. She found that obstetricians and gynaecologists were not prepared to talk to her about patient advocacy. She also detected resentment about anything relating to Cartwright, including her appointment. She cited gynaecologist Dr Tony Baird, who had written to the Director-General of Health that, while there was general support for advocacy, there was dislike of the term itself as 'adversarial'.[37]

National Women's was described by Williams as 'unremittingly sexist'. She found charge nurses, midwives, senior consultants and junior doctors equally culpable, explaining that none of them would give her time to explain the problem before launching a personal attack on the person she was advocating for. She would patiently listen to this diatribe before getting to the heart of the problem and she said that most of the time she coped reasonably well with that process and achieved the desired effect for the woman concerned. 'However', she added, 'there were a few encounters with medical staff that tested every ounce of control and restraint I had as they made their unbelievably sexist and patronising attitudes towards women patently clear.'[38]

Williams recounted how she persuaded the Auckland Area Health Board to undertake a survey of the patient advocacy system in December 1990.[39] Phillida Bunkle referred to this survey, released in July 1991, as showing a very high level of satisfaction among clients who had used the service, and a clear desire to use the service among patients who had not known about it. Bunkle commented, however, that the study also revealed open hostility to the advocacy service and the advocate personally within the hospital, and that the advocate operated in a difficult and unpredictable climate. Crucially, she wrote, the effectiveness was limited by the absence of any process for independent investigation or complaint resolution procedures.[40]

Eventually the government did enact provision for such independent investigation. In his submission to the Inquiry, Peter Davis, senior lecturer in medical sociology in the Auckland Medical School Department of Community Health, had suggested the establishment of a 'health services commissioner answerable to Parliament'.[41] Cartwright endorsed this proposal. Labour Health Minister Helen Clark introduced a Health Commissioner Bill to Parliament in 1990,[42] but the Health and Disability Act did not pass into law until 1994 under a National government. The Act aimed 'to promote and protect the rights of health and disability consumers, and to facilitate the fair, simple, speedy and efficient resolution of complaints.' The government appointed the first commissioner, Robyn Stent, in December 1994.[43]

The University of Auckland

The Council of the University of Auckland also set up a committee to oversee the implementation of the Cartwright Report in 1988, chaired by Pro-Chancellor Helen Ryburn. She co-opted three women, one from the academic staff, one from the general staff and a sixth-year medical student. Although various community consumer groups wrote to Ryburn offering to serve on the committee, none was appointed, much to Coney's disappointment.[44] When Lynda Williams sought information on the composition of the committee, Medical School dean David Cole assured her that there were no medical members.[45]

The Ryburn Committee reported in February 1989. While it noted that ethical and communication skills occupied a greater part of the Medical School syllabus than the Cartwright Report had acknowledged, it considered how this could be enhanced. The report claimed that none of the Medical School doctors disputed 'the paramount importance of patient welfare', but admitted, 'The problem for the Medical School is to satisfy the public that this is, in fact, the basic principle guiding all their work.' It suggested two ways of achieving this. The first was, 'To be seen to be accepting and implementing the Cartwright recommendations', and the second was to make the Cartwright Report required reading for 'all O & G students'. The committee recommended endorsing the Cartwright Report to help restore the school's public image.[46]

During the discussion of teaching on cervical screening, the Medical Students Association representative stated that the teaching given to students was adequate, 'but in view of the strong comments in the Cartwright Report, the Committee still felt this should be looked at'.[47] Dr Jeffrey Robinson, professor of obstetrics and gynaecology at the University of Adelaide, Australia, was invited to survey this teaching and found it adequate.[48] The committee commented on the shortage of women gynaecologists, which it attributed to the practice at National Women's of insisting that all postgraduate study be done on a full-time basis, and recommended that the 'O & G Department' restructure the Diploma course so that it could be done part time.[49] In his response, the dean of medicine Derek North pointed out that the Diploma of Obstetrics and Gynaecology was a short formal course of one month, associated with experiential training over a period of 6–9 months. The latter could be taken on a part-time basis. North also pointed out that the Department of Obstetrics and Gynaecology had, for some years, been willing to accept job-sharing for women doctors in the course of speciality training.[50] Under Bonham's leadership, the department had a long tradition of encouraging women to train, as discussed in chapter 9, which now appeared to have been forgotten.

North also stated that the Postgraduate Committee was planning to bring a British feminist sociologist Dr Ann Oakley to Auckland as the 1989 ASB Bank Visiting Professor in Women's Health. Oakley was well known for her 1984 book, *The Captured Womb: A History of the Medical Care of Pregnant Women*, and various other books on maternity and other women's issues.[51] Oakley's remit in Auckland was to speak on women's health issues to both professional and lay groups during her three-week visit to the School of Medicine. North noted that this visit had been planned some months before the release of the Cartwright Report.[52]

Derek North invited Sandra Coney to address the Medical School in October 1988 on the implications of the Inquiry for the medical profession.[53] Joan Donley later referred to a meeting between fourteen members of the Auckland Women's Health Council and seventeen deans and professors of the Medical School. She wrote, 'Our spokesperson, Marin Adams explained to them how they were in the same position as the priests in the Middle Ages, they interpreted God for the people UNTIL Martin Luther nailed his thesis to the door of the church, then the people began to understand that they could interpret God for themselves!'[54]

In 1989 the Medical School announced a scheme to recruit up to eight women, paying them $60 for each two-hour session, to enable medical students to perform vaginal examinations. The press reported that women's groups, led by the Auckland Women's Health Council, regarded this to be exploitative and dubbed it the 'rent a vagina' row. They wanted the examinations integrated with other topics in women's health, taught by lay women who were paid a lecturer's salary, and monitored jointly by medical staff and consumers. North replied that he would not accede to a 'pistol at the head approach', and maintained that control must remain with the professionals.[55] Three years later, in 1992, Dr Helen Roberts, a newly appointed lecturer in women's health at the Medical School, established a scheme whereby women volunteers (gynaecology teaching associates or 'GTAs') were trained to instruct medical students in how to perform gynaecological examinations. The GTAs were generally health professionals and educators and were recruited through the Family Planning Association.[56]

Medical education was changing in the light of social change, and probably would have done so without the aid of the Cartwright Report. The report, however, gave the changes in medical education a much higher public profile through the media than they otherwise would have had, and put the medical profession on the defensive.

Bonham's Successor – The Postgraduate Chair

The replacement of Bonham as professor of obstetrics and gynaecology commanded a level of public interest unusual for an academic chair. Joan Donley told a friend in August 1988 that in her view, 'no consultant who has been trained in the environment provided by Bonham and Co since 1963 should be allowed to step into that position'. She believed that, since obstetrics and gynaecology dealt with women, it was logical that a woman should head the department, though she warned that 'a woman O&G can often be (and often is) worse than any man is since women are cloned into the authoritarian medical mould and to achieve in that hierarchy they have to be "better" than the males'. However, she added, 'one woman who is acceptable is Wendy Savage and she has indicated a willingness to apply'.[57]

British obstetrician and gynaecologist Dr Wendy Savage had qualified in medicine in 1957, gained membership of the RCOG in 1969, and was at that time senior lecturer and honorary consultant in obstetrics and gynaecology at the London Hospital. She had been suspended from her clinical work there in 1985 when she was accused of mismanagement of five cases. She was said to have put women and their babies at risk by allowing labour to proceed without intervention, in circumstances where the clinical protocols would have indicated Casearean section. However, she was cleared of all charges at a public inquiry and reinstated.[58]

Savage had prior experience of working as an obstetrician and gynaecologist in New Zealand. She had practised in Gisborne from 1973 to 1976 and, according to a later report, sent 'shockwaves' through the local hospital when she set up an abortion service and family planning and venereology clinics. She had unsuccessfully applied for the chair in obstetrics and gynaecology at Wellington in 1983.[59]

Donley started a public campaign to get Wendy Savage appointed to the chair. She wrote to the *Auckland Star* in September 1988 that it was important to get not just any woman consultant obstetrician, but one who could demonstrate not only her concern for women and their right to control their own bodies, but also her ability to challenge the male domination of this specialty; such a person, she said, was Wendy Savage of London.[60] Other letters followed, agreeing with Donley. One urged the appointment of Savage, who had 'proved her concern and fought for the right for women to have control over their own bodies'.[61] Another appealed urgently for a woman of Wendy Savage's calibre 'to help us to fight for our rights in New Zealand'.[62] Yet another opined that Savage would be one of the few medical professionals whom New Zealand women could trust to deliver the medical care they wanted. This writer pointed out that Savage had already survived a determined effort by her conservative colleagues to strip her of her teaching role and clinical duties at a London hospital 'for daring to practise medicine as it is recommended by the Cartwright inquiry'.[63] A fourth felt that males were poorly equipped 'to understand, let alone implement, the Cartwright recommendations'. She felt that women's communication skills, caring and outlook on life were necessary for the top positions of the medical world, and argued that Savage had 'all the experience, qualifications and philosophy the women of

Auckland could ask for'.[64] The *Sunday Star* profiled Savage in October 1988 and noted a groundswell of support among New Zealand women who felt she would restore public faith in National Women's.[65]

Donley also wrote to Colin Mantell, the new head of the Department of Obstetrics and Gynaecology at the University of Auckland. She began her letter by saying that she appreciated that male obstetricians found it very difficult to cede or share power to female colleagues. She reiterated that the Cartwright Report had highlighted the defensiveness and arrogance of the medical profession, and reminded Mantell that Cartwright had concluded that she could not leave the encouragement of new habits and practices to the medical profession alone. She suggested that Mantell approach Savage as a possible replacement for Bonham, as in doing so, 'your Department would be demonstrating, both nationally and internationally, that the criticisms and recommendations of the Cartwright Report are being taken seriously and that there IS a will to confront and resolve old habits and attitudes'.[66]

Mantell was also canvassed by others, notably Maternity Action, which had been set up in 1984 (see chapter 7). They told him that an appointment made from within New Zealand would be detrimental to women's health as it would perpetuate the teaching of Bonham and Green. They too supported Savage, explaining that she had shown that she was not threatened by the reallocation of power to women and had experience in challenging medical authority.[67] Karen Guilliland wrote that women would trust Savage, which was necessary to regain faith in their health providers. She thought Savage would provide a role model for doctors, midwives and nurses, and restore the morale of New Zealand's maternal and child health services.[68] The university's own Academic Women's Group declared that the chair needed to be filled by 'someone not associated with the Auckland university medical school and preferably a woman'. They explained that, 'People associated with the Auckland medical school at any time in the past 20 years are tainted by the failure of the medical community here to stop or even to publicly condemn Professor Green's research project'.[69]

Savage visited New Zealand over New Year 1988–89. One record of her visit, found in Joan Donley's collected papers, was the transcript of a talk she gave to the Whangarei Home Birth Support Group, printed in the group's newsletter. She told her audience that while the 1958 survey (which Bonham had been involved with) had supported hospitalised childbirth, home births were safer, and that births should be in the hands of independent midwives. She stated, 'We have a system that is politically and medically male orientated. . . . we are victims of our socialisation system and stereotyping which depicts women as frail beings who need to be controlled and looked after.' She believed it was doctors who benefited from the current system and that 'women were often too busy being kind to men and not wanting to hurt them. Women must rise above this and become a real nuisance to get their needs met'.[70]

While Savage applied for the chair, she subsequently withdrew her application. Yet the story of her application and strong support by feminist lobbyists in New Zealand is indicative of the mood of the time. When Dr Judith Lumley from Australia turned down the job offer in November 1989, Donley still hoped that Savage might be appointed. The Auckland Women's Health Council felt that it should be represented on the interview panel.[71] When there was still no replacement for Bonham by mid–1990, the Medical School decided to offer temporary visiting professorships to distinguished academics,[72] one of whom was Dr Allan McLean, a New Zealander then working in London; he was disinclined to stay in Auckland after assessing the mood as 'anti-male'. He also thought that he would be hindered in his research into cancer of the vulva, as it was too close to the research under fire. McLean was subsequently appointed to a chair at the Royal Free Hospital in London.[73]

In early 1992, the post was readvertised and the *Sunday Star* reported, 'Two compete for varsity's key health job'. The article included a picture of Wendy Savage with caption 'popular', but noted that she was not one of the applicants. Savage was reported as saying that she did not believe there to have been any obvious change in attitudes by doctors at National Women's Hospital since the Cartwright Inquiry, adding, 'I understand the changes in the status of midwives (which gave midwives more autonomy in hospital deliveries) have not been entirely welcomed by doctors.'[74]

The position went to Dr Gillian Turner, a 52-year-old obstetrician-gynaecologist from Bristol, England. On her appointment in 1992 she told the press that most obstetricians in Britain had been brought up knowing about National Women's Hospital: 'it was a household name in Britain'.[75] Profiled in the *New Zealand Herald* later that year, she said that when she undertook obstetrical and gynaecological training at London's Hammersmith Hospital, everyone knew of National Women's. She said it had 'an enormously high international profile', and that its 'fall from grace', which Turner referred to as 'that whole episode', had had 'a huge impact on medicine generally', and particularly on gynaecology, as hospitals throughout the Western world revised their attitudes and procedures in the wake of what happened at National Women's.[76] She found the hospital a very demoralised place when she arrived to take up the chair, and considered the atmosphere in Auckland fraught.[77] She remained in the post for five years before returning to Britain.

Medical Research and Ethics Committees

One of the recommendations of the Cartwright Report related to informed consent for medical research. Cartwright concluded that Green's patients had not known other than intuitively that they were part of a research programme. However, it was not only Green who was regarded as culpable. Cartwright reviewed ten years of the National Women's Hospital Ethics Committee proceedings (1977–87) and concluded that the committee lacked independence and impartiality (it was chaired by Bonham) and had a poor record on patients' rights.[78] The committee was disbanded immediately after the Inquiry, as Cartwright suggested.[79]

National Women's had set up an Ethics Committee in 1973, as discussed earlier, the same year that the Auckland Hospital set up its Ethics Committee. The timing of its establishment was in line with international trends, or at least those in Britain where many of New Zealand's doctors trained. For instance, Glasgow Royal Maternity and Women's Hospital set up an Ethical Advisory Committee in 1972.[80] Kolstad recalled in 1987 that Scandinavian hospitals had set up such committees in the mid–1970s.[81]

In 1975 National Women's appointed a lay member to its Ethics Committee, Harry Israel, who was still the only lay member at the time of the Inquiry. One of Cartwright's recommendations was for 50 per cent of the membership of ethics committees to be non-medical.[82] This was indicative of a growing distrust of doctors to protect the interests of patients. The Health Department produced a national standard for hospital and area health board ethics committees in 1988. This required half the members of an ethics committee to be non-medical. In May 1989 the Auckland Area Health Board set up a working party on ethics committees to organise the implementation of the standard in Auckland, and by 1990 there were two ethics committees in Auckland to cope with the workload.[83] The Auckland Women's Health Council kept a watching brief on both, sending representatives to meetings.[84] Judi Strid, coordinator of the council, later recounted its attempts to have all research proposals discussed in the open part of the meeting rather than 'in committee'. By 1993 she believed that patients' interests were still not adequately protected, pointing out that a report commissioned by the Auckland Area Health Board that year 'overlooked a major consumer concern: that in trying to foster research, the committees often lose their prime focus, the protection of research participants from harm'. She accepted, however, that the Health Department's decision to set up a National Advisory Committee on Health and Disability Services Ethics was a move in the right direction.[85]

In a 1995 review of National Women's Hospital, Professor Jeffrey Robinson from Adelaide, Australia, commented on a new diffidence in initiating medical research at the hospital, pointing to 'the absence of a strong academic presence in gynaecology and midwifery'. He recommended that the hospital 'should seek to regain and maintain its pre-eminence in research'.[86] Various ventures such as the Liggins Institute were subsequently set up with that goal in mind, although this was concerned with fetal and perinatal health and medicine rather than with obstetrics and gynaecology.

Cervical Screening

Sandra Coney declared that the National Cervical Screening Programme launched in 1990 was one of the few tangible outcomes of the Cervical Cancer Inquiry.[87] The Ministry of Women's Affairs' submission had drawn attention to the importance of cervical screening as 'a major women's health issue' and maintained that the level of concern among women in the community about the lack of a screening programme was 'very high'. Women's Affairs pointed out that it had devoted considerable time and energy to this issue since its establishment two years previously.[88]

The Health Department had begun planning a national screening programme before the Cartwright Inquiry. Professor David Skegg published guidelines in 1985 and the Health Department and Cancer Society of New Zealand held a conference in 1985 to discuss implementation of those guidelines. Coney claimed that 'as the Post-Graduate School [of Obstetrics and Gynaecology, the University of Auckland] was highly influential among general practitioners, its opposition could effectively undermine any attempt to institute a programme'.[89] Despite her concerns, planning had proceeded apace. In early 1987, before Coney and Bunkle's *Metro* article, the Health Department held a series of meetings, in which the Ministry of Women's Affairs participated, to discuss pilot projects for cervical screening aimed at Maori women and lower socio-economic status women.[90] The Minister of Health on 2 September 1987 announced funding for three pilot schemes – two in Maori areas and another in a lower socio-economic area – and invited local groups to apply to manage the schemes. Twenty-one local women's health collectives or community groups applied and the first two successful bids, for Nelson and Kawerau, were announced in March 1988.[91] Later that year the Health Department funded schemes in Whanganui and Kaikohe.[92]

WIN (Women in Nelson Inc), which organised the Nelson scheme, was a women's organisation with its own women's health centre. Some local general practitioners questioned its ability to organise a screening programme, but most doctors appeared to accept that cervical screening programmes should be tied to broader programmes promoting women's health and should be run by women themselves.[93] While the Health Department had been preparing to extend cervical screening prior to the Cartwright Inquiry, the Inquiry increased public awareness and effectively quashed any remaining opposition. Sociologist Pamela Hyde explained that, as a result of the Inquiry, those members of the medical profession who questioned screening now had to oppose not only their peers but also the conclusions of a formal inquiry, and had to do so in a climate in which the medical profession had been found wanting as guardians of women's bodies: 'To be seen to be against screening was to be seen to be on the side of patriarchal medicine.'[94]

In December 1988 the Health Department held a national workshop on cervical screening in Porirua with about a hundred participants representing consumers and health professionals. Its purpose was to consult as widely as possible, based on the belief that it was no longer appropriate to leave decision-making to health professionals. In September 1989 Health Minister Helen Clark set up a ministerial review of the cervical screening programme because of wide-spread concern about delays in implementation. The review committee included Sandra Coney and Waireti Walters from the Maori Women's Welfare League. The review recommended that an 'expert group' be set up to oversee the implementa-tion of the programme. This body, which first met in December 1989, included Joy Bickley who had written a critical review of screening in August 1987,[95] and Sandra Coney.[96] When the new National government disbanded the group at the beginning of 1991, Clark called this 'a triumph of bureaucracy over the interests of the health of women' and Coney complained that it excluded women's voices and was 'a retreat into a medical model'.[97]

When the government first announced its national cervical screening pro-gramme in mid–1989, it was to be based on a voluntary register of women. Coney believed this to be 'fraught with problems', since 'women at risk may be the very people who do not want to register for various reasons'.[98] She considered the written informed consent which accompanied the voluntary register to be prob-lematic: 'Written consent is usually associated with major procedures, involving general anaesthetics and operative interventions. The requirement of written consent gave out an erroneous signal; it implied that enrolling in the programme was a major decision, with serious implications. In fact, the only reason signed consent was needed was to enable the laboratories to legally pass the informa-tion to the register'.[99] Professor David Skegg also opposed the voluntary register, commenting that 'recent proposals for requiring written consent can only be described as dotty'.[100] As discussed earlier, submitting to a smear could be con-sidered a significant decision for women, which might lead to further invasive tests. In the United Kingdom, an advisor on screening maintained that it was important for women to have full information on the meanings of the results and to realise the possible implications and risks of undergoing a smear test.[101]

Responding to lobbying by feminists and community health physicians, in 1993 the government legislated for an 'opt-off' register. Section 74A, an amend-ment to the 1956 Health Act, required that the smear results of all women be forwarded to the programme register, unless they 'opted-off', that is, asked not to have their results included.

Feminist sociologist Pamela Hyde argued that by opposing the voluntary or 'opt-on' register, these feminists had in effect suggested that women could not be relied upon to make rational and informed decisions. By claiming that women were unable to distinguish between informed consent for a major surgical pro-cedure and informed consent for the purposes of having their smear results transferred from the laboratory to a register, they appeared to suggest that women were naive in medical matters. Hyde maintained that feminists had co-opted existing medical values and standpoints which assumed that women were in need of management and control. Moreover, she argued, 'Feminists found themselves in a position of supporting a paternalistic approach to medicine and contributing to the increasing medicalisation of women's bodies', and, 'removing from women the autonomy of being able to choose whether they accessed that health service in the first place'.[102] Ironically, despite her endorsement of the screening programme, Coney herself articulated the feminist interpretation of screening in her book on the menopause published in 1994. In a section entitled 'Medicalisation gone mad', she wrote of screening, though without mentioning cervical screening:

> It is no longer possible for an individual to simply regard herself as a healthy woman in the prime of life. This can only be stated with authority if medicine has confirmed that by 'screening' various parts of her body. She should submit to 'monitoring' by periodic testing . . . .. Her body becomes a machine that must be given its 'warrant of fitness' to be deemed to be in proper running order. The woman who eschews protecting herself in this way is seen as inexplicably reckless and careless of herself.[103]

Maori women had long been identified as the group least likely to seek smear tests and also that with the highest rates of cervical cancer. The Health Department planning meetings in early 1987 were informed that the risk of cervical cancer among Maori women was almost three times that for non-Maori women, while one study showed that proportionately over three times as many Maori women as Pakeha women had never been screened.[104] Green and Bonham had recognised and commented upon this, and Bonham had previously suggested to the Cancer Society that a pilot programme be set up in Northland, an area of New Zealand with a predominantly Maori population.[105]

During the Inquiry, Te Ohu Whakatupu and the Department of Health's Maori Health Project Group convened a Cervical Screening Working Group.[106] Te Ohu Whakatupu stated that unless Maori women were provided with primary health care based on Maori health values and made available at culturally appropriate sites, Maori women would continue to be alienated and excluded from any screening programme.[107] Once the 'opt-off' register was established, Maori women lobbied to have the Maori register kept separate and protected by kaitiaki (guardians), and they succeeded in this.[108] The concerns of the women at the time included a desire for Maori control over their data, control over the possible negative portrayal of Maori women and ensuring that the data be used to benefit Maori women.[109]

Screening was generally viewed as a positive outcome of the Inquiry, although the Health Department had initiated plans for a screening programme before the Cartwright Inquiry and launched some pilot schemes. While the Cartwright Inquiry helped ensure government funding and silenced any further debate, the timing was similar to other Western countries: the UK introduced a national screening programme in 1990 and Australia in 1992. Green and some of his colleagues had been concerned that a national screening programme would primarily benefit white middle-class women, and these concerns appeared to come to pass. In 2007 Dr Ron Jones reported that, 'even after 17 years the Programme has still not quite achieved its target of 75% coverage for the entire population – and only 50% for Maori and Pacific Island women.'[110]

Conclusion

In 1990 Helen Clark, then Minister of Health and later Prime Minister, reflected on the effects of the Cartwright Inquiry when she was interviewed for *Time* magazine. Clark stated that the 'awareness of the need to question had been the main result of the Cartwright report' and explained, 'Of course, the doctors see that very negatively. They see it as building distrust. I don't see it that way. I think people are entitled to be participants in the decisions about their care.' Yet the reporter went on to comment, 'For New Zealand women who do use the hospital system, trust will be a long time returning. Many are not just questioning of doctors; they are openly hostile.' An Auckland gynaecologist was cited as saying, 'It's going to take a long time for women to have confidence again in professional medical people.'[111]

Clark commented that the Cartwright Inquiry had politicised women's health issues generally, and referred specifically to the 1990 Nurses Amendment Act. As she explained, this Act effectively made midwives equal to doctors in maternity care, giving them the right to order blood tests, prescribe medicines and deliver babies without a doctor, and had provided for pay equity. Clark saw this as an important move towards curbing over-intervention in childbirth and encouraging a questioning of the medical model. To pinpoint this as a result of the Cartwright Inquiry could be considered ironic as the Inquiry itself did not question but rather supported 'the medical model' in relation to CIS. What the Inquiry did do, however, was question the reliability of the medical profession. People retained their faith in the ability of medical science to effect a cure but not their faith in those who practised it. Joan Donley, who contributed to the *Time* magazine article, explained that, 'the Cartwright inquiry changed the climate because it undermined the status and credibility of the obstetricians and gynaecologists and made it easier for this [Nurses Amendment Act] to happen.'[112]

The women's health movement saw in the Inquiry an opportunity to change power relationships within the health system and to challenge male medical dominance, one of its long-standing goals. The publicity surrounding the Inquiry gave the movement more authority and forced the medical profession and the Health Department to take it seriously. The balance of power had shifted. Did it build distrust or was it a move towards a new and better world? Both cases can be argued, although many of the changes that emerged post-Cartwright were part of a long-term trend and not a dramatic break with the past, generally related to the wider social climate and in keeping with social movements occurring in the Western world generally.

British feminist sociologist Ann Oakley had found herself constantly drawn into discussions of the Cartwright Inquiry during her 1989 visit to Auckland. She believed that the issues brought to the forefront of public and professional debate in New Zealand were common to the health care scene in many countries at that time. To illustrate this fact, she noted that during her absence from the UK in 1989 a new consumer organisation had been set up – CERES, or Consumers for Ethics in Research – which aimed to 'speak for consumers' on the question of how medical research should be done. She correctly identified a breakdown in the traditional relationship between the public and the professional providers of medical care, a bond founded on patient trust in the therapeutic relationship, which was paternalistic rather than democratic.[113] American historian David Rothman argued that the new women's health movement in America in the 1970s in particular saw the transformation of the adage 'never trust a stranger' to 'never trust a doctor'.[114] Two other American historians in an article on the development of modern medicine also identified a crisis in trust between doctors and patients occurring in the 1970s and 1980s.[115] And yet, while the changes were not unique to New Zealand, they had assumed a particularly dramatic form in the course of the Cartwright Inquiry and its implementation.

11.

The Aftermath
Public Perceptions of Unethical Practice

In 1990 Sandra Coney reported Graeme Overton's 'alarming statement' that at National Women's Hospital they were 'still doing mainly what Herb did' in the treatment of CIS.[1] What she failed to appreciate was that National Women's was far from alone in this. Professor Malcolm Symonds from the Department of Obstetrics and Gynaecology at the University of Nottingham in the UK wrote the same year, 'The worrying thing about all of this [the findings of the Inquiry] is that what is actually suggested in the management protocol [in 1966] really looks very little different to the standard method of managing in situ carcinoma at the present time.'[2] Auckland University professor and later head of the Department of Obstetrics and Gynaecology, John France, later confirmed that, 'You would go out to a meeting somewhere else, internationally the support was all for us, colleagues overseas were strongly supportive. They were doing similar sort of things.'[3] And yet a view that unethical practice had occurred at National Women's, which was aired in the press during the Inquiry, became increasingly entrenched in the literature and in the public mind, both locally and overseas.

The *New Zealand Herald* report on the contribution of the overseas witness from Norway, Professor Per Kolstad, declared, 'Some women diagnosed at National Women's Hospital as having cervical carcinoma in situ had "severe" and "terrifying mismanagement", a Norwegian authority told the cancer inquiry in Auckland yesterday.'[4] This damning statement, made by Kolstad in the evidence he read out at the Inquiry, did not stand up to subsequent cross examination by David Collins, which was not reported in the press. Kolstad had used the phrase 'terrifying mismanagement' in relation to Clare Matheson's treatment, but admitted to Collins that he had not seen her complete file, and had not realised that she had five Grade 1 (negative) smears and one equivocal report of probably carcinoma in situ in the four years preceding her discharge from National Women's in 1979; he also admitted that his main source of information had been the *Metro* article. In response to other questions, he explained that he had reviewed the six files upon which he based his evidence after a 55-hour journey from Norway. He had been unaware that one of the cases he was critiquing had been under

McIndoe's charge, and that another where he had specifically criticised Green had been seen by Green on only seven of the 31 times she visited the hospital.[5] Such nuances were not reported in the press.

Following the Inquiry, Green's work at National Women's Hospital was increasingly referred to as research into untreated cervical cancer. An article on the Inquiry in the British newspaper *The Guardian* – entitled 'Death Watch' – in August 1988 cited Clare Matheson's claim that she felt as if she had been in Auschwitz. The article referred to McIndoe's Group 1 patients being given 'conventional treatment' and to Group 2 being 25 times more likely to develop invasive cancer. Death rates and actual figures were given: 0.5 per cent in Group 1 (four deaths) and 6 per cent in Group 2 (eight deaths). While twelve deaths from cervical cancer between 1955 and 1976 out of nearly 1000 patients with CIS did not warrant comparison with Auschwitz, the article called it a 'scandal' and blamed the no-fault compensation system in New Zealand (even though the scheme had not been introduced until 1974). The article concluded that doctors should be made accountable for their mistakes.[6]

Intriguingly, Malcolm Coppleson, who had been such an important influence on Green, attacked him in 1989 for his 'flat-earth gynaecology' and referred to 'the eccentric Auckland experiment'. However, the primary aim of Coppleson's leading article in the *Medical Journal of Australia*, written with law professor Paul Gerber, appeared to be to absolve overseas specialists from any responsibility for what had happened at Auckland: 'we reject any suggestion of collective responsibility on the part of the world's medical practitioners'. According to Coppleson, they simply had not realised what was happening in the face of National Women's Hospital's 'artfully-contrived damage control'. In his view Green's findings had been 'sanitised' to conceal the magnitude of the problem. Coppleson stated that, 'It is true to say that Green's opinion that CIS was being treated too radically was well known throughout the world of cervical neoplasia – and was scoffed at . . . . That is a far cry from concluding that it was known to anyone abroad – or that it reasonably could have been suspected – that there were patients in Auckland with untreated invasive cervical cancer.'[7] This accusation of secrecy was refuted by Dr Joe Jordan, an overseas medical witness to the Inquiry who had access to patients' files. He noted in his submission, 'In fairness to him [Green], he has always written about his work and nothing has been kept secret from other workers in the field.'[8] The claim that Green's views were 'scoffed at' could also be challenged. As recently as 1981, Coppleson himself had claimed that 'many thinking gynaecologists' were questioning the indispensability of cone biopsies and hysterectomy, as a result of the success of minimal interference with patients who had abnormal smears and pathology.[9] Coppleson's negative view in 1989 of the conservative treatment practised at National Women's could be at least partly explained by the acknowledgement at the end of the article. He and co-author Paul Gerber thanked Charlotte Paul 'who had discussions with Gerber', and stated that 'her clear description of the Inquiry in the British Medical Journal has enabled us to see the background to this cause célèbre in much clearer focus than otherwise would have been possible'.[10]

The same issue of the *Medical Journal of Australia* contained an article by medical ethicist Paul McNeill who also acknowledged Paul's assistance. The article opened with the following statement: 'The New Zealand Cervical Cancer Inquiry established that patients with carcinoma-in-situ were left untreated in order to observe the natural history of their disease.' McNeill commented on the 'dangerous practices' at National Women's and Green's 'total disregard for patients' welfare'. Of 'those responsible for the 1966 trial', by which he presumably meant Green, he wrote, 'Their attitude to patients was found to be punitive on occasions and their communication was uncaring and inadequate.' This conclusion was at odds with the Cartwright Report itself, which stated that, 'During the course of the Inquiry there was no serious suggestion that he [Green] had anything other than a benevolent attitude towards his patients.'[11] McNeill suggested that Green's 'lack of concern' could be partly explained by the fact that the majority of his patients were from working-class backgrounds and included a disproportionate number of Maori women. His source for the latter conclusion was Sandra Coney's book in which she commented that 15 per cent of the hospital files they looked at were those of Maori women.[12] McNeill noted, 'the attitude of male doctors to female patients from different backgrounds may have been an element in this lack of concern'.[13] He repeated this suggestion in his 1993 book, *The Ethics and Politics of Human Representation*.[14] This was picked up by other writers, such as the author of a 1995 article in an Australian feminist journal, *Hecate*, which referred to a 'discredited, highly unethical study in Auckland, in which cervical cancer was detected and allowed to progress under observation without treatment, for data on the biology of the cancer'. The author claimed that the low social status of many patients was a factor in the 'disregard shown by doctors for the welfare of the women'.[15]

The 'low social status' did not apply to the patient who had helped spark the Inquiry, Clare Matheson. Her husband was a civil servant and she herself was a secondary school teacher and deputy school principal by 1985.[16] it is also the case that some of the patients included in the CIS statistics were private patients at the hospital, or patients who moved between the two sectors, as Matheson did. There was no stigma in New Zealand in attending a public hospital in this period.[17] Nor could the 15 per cent Maori patients be regarded as 'disproportionally high'. National Women's catchment area included Northland, which was one of the main centres of Maori population; nationally, Maori comprised more than 10 per cent of New Zealand's population in the 1970s, and the rate of cervical cancer among Maori women was three times higher than among non-Maori.[18]

The perception that Green's work involved research into untreated cervical cancer has persisted. A 1998 study of the ethics of biomedical research internationally referred to the Inquiry, commenting that unfortunately abuse of subjects was an international phenomenon: 'In New Zealand investigations in the 1980s focused on research in the 1960s and 1970s in which women with cervical cancer in situ were left untreated to study the natural history of the disease. As was expected, many developed invasive carcinoma from which some died.'[19] In 2005 a World Health Organization publication on cervical cancer screening referred to McIndoe's 1984 study and claimed that in Auckland, 'a total of 948 women with histologically confirmed CIS diagnosed on punch biopsy but untreated were followed during 5–28 years.'[20] Not just Group 2, but now all patients were represented as 'untreated'.

Feminist historian and sociologist Ann Oakley, who had been the ASB Visiting Professor to the University of Auckland School of Medicine in 1989, told a 2006 Wellcome Trust Witness Seminar in London that she happened to know 'a little bit about the history of National Women's Hospital in Auckland and it doesn't have a very good history itself in terms of ethics of trials'. She stated that, 'From the late 1950s for some 20 years staff at the National Women's Hospital carried out an uncontrolled experiment examining the natural history of untreated cervical cancer. Some women with abnormal smears . . . were left untreated, and outcomes in this group were compared with those in treated women .. . . The experiment lacked a scientific research design since there was no proper control group, and there was no provision for informed consent.'[21] The enduring but false impression was of 'untreated cervical cancer'.

In 2008 an article in the *New Zealand Herald,* entitled 'Unfortunate, But True', referred to a recent article in *The Lancet Oncology* that 'should settle the argument for all time', i.e. that CIS was a precursor for cancer, and that Green 'under-treated and played with the lives of hundreds of New Zealand women'.[22] The authors of the *Lancet* article (David Skegg, Charlotte Paul, Ron Jones and four others) had analysed the records of patients treated for CIS at National Women's Hospital between 1955 and 1976 and concluded that those who had had less than cone biopsy as initial treatment, and continued to have a positive smear, had a 50 per cent chance of developing cancer over a 30-year period: 'All 31 cancers occurred in women diagnosed with CIN3 in 1965–74 and who had persistent disease.'[23] The significance of this retrospective review and of the omission of those whose smear test reverted to normal within two years of initial diagnosis has already been addressed (see chapter 3).

In many respects the 2008 *Lancet* article appeared to be a continuation of accusations levelled at Green 20 years previously. It referred to his 'premise that CIS was not a precursor of invasive cancer', citing his 1966 contribution to the *American Journal of Obstetrics and Gynecology.* In that article Green had concluded that, 'These then are still the two uncertain factors – the length of the preinvasive phase and the proportion going on to invasion. Clinical evidence is tending to show, but cannot prove, that the latter is small – probably much less than 10 per cent.' Discussing cases at National Women's, he wrote that the conclusion from the results of treating 446 cases was that 'it is safe to treat carcinoma in situ conservatively, provided that the original biopsy is adequate and excludes invasive cancer and that follow-up is thorough'. Had he believed that CIS was never a precursor to invasive cancer, he would not have suggested such a careful follow-up protocol. In that article he referred to 40 cases of CIS with microinvasion which had been treated conservatively. Only three of the 40 showed subsequent positive smears (after an interval of one to four years); two of these then had hysterectomies performed and a third a repeat conisation.[24] The 2008 article stated that, under Green, 'treatment of curative intent was withheld from some women'. This reiterated Sandra Coney's claim that Green had 'no intention to cure some of his patients,'[25] which had been emphatically rebutted by Judge Cartwright: 'I do not accept that Dr Green had no intention to cure his patients.'[26] Indeed, Cartwright commented that 'many [patients] took the opportunity of this Inquiry to make sure that I realised that Dr Green was a person who cared intensely about his patients.'[27] The 2008 article also reported that some women continued without curative treatment after 1975 since the study was 'never formally ended'. Yet the Medical Council had accepted that the 1978 protocols laid down by the Hospital Medical Committee spelt the end of the 'trial', and the International Federation of Gynecology and Obstetrics published statistics from the late 1970s which showed that the hospital ranked among the best institutions in the world in regard to five-year survival rates for patients with carcinoma of the cervix.[28] If women were inadequately treated, this was not reflected in the hospital statistics.

Other statements in the 2008 *Lancet* article are also open to question. It cited an 'independent analysis by McIndoe and colleagues in 1984'; this cannot be described as independent since McIndoe and McLean had worked closely with Green for years. The article claimed that 'no records exist of which women were chosen for that study', but, as discussed in chapter 3, McIndoe had explained that 'there was no choosing of women, nor did they choose to be in any particular group. The two groups we discuss result from a method we have applied to analyse data.'[29] Finally, the 2008 *Lancet* article mentioned '69 [women] for whom concern remained' following the Cartwright Inquiry. The reference for this figure was to the Cartwright Report and Charlotte Paul's 1988 *British Medical Journal* article. Early in the Inquiry, Cartwright had ordered a follow-up of 139 women who had attended National Women's with a positive cervical smear. This follow-up was carried out without finding any fault in treatment (see chapter 10). Paul did not mention '69 women' in her 1988 *BMJ* article, but stated that, 'A review was undertaken for Judge Cartwright of the case notes of patients with carcinoma in situ treated at the hospital. Several women were recommended for further advice or treatment.'[30]

In the decades following the Inquiry, the view of Green as an unscrupulous experimenter came to be accepted unreservedly, not only overseas but also within

New Zealand. In September 2007 Auckland obstetrician Anil Sharma referred in passing to the 'dreadful "unfortunate experiment" in Auckland', which, he explained, had conducted 'unethical research on women with CIN'.[31] This was repeated in the media reports of the twentieth anniversary of the publication of the report in 2008. The cover of *New Zealand Doctor* for 10 September 2008 featured a full-page picture of a group of doctors examining a child in the late nineteenth century with the caption 'How Cartwright changed it all . . .'. The opening statement of the feature article under that name read: 'It was a betrayal so profound it shook the bedrock of doctor–patient relations and altered the health-care landscape irrevocably.' The article referred to Green's '1955–76 experiment', stating that of more than 100 women involved (presumably the 131 in McIndoe's Group 2), 21 died. How that figure was reached remains unknown as there were no references; it was simply presented as fact.[32]

The article was reflective of the way in which the narrative had been simplified over the years. Contrary views about what had taken place at National Women's which were presented to the Inquiry and the debates which occurred immediately following the Inquiry were forgotten, as the belief that unethical practices had occurred at National Women's came to be the accepted wisdom.

12.

Conclusion
An 'Unfortunate Experiment'?

The programme of conservative treatment advocated by National Women's Hospital
was not unfortunate; for many patients it was a very fortunate programme as
unnecessary surgery was avoided, as were the traumatic consequences of surgery that
opponents of National Women's Hospital would advocate should have been performed.
I am, of course, acutely conscious of the fact that some patients who were managed
died. It cannot be said, however, that they died as a consequence of the treatment
they received at National Women's Hospital. The policy adopted at National Women's
Hospital was no more an experiment than any medical decision made every minute,
every day, by almost every doctor in New Zealand. In one sense, no doctor prescribes a
cure knowing absolutely that the consequences will be as expected. To that extent there
is an element of the unknown in every medical decision. That does not mean to say that
doctors 'experiment'. In the same sense, those of us at National Women's Hospital who
advocated the conservative treatment programme could not be absolutely certain about
the consequences of our methods of treatment. Neither could we be certain about the
consequences if hysterectomies had been performed on every patient with carcinoma in
situ. It is both emotive and misleading for the conservative management programme at
National Women's Hospital to be labelled 'unfortunate' and/or 'an experiment'.
GREEN, TRANSCRIPTS OF PUBLIC HEARINGS, DAY 11, 24 AUGUST 1987, P. 726, BAGC
A638/3A 101B, ARCHIVES NEW ZEALAND AUCKLAND.

Despite Associate Professor Herb Green's rebuttal, the concept of the 'unfortunate experiment', first coined by epidemiologist Professor David Skegg in 1986,
persisted in the following two decades, most recently in the press coverage of the
2008 *Lancet Oncology* article.[1] The *Lancet* article stated that unethical practices
had occurred at National Women's,[2] but did not convincingly prove that Green
had no intent to cure his patients. Nor did it recognise the complexities and
uncertainties attached to past medical decisions. Green advocated conservative
treatment in response to CIS, a minimal or a wait-and-watch approach, based on
the belief that most lesions of the cervix would not become invasive cancer, and
that early radical treatment could be worse than the disease itself, which was after

all based on a cytopathologic and not a clinical diagnosis. He was not alone in this belief; many other gynaecologists worldwide advocated a similar approach. The view that he was out on a limb, propagated by Coney and others at the time of the Inquiry, is not supported by a review of the international medical literature. For instance, the 1979 edition of *Recent Advances in Obstetrics and Gynaecology*, published just two years before Green retired, described the 'discrepancy of opinion as to whether or not dysplasia of the cervix represents a separate disease process which seldom, if ever, progresses to invasive carcinoma'.[3] Nor could one brush this off with the idea that 'dysplasia' was of lesser significance than CIS; as Ralph Richart said in 1981, 'It is a dictum among pathologists that one man's dysplasia is another's carcinoma in situ.'[4]

Green saw himself first and foremost as a clinician. One of his former patients wrote to *Metro* magazine that she was puzzled by the depiction of Green as 'a fanatical egoist hell-bent on proving his theories on carcinoma in situ at the expense of his patients'. She wrote, 'Such a caring, compassionate and obviously dedicated doctor could, in my opinion, not act in the manner he was accused of doing at the "trial" – it would have been completely out of character.'[5] She was not the only patient to spring to Green's defence.

It is important, as Coney advised, in telling the story of the 'unfortunate experiment' that the women at the centre – the patients – do not become invisible.[6] Coney and her colleagues claimed to speak for all New Zealand women, and yet this research has shown that there was no unified women's voice in relation to National Women's Hospital and the Cartwright Inquiry. Rather, there were multiple voices. Coney and her feminist colleagues did not speak for all New Zealand women, any more than the New Zealand Nurses' Association spoke for all nurses or Te Ohu Whakatupu spoke for all Maori women. Recent historical scholarship has questioned the tendency among feminist historians to view women of the past as victims of male patriarchy. Canadian feminist medical historian Wendy Mitchinson has argued that women of the past should not be regarded as victims in the medical encounter, that they participated in building their relationship with physicians through various demands, negotiation and renegotiation; 'that the patient–physician dynamic was composed of both sites of resistance and sites of compliance, that women as individual patients did have some agency'. Mitchinson included Coney's book among those she listed as feminist histories which had falsely portrayed women as victimised by the medical establishment.[7] This study has contributed to that recent historiography by moving beyond such victimisation theories to consider the multiple responses of those involved with National Women's Hospital.

In understanding medical history, not only do the voices of women as patients need to be heard, so too do those of their practitioners, the other half in the medical encounter. Balancing the risks and benefits of treatment is a highly subjective exercise, as noted by medical historian Thomas Schlick: 'despite their appearance of objectivity and neutrality, processes of risk assessment and management always involve value judgements'.[8] The Cartwright Inquiry revealed a profession divided about the benefits of modern medical intervention. These debates need to be contextualised within a wider intellectual framework. From the 1970s there was widespread disagreement within and beyond the medical profession about high-tech medicine. The intellectual movement was led by people such as the philosopher Ivan Illich and professor of social medicine Thomas McKeown. Green and some of his colleagues drew upon this intellectual movement. Dr W. J. Ramsay of Christchurch belonged to the same generation as Green and had trained in London in the immediate post-World War II period. He wrote to Judge Cartwright explaining how he opposed the high rate of caesarean sections, 20 per cent in New Zealand as opposed to 7 per cent in England, adding that any questioning of this was 'a threat to the reputation and income of some unscrupulous or unskilled practitioners'. He regretted that New Zealand had followed the American pattern in medicine. He told Cartwright that Green and his colleagues had made a 'creditable stand against exploitation of the vulnerability of women'.[9]

In common with health professionals in many other countries, the New Zealand medical profession was not homogeneous in its views. These disputes were settled in the case of New Zealand, however, in a non-medical arena, through an official inquiry sparked by two feminists. Sandra Coney and Phillida Bunkle latched onto these internal divisions to pursue their wider goal of changing the power relationships within medicine. While they aligned themselves with one of those factions (the anti-Green lobby), it is clear from Coney's pronouncements that they had a broader agenda. Coney argued that the inadequacies identified at National Women's Hospital applied to many doctors, in fact most New Zealand obstetricians and gynecologists, and that the issues were current and not historical, even though Green had retired in 1982. She stated that the Cartwright Report 'revealed those doctors (particular individuals, groups of doctors and the medical professional bodies), the university, health administrators, and the Department of Health had all "failed in their basic duty to the patients"'.[10] Coney's agenda can only be understood in the light of the politics of health – she and others in the women's health movement had already clashed with Drs Dennis Bonham, Ron Jones, Colin Mantell, Bruce Faris, Richard Seddon, Mont Liggins and Herb Green, who were all based at National Women's. In one sense, what happened at the Inquiry was a continuation of the struggle between these feminists and the profession of obstetricians and gynaecologists. The Cartwright Inquiry was not just about the research of one individual and his approach to his female patients; it was the culmination of a clash of ideologies and approaches to medicine, with women's bodies having become highly politicised.

Mistrust was the key element that saw Coney and Bunkle's *Metro* article escalate into a full-blown government inquiry and, ultimately, an interrogation of the entire medical profession. And yet while mistrust of doctors was at an all-time high, there appeared to be at the same time faith in the powers of medical science and medical treatment to effect a cure. The uncertainties or shortcomings of modern medical science and technology in relation to CIS and cervical cancer were unacknowledged in the public debates. Likewise in America, medical historian Elizabeth Siegel Watkins wrote of the 1980s that while there was much scepticism relating to medical ethics, medical science still commanded respect.[11]

Public confidence in the medical profession in New Zealand was at an all-time low immediately following the Cartwright Inquiry and the publicity surrounding it. A survey conducted by the Heylen Research Centre in 1989 showed that only 37 per cent of people had 'full trust and confidence in doctors', compared with 52 per cent in 1985; this was a greater drop than that suffered by any other profession.[12] The American equivalent in the late 1980s was 42 per cent, but this was a level it had been at for more than a decade.[13] In New Zealand the anti-male doctor feeling reached a feverish pitch just after the Inquiry. For instance, an article in the *Northern Advocate* regretted that the 'lessons' of the Cartwright Report had not been learnt. The author explained that a group of Northland women were marshalling forces to protest against the possibility of a male doctor being appointed as women's health service adviser for the Northland Area Heath Board. A representative of the women claimed that in the aftermath of Cartwright, 'it is appalling that the board could even think about appointing a man'. The report, she said, showed clearly that male medical professionals could not be trusted to act in a way that safeguarded women's health.[14]

Coney's view that the medical profession adopted a siege mentality was repeated in a 2008 *Herald* article on the Inquiry, reinforcing an enduring negative image of the profession.[15] Yet it is clear from the debates following the Inquiry that the profession was severely torn. This involved not only ideological differences discussed above but also personalities – Green and Bonham had forceful personalities and were not universally liked by their colleagues. There was also sympathy for the feminist cause, regardless of the details of the specific case, and a generational divide. Janet McCalman gave a graphic account of the post-World War II generation of doctors in her history of the Melbourne Women's Hospital, describing them as extremely dedicated professionals: 'Their wives and children saw little of them; they were for the most part people obsessed with medicine.'[16] This would also be a fair assessment of Bonham and Green. By the 1980s their generation was approaching or past retirement age and the world around them was changing. Consumerism and the women's health movement were new forces to be reckoned with in the 1980s, women's bodies had become politicised and, for some at least, progress seemed to be dependent on castigating these older doctors who were regarded as being inevitably paternalistic.

Changes occurring within medical and hospital practice were graphically illustrated in the Cartwright Inquiry and the publicity surrounding it – the Inquiry was both a product of, and contributed to, those social changes. The sheer volume of material generated by the Inquiry, both published and unpublished, provides a unique opportunity to investigate not medical wrong-doing, as many would have it at the time and subsequently, but rather social relations in medicine in the late twentieth century.

Notes

1 Introduction

1   Allan M. Brandt and Martha Gardner, 'The Golden Age of Medicine?', in Roger Cooter and John Pickstone (eds), *Companion to Medicine in the Twentieth Century,* Routledge, London, 2003, pp. 21–37.

2   *The Report of the Committee of Inquiry into Allegations Concerning the Treatment of Cervical Cancer at National Women's Hospital and into Other Related Matters (Report of the Cervical Cancer Inquiry,* CCR), Government Printing Office, Auckland, 1988, p. 70.

3   *Journal of General Practice,* 4, August 1987, p. 1.

4   Julie Roberts, 'We Had To Speak Out About What We Knew', *New Zealand Woman's Weekly (NZWW),* 8 August 1988, p. 16.

5   Rosemary McLeod, 'The Importance of Being Sandra Coney', *North and South,* July 1988, p. 56; this was repeated by Patricia Sargison, *Notable Women in New Zealand Health: Te Hauora ki Aotearoa: Ona Wāhine Rongonui,* Longman Paul, Auckland, 1993, p. 83.

6   *New Zealand Herald (NZH),* 10 May 2008.

7   Sandra Coney and Phillida Bunkle, 'An Unfortunate Experiment at National Women's', *Metro,* June 1987, p. 50.

8   'Ruth' later identified herself as Clare (Hallas) Matheson.

9   Sandra Coney, *The Unfortunate Experiment: The Full Story Behind the Inquiry into Cervical Cancer Treatment,* Penguin Books, Auckland, 1988, p. 11. The statistician denied making any such claim – see chapter 3.

10 CCR, p. 131.

11 *Ibid.,* p. 70.

12 *Central Leader,* 16 August 1988.

13 Alan Gray, 'Sandra Coney and the National Women's Hospital Affair', Correspondence, *New Zealand Medical Journal (NZMJ),* vol. 103, 1990, p. 378.

14 Paul Gerber and Malcolm Coppleson, 'Leading Article: Clinical Research after Auckland', *The Medical Journal of Australia,* vol. 150, 1989, pp. 230–3. This quotation was also cited in Clare Matheson, *Fate Cries Enough,* Sceptre NZ, Auckland, 1989, pp. 236–7.

15 Jocelyn Keith, 'Bad Blood: Another Unfortunate Experiment', *New Zealand Nursing Journal (NZNJ),* December/January 1989, pp. 20–21; ref. to James H. Jones, *Bad Blood: The Tuskegee Syphilis Experiment: A Tragedy of Race and Medicine,* Free Press, New York, 1981.

16 *NZNJ,* December/January 1989, p. 15.

17 Fertility Action, Submission to Inquiry, p. 253; reported in *Auckland Star,* 27 January 1988; and *NZH,* 27 January 1988.

18 *Auckland Star,* 11 August 1988.

19 Linda Bryder, *A Voice for Mothers: The Plunket Society and Infant Welfare 1907–2000,* Auckland University Press, Auckland, 2003.

20 CCR, pp. 95, 99.

21 Coney, *The Unfortunate Experiment,* p. 135.

22 CCR, p. 33.

23 *Ibid.,* pp. 198–9.

24 Sandra Coney, 'Cartwright Ten Years On: Access to Health Care: The Over-Riding Issue', *Women's Health Watch,* 42, November 1997 (http://www.womens-health.org.nz/publica-tions/WHW/whwnov97.htm#cartten).

25 Coney, *The Unfortunate Experiment,* p. 16.

26 David J. Rothman, *Strangers at the Bedside: A History of How Law and Bioethics Transformed Medical Decision Making,* Basic Books, New York, 1991, pp. 100, 142–3.

27 Kathy Davis, *The Making of Our Bodies, Ourselves: How Feminism Travels across Borders,* Duke University Press, Durham and London, 2007, pp. 163, 164.

2 Carcinoma in Situ

1   CCR, p. 38.

2   *Ibid.,* p. 33.

3   McIndoe, cited in Ronald Jones and Norman Fitzgerald, 'The Development of Cervical Cytology and Colposcopy in New Zealand: 50 Years since the First Cytology Screening Laboratory at National Women's Hospital', *NZMJ,* vol. 117, 2004, pp. 1179–88 (URL: http://www.nzma.org.nz.ezproxy.auckland.ac.nz/journal/117-1206/1179); Barbara Heslop, '"All About Research": Looking Back at the 1987 Cervical Cancer Inquiry', *NZMJ,* vol. 117, 2004, p. 1199 (URL: http://www.nzma.org.nz.ezproxy.auckland.ac.nz/journal/117-1199/1000).

4   Barron H. Lerner, *The Breast Cancer Wars: Fear, Hope and the Pursuit of a Cure in Twentieth-century America,* Oxford University Press, Oxford, 2001, pp. 5, 29, 30, 51; Walter Sanford Ross, *Crusade: The Official History of the American Cancer Society,* Arbor House, New York, 1987, pp. 76, 86; Monica J. Casper and Adele E. Clarke, 'Making the Pap Smear into the "Right Tool" for the Job: Cervical Cancer Screening in the USA, circa 1940–95', *Social Studies of Science,* vol. 28, 2, 1998, pp. 255-90.

5   Lerner, *Breast Cancer Wars*, pp. 29, 30.
6   Patrice Pinell, 'Cancer', in Roger Cooter and
    John Pickstone (eds), *Companion to Medicine
    in the Twentieth Century*, Routledge, London,
    2003, p. 676.
7   Ross, *Crusade*, p. 76.
8   George N. Papanicolaou and Herbert F. Traut,
    'The Diagnostic Value of Vaginal Smears in
    Carcinoma of the Uterus', *American Journal
    of Obstetrics and Gynecology (AJOG)*, vol. 42,
    1941, pp. 193–206; and G. N. Papanicolaou and
    H. F. Traut, *Diagnosis of Uterine Cancer by
    Vaginal Smear*, Commonwealth Fund, Oxford
    University Press, Oxford, 1943.
9   Casper and Clarke, 'Making the Pap Smear
    into the "Right Tool"'.
10 Ross, *Crusade*, p. 86. 'Roswell Park' was a
    Research Institute at the University of Buffalo.
11 Casper and Clarke, 'Making the Pap Smear
    into the "Right Tool"', p. 261.
12 Lerner, *Breast Cancer Wars*: he writes that it
    was pathologist Albert C. Broders from the
    Mayo Clinic in Minnesota who first named
    it, p. 197. James S. Olson claims it was Paul A.
    Younge from Harvard: James S. Olson (com-
    piler), *The History of Cancer: An Annotated
    Bibliography*, Greenwood Press, New York,
    1989, p. 142.
13 Charles E. Rosenberg, 'Disease in History:
    Frames and Framers', *Millbank Quarterly*, vol.
    67 (Suppl. 1), 1989, pp. 1–15.
14 W. A. McIndoe, 'Diagnosis of Pre-clinical
    Carcinoma of the Cervix with Particular
    Reference to the Place of Cytology', November
    1960, National Women's Hospital Medical
    Committee (NWHMC) minutes, 17
    September 1962, BAGC A638 22b, Archives
    New Zealand Auckland (ANZA).
15 Cited in Ross, *Crusade*, pp. 87–88.
16 Casper and Clarke, 'Making the Pap Smear
    into the "Right Tool"', p. 261.
17 Ross, *Crusade*, p. 89.
18 Lerner, *Breast Cancer Wars*, p. 51.
19 *Ibid.*, p. 51.
20 *Ibid.*, p. 51.
21 Tina Posner, 'What's in a Smear? Cervical
    Screening, Medical Signs and Metaphors',
    Science as Culture, vol. 2, part 2, 11, 1991, p. 167.
22 Arthur T. Hertig and Paul A. Younge, 'A
    Debate: What is Cancer in Situ of the Cervix?
    Is it the Preinvasive Form of True Carcinoma?'
    (Presented at the 75th Annual Meeting of the
    American Gynecological Society 1952), *AJOG*,
    vol. 64, 4, 1952, pp. 807–15.
23 John L. McKelvey, 'Carcinoma In Situ of the
    Cervix: A General Consideration' (Presented
    at the 75th Annual Meeting of the American
    Gynecological Society 1952), *AJOG*, vol. 64,
    4, 1952, p. 817. McKelvey lectured at National
    Women's in 1958 and 1963: Green to Registrar,

16 April 1963, Box 314, University of Auckland
    Archives (UOAA).
24 Howard C. Taylor, 'Discussion', *AJOG*, vol. 64,
    4, 1952, p. 827.
25 Richard W. TeLinde, 'Discussion', *AJOG*, vol.
    64, 4, 1952, p. 829; see also Howard W. Jones,
    Georgeanna S. Jones and William E. Ticknor,
    Richard Wesley TeLinde, Williams and
    Wilkins, Baltimore, 1986, pp. 45–46.
26 Curtis J. Lund, 'An Epitaph for Cervical
    Carcinoma', Chairman's address, section
    of Obstetrics and Gynecology at the 109th
    Annual Meeting of the American Medical
    Association, *Journal of the American Medical
    Association*, vol. 175, 2, 1961, pp. 98–99.
27 D. A. Boyes, H. K. Fidler and D. R. Lock,
    'Significance of In Situ Carcinoma of the
    Uterine Cervix', *British Medical Journal (BMJ)*,
    1962, vol. 1, p. 203.
28 Hugh C. McLaren, *The Prevention of Cervical
    Cancer*, The English Universities Press,
    London, 1963, pp. 3, 96.
29 John Stallworthy and Gordon Bourne
    (eds), *Recent Advances in Obstetrics and
    Gynaecology*, 11th edn, Churchill, London,
    1966, p. 338.
30 Jose Shelley, 'The Smear', *SHE*, June 1964.
31 Marie P. S. Grant, 'Cytology in Prevention of
    Cancer of Cervix', *BMJ*, (22 June) 1963, p. 1640.
32 Report of meeting held 23 February 1965,
    SA/NWF/F13, 56, Wellcome Library, London,
    UK.
33 Cervical Cancer Prevention Campaign, 23
    February 1965, SA/NWF/F13, 56, Wellcome
    Library, London, UK.
34 Emerson Day, 'The 24-hour Cancer Cure', in
    Walter Sanford Ross, *The Climate is Hope:
    How They Triumphed over Cancer*, Prentice-
    Hall, New Jersey, 1965; reprinted in New
    Zealand, A. H. and A. W. Reed, Wellington,
    1967, p. 28.
35 J. B. Graham, L. S. J. Sotto and F. P. Paloucek,
    Carcinoma of the Cervix, W. B. Saunders Co.,
    Philadelphia and London, 1962, p. 12.
36 *Ibid.*, p. 432.
37 Leopold G. Koss, Fred W. Stewart, Michael J.
    Jordan, Frank W. Foote, Genevieve M. Baker
    and Emerson Day, 'Some Histological Aspects
    of Behavior of Epidermoid Carcinoma In Situ
    and Related Lesions of the Uterine Cervix: A
    Long-term Prospective Study', *Cancer*, vol. 16,
    2, 9, 1963, pp. 1160–211. On Koss, see Leopold
    G. Koss, interviewed by Stephen G. Silverberg,
    M.D., Vancouver, March 2004, *International
    Journal of Gynecological Pathology* (http://
    ovidsp.tx.ovid.com.ezproxy.auckland.ac.nz/
    spb/ovidweb.cgi).
38 Malcolm Coppleson and Bevan Reid, with
    the assistance of Ellis Pixley, *Preclinical
    Carcinoma of the Cervix Uteri: Its Nature,*

Origin and Management, Pergamon Press, Oxford, 1967, p. 152.

39 *Ibid.*, p. 167.

40 Graham, Sotto and Paloucek, *Carcinoma of the Cervix*, p. 92; Coppleson and Reid, *Preclinical Carcinoma of the Cervix Uteri*, p. 167.

41 Coppleson and Reid, *Preclinical Carcinoma of the Cervix Uteri*, p. 167.

42 *Ibid.*, p. 7.

43 *Ibid.*, Preface by Stallworthy, pp. vii-viii.

44 Eamonn De Valera, Discussion to paper by James S. Krieger and Lawrence J. McCormack (Cleveland, Ohio), presented at the 35th Annual Meeting of the Central Association of Obstetricians and Gynecologists, 1967, 'Graded Treatment of In Situ Carcinoma of the Uterine Cervix', *AJOG*, vol. 101, 2, 1968, pp. 171-9, 181.

45 Angela Raffle and Muir Gray, *Screening: Evidence and Practice*, Oxford University Press, Oxford, 2007, p. 10.

46 Simon Szreter, 'The Importance of Social Intervention in Britain's Mortality Decline c.1850–1914: A Re-interpretation of the Role of Public Health', *Social History of Medicine*, vol. 1, 1, 1988, p. 2.

47 E. G. Knox, 'Cervical Cytology: A Scrutiny of the Evidence', in Gordon McLachlan (ed.), *Problems and Progress in Medical Care, Essays on Current Research, Second Series*, Nuffield Provincial Hospitals Trust, Oxford University Press, London, 1966; E. G. Knox, 'Cervical Cancer', in Thomas McKeown (ed.), *Screening in Medical Care: Reviewing the Evidence: A Collection of Essays with a Preface by Lord Cohen of Birkenhead*, Nuffield Provincial Hospitals Trust, Oxford University Press, London, 1968, pp. 43-54.

48 Knox, 'Cervical Cytology', p. 277.

49 See, for example, Editorial, 'Preventing Cancer of Uterine Cervix', *BMJ*, vol. 1, 1962, pp. 1817–18.

50 Coney, *The Unfortunate Experiment*, p. 50; see also CCR, p. 139.

51 Knox, 'Cervical Cytology', pp. 277–307.

52 *Ibid.*, p. 293.

53 Ralph Richart, 'Natural History of Cervical Intraepithelial Neoplasia', *Clinical Obstetrics and Gynecology*, vol. 10, 1967, pp. 748–94.

54 A. Stafl and E. J. Wilkinson, 'Cervical and Vaginal Intra-epithelial Neoplasia', in John Stallworthy and Gordon Bourne, *Recent Advances in Obstetrics and Gynaecology*, 13th edn, Churchill Livingstone, Edinburgh, 1979, p. 258.

55 Knox, 'Cervical Cytology', p. 307.

56 Knox, 'Cervical Cancer', p. 54.

57 Thomas McKeown and E. G. Knox, 'The Framework Required for Validation of Prescriptive Screening', in McKeown (ed.), *Screening in Medical Care*, p. 161.

58 Ornella Moscucci, *The Science of Woman: Gynaecology and Gender in England 1800–1929*, Cambridge University Press, Cambridge, 1990, pp. 181–206.

59 G. H. Green, 'The Progression of Pre-invasive Lesions of the Cervix to Invasion', *NZMJ*, vol. 80, 1974, p. 282.

60 *New Zealand Official Year-book, 1951–52*, 57th issue, compiled in the Census and Statistics Department New Zealand, Government Printer, Wellington, 1952, pp. 94–95.

61 McLaren, *Prevention of Cervical Cancer*, p. 1.

62 NWHMC minutes, 4 July 1949, YCBZ 15492 1a, ANZA.

63 See, for example, A. B. Miller, J. Lindsay and G. B. Hill, 'Mortality from Cancer of the Uterus in Canada and its Relationship to Screening for Cancer of the Cervix', *International Journal of Cancer*, vol. 17, 5, 1976, pp. 602–12. Miller was later Professor, Department of Preventive Medicine and Biostatistics, University of Toronto, Ontario, Canada.

64 Memo presented to NWHMC, 28 September 1954: NWHMC minutes, 12 October 1954, BAGC A638 37a, ANZA.

65 NWHMC minutes, 30 April 1957, BAGC A638 38a, ANZA.

66 Jones and Fitzgerald, 'The Development of Cervical Cytology and Colposcopy'.

67 H. M. Carey and S. E. Williams, 'Cytological Diagnosis of Pre-Clinical Carcinoma of the Cervix', *NZMJ*, vol. 57, 1958, pp. 227–35.

68 Carey to Sec. BECC, 26 May 1961, NWHMC minutes, 20 June 1961, BAGC A638 39a, ANZA.

69 W. A. McIndoe, 'A Cervical Cytology Screening Programme in the Thames Area: Second and Third Years of Study', *NZMJ*, vol. 65, 1966, p. 648.

70 J. D. Baeyertz, 'Results of a Cervical Smear Campaign in Wanganui', *NZMJ*, vol. 64, 1965, pp. 618–25.

71 E. J. Marshall, 'Cervical Smear Survey', *NZMJ*, Supplement, vol. 63, 1964, pp. 18–22.

72 R. E. W. Darby and S. E. Williams, 'The Cytological Diagnosis of Carcinoma of the Cervix', *NZMJ*, vol. 64, 1965, pp. 98–102.

73 Shelley, 'The Smear'.

74 NWHMC minutes, 10 May 1955, BAGC A638 37a, ANZA.

75 Under the medical benefits of the 1938 Social Security Act, public hospitals such as National Women's provided free treatment, but private fee-paying hospitals were allowed to remain in existence and doctors could also bring their own fee-paying patients into public hospitals. On the history of hospital services, see Iain Hay, *The Caring Commodity: The Provision*

of Health Care in New Zealand, Oxford University Press, Auckland, 1989.

76   'Obituary: Sydney Wallace Jefcoate Harbutt', *NZMJ*, vol. 111, 1998, pp. 60–61.

77   Jefcoate Harbutt, 'New Concepts in the Treatment of Carcinoma of the Cervix', *NZMJ*, vol. 54, 1955, p. 359.

78   S. W. J. Harbutt to Medical Director, 13 June 1955, NWHMC minutes, 14 June 1955, BAGC A638 37a, ANZA.

79   Green, Evidence to Cervical Cancer Inquiry, p. 3, UOAA.

80   Green, Transcripts of Public Hearings, Commission of Inquiry into Cervical Cancer, National Women's Hospital, Day 20, 4 September 1987, p.1506, BAGC A638/5a 101D, ANZA.

81   Dawson to Honorary Secretary, RCOG, 17 December 1948, B9/4/1, NZ Regional Council, 1947–50, Archives of the RCOG, London, England.

82   W. A. McIndoe, 'The Diagnosis of Pre-clinical Carcinoma of the Cervix with Particular Reference to the Place of Cytology', November 1960, NWHMC minutes, 17 September 1962, BAGC A638 22b, ANZA.

83   Auckland Hospital Board, Obstetrical and Gynaecological Unit, Cornwall Hospital, Auckland, *First Clinical Report, for the Year Ended 31st March 1949*, prepared by the registrars F. L. Clark and G. H. Green, 1950; Auckland Hospital Board, Obstetrical and Gynaecological Unit, Cornwall Hospital, Auckland, *Second Clinical Report, for the Year Ended 31st March 1950*, prepared by G. H. Green, 1951.

84   Green, Evidence to Cervical Cancer Inquiry, p. 6, UOAA.

85   On Liley, see L. Bryder, 'Liley, Albert William', in W. F. Bynum and Helen Bynum, *Dictionary of Medical Biography*, vol. 3, Greenwood Publishing Group, Westport CT, 2007, p. 793.

86   Sir Graham (Mont) Liggins, Interview by Jenny Carlyon, 12 May 2004.

87   'Readers Respond', *Metro*, August 1990, pp. 124–5. Like CIS, microinvasion was a histological diagnosis. As Jeffcoate explained in 1967: 'The condition [CIS] shows all the factors of malignancy except that the stroma underlying the epithelium is not invaded at any point. Sometimes a few cells break through the basement membrane but progress no further; this is described as micro-invasion and is not necessarily ominous': T. N. A. Jeffcoate, *Principles of Gynaecology*, 3rd edn, Butterworths, London, 1967, p. 509.

88   NWHMC minutes, 30 July 1957, BAGC A638 38a, ANZA.

89   Green, evidence to Cervical Cancer Inquiry, pp. 7, 78, UOAA.

90   NWHMC minutes, 4 February 1958, BAGC A638 38a, ANZA.

91   NWHMC minutes, 5 May 1958, BAGC A638 38a, ANZA.

92   NWHMC minutes, 21 July 1958, BAGC A638 37a, ANZA.

93   NWHMC minutes, 21 September 1961, BAGC A638 39a, ANZA.

94   NWHMC minutes, 19 August 1963, BAGC A638 22b, ANZA.

95   NWHMC minutes, 17 September 1962, BAGC A638 22b, ANZA.

96   Coppleson and Reid, *Preclinical Carcinoma of the Cervix Uteri*, p. x.

97   *Ibid.*, p. 16. Adolf Stafl, Eduart G. Friedrick Jnr and Richard F. Mattingly, 'Detection of Cervical Neoplasia: Reducing the Risk of Error', *CA: A Cancer Journal for Clinicians*, vol. 24, 1974, p. 26; Stafl and Wilkinson, 'Cervical and Vaginal Intra-epithelial Neoplasia', p. 264.

98   Jones and Fitzgerald, 'The Development of Cervical Cytology and Colposcopy', p. 1186.

99   NWHMC minutes, 17 September 1962, BAGC A638 22b, ANZA.

100   NWHMC minutes, 15 June 1964, BAGC A638 39b, ANZA.

101   NWHMC minutes, 17 August 1964, BAGC A638 39b, ANZA.

102   NWHMC minutes, 19 October 1964, BAGC A638 39b, ANZA.

103   NWHMC minutes, 26 May 1965, BAGC A638 39b; NWHMC minutes, 30 August 1965, BAGC A638 39b, ANZA.

104   'Colposcope Used in Detection of Cancer', *Auckland Star*, 3 October 1968.

105   NWHMC minutes, 19 October 1964, BAGC A638 39b, ANZA.

106   Matheson, *Fate Cries Enough*, p. 31.

107   Coppleson and Reid, *Preclinical Carcinoma of the Cervix Uteri*, pp. 7, 190.

108   Robert Baker, 'Transcultural Medical Ethics and Human Rights', in Ulrich Tröhler and Stella Reiter-Theil in cooperation with Eckhard Herych (eds), *Ethics Codes in Medicine: Foundations and Achievements of Codification since 1947*, Ashgate, Aldershot, 1998, pp. 322–4; Pascal Arnold and Dominique Sprumont, 'The "Nuremberg Code": Rules of Public International Law', in Tröhler and Reiter-Theil (eds), *Ethics Codes in Medicine*, pp. 91, 127; William J. Winslade and Todd L. Krause, 'The Nuremberg Code Turns Fifty', in Tröhler and Reiter-Theil (eds), *Ethics Codes in Medicine*, pp. 147–8, 156.

109   Green, Transcripts of Public Hearings, Day 21, 7 September 1987, p. 1592, BAGC A638/5a 101D, ANZA.

110   Heslop, '"All About Research"', 2004.

111   NWHMC minutes, 20 June 1966, BAGC A638 40a, ANZA; reprinted in CCR, pp. 21–22.

112 Per Kolstad, 'Diagnosis and Management of Precancerous Lesions of the Cervix Uteri', *International Journal of Gynaecology and Obstetrics*, vol. 8, 4, 2, 1970, p. 559.

113 M. Coppleson, 'Cervical Intraepithelial Neoplasia: Clinical Features and Management', in Malcolm Coppleson (ed.), *Gynecologic Oncology. Fundamental Principles and Clinical Practice*, vol. 1, Churchill Livingstone, New York, 1981, p. 416.

114 CCR, p. 21.

115 Green, Transcripts of Public Hearings, Day 11, 24 August 1987, p. 680, BAGC A638/3a 101B, ANZA.

116 Green, 'The Early Diagnosis of Gynaecological Malignancy', Refresher Course in Obstetrics, February–March 1967, Postgraduate School of Obstetrics and Gynaecology, UOAA. The latter point was also made by Dr Eamonn De Valera from Dublin, Ireland, 'Discussion', *AJOG*, vol. 101, 1, 1968, p. 181. An article in an international journal in 1966 also suggested the possibility of two forms of the disease: David J. B. Ashley, 'Evidence for the Existence of Two Forms of Cervical Carcinoma', *Journal of Obstetrics and Gynaecology of the British Commonwealth*, vol. 73, 1966, pp. 382–9.

117 G. H. Green, 'Invasive Potentiality of Cervical Carcinoma in Situ', *International Journal of Obstetrics and Gynaecology*, vol. 7, 4, 1969, pp. 168–9; see also Coppleson, 'Cervical Intraepithelial Neoplasia', p. 409.

118 G. H. Green and J. W. Donovan, 'The Natural History of Cervical Carcinoma in Situ', *Journal of Obstetrics and Gynaecology of the British Commonwealth,* vol. 77, 1, 1970, p. 1.

119 Memo, G. H. Green to Medical Superintendent (through Bonham), NWH, 7 November 1973, Cartwright Inquiry Exhibit 3b, UOAA.

120 Lerner, *Breast Cancer Wars*, pp. 10, 93, 208.

121 Sir Richard Doll, 'Concluding Remarks' (Given at the Medical Research Council Epidemiology Symposium, Green Lane Hospital, Auckland, 2 November 1973), *NZMJ*, vol. 80, 1974, p. 403. Doll continued to support NWH research proposals, including work on the data on CIS into the 1980s: MRC of NZ, Referee Report on Grant Application, Form MRC/4, 1981, MRC folder, Bonham, Epidemiology and Data Processing in Obstetrics, Neonatology and Gynaecology, 82/137, UOAA.

122 CCR, p. 82.

123 Green, 'The Progression of Pre-invasive Lesions of the Cervix to Invasion'.

124 Knox, 'Cervical Cytology', p. 293.

125 CCR, p. 38.

126 A. R. Chang, 'Health Screening: An Analysis of Abnormal Cervical Smears at Dunedin Hospital 1963–82', *NZMJ*, vol. 98, 1985, p. 106.

127 Leading article, 'Screening for Cervical Cancer', *BMJ*, vol. 2, 1976, pp. 659–60.

128 A. L. Cochrane, *Effectiveness and Efficiency: Random Reflections on Health Services,* The Nuffield Hospitals Trust, London, 1972, pp. 26–27.

129 A. L. Cochrane and W. W. Holland, 'Validation of Screening Procedures', *British Medical Bulletin*, vol. 27, 1, 1971, pp. 3–8. On Cochrane's career and influence, see Iain Chalmers, 'Cochrane, Archibald Leman', in Bynum and Bynum, *Dictionary of Medical Biography*, vol. 2, pp. 353–5.

130 A. B. Miller, 'Control of Carcinoma In Situ by Exfoliative Cytology Screening', in Coppleson (ed.), *Gynecologic Oncology*, 1981, pp. 382–3.

131 Edward W. Savage, 'Current Developments: Microinvasive Carcinoma of the Cervix', *AJOG*, vol. 113, 5, 1972, pp. 708–17.

132 Editorial, 'Cervical Epithelial Dysplasia', *BMJ*, vol. 1, 1975, pp. 294–5.

133 A. Singer, 'Cervical Epithelial Dysplasia', Correspondence, *BMJ*, vol. 1, 1975, pp. 679–80; Singer was later to be invited to New Zealand in 1990 by the Medical Council to review the evidence for Bonham's hearing and his responsibility for Green's research. (On Bonham's hearing before the Medical Council, see ch. 9.)

134 R. J. Walton, 'Cervical Cancer Screening Programs: Report of the Task Force Appointed by the Conference of Deputy Minister of Health in December 1973 and Submitted to the Conference March 16 and 17 1976: Part I: Epidemiology and Natural History of Carcinoma of the Cervix', *Canadian Medical Association Journal*, vol. 114, 1976, p. 1010.

135 Hertig and Younge, 'A Debate', p. 807.

136 D. R. Popkin, 'Editorial: Cervical Cancer Screening Programs: A Gynecologist's Viewpoint', *Canadian Medical Association Journal*, vol. 114, 1976, pp. 982–3; L. G. Koss, 'Concept of Genesis and Development of Carcinoma of the Cervix', *Obstetrical and Gynecological Survey*, vol. 24, 1969, p. 851.

137 Leopold G. Koss, 'The Papanicolaou Test for Cervical Cancer Detection: A Triumph and a Tragedy', *Journal of the American Medical Association*, vol. 261, 5, 1989, pp. 737, 740.

138 L. W. Coppleson and B. W. Brown, 'Control of Carcinoma of Cervix: Role of the Mathematical Model', in Coppleson (ed.) *Gynecologic Oncology*, 1981, p. 390.

139 James S. McCormick, 'Cervical Smears: A Questionable Practice?', *The Lancet,* vol. 1, 1989, pp. 207–9.

140 Anon, 'Cancer of the Cervix: Death by Incompetence', *The Lancet*, vol. 2, 1985, pp. 363–4.

141 Green told Coney and Bunkle it was about 5

per cent: Sandra Coney and Phillida Bunkle, 'Interview notes with Professor Herbert Green', 27 November 1986, p. 8; Exhibit 3a, UAAA; Green, Transcripts of Public Hearings, Day 13, pp. 850–2, BAGC A638/3a 101B, ANZA.

142 Per Kolstad and Klem Valborg, 'Long-term Follow-up of 1121 Cases of Carcinoma in Situ', *Obstetrics and Gynecology, Journal of the American College of Obstetricians and Gynecologists (O&G)*, vol. 48, 2, 1976, p. 128.

143 Ralph M. Richart and John J. Sciarra, 'Treatment of Cervical Dysplasia by Outpatient Electrocauterization', *AJOG*, vol. 101, 2, 1968, pp. 200–5.

144 *Auckland Star*, 28 August 1987.

145 Coney, *The Unfortunate Experiment*, pp. 167, 168.

146 J. A. Jordan, 'Minor Degrees of Cervical Intraepithelial Neoplasia: Time to Establish a Multicentre Prospective Study to Resolve the Question', *BMJ*, vol. 297, 1988, p. 6.

147 Andrew G. Östör, 'Review: Natural History of Cervical Intraepithelial Neoplasia: A Critical Review', *International Journal of Gynecological Pathology*, vol. 12, 2, 1993, pp. 186–92.

148 Charlotte Paul, 'Letter from New Zealand: The New Zealand Cervical Cancer Study: Could It Happen Again?', *BMJ*, vol. 297, 1988, pp. 533–9.

149 Paul, 'Letter from New Zealand'; David Slater, 'New Zealand Cervical Cancer Study: Could It Happen Again?', Correspondence, *BMJ*, vol. 297, 1988, p. 918.

150 R. M. Richart, Yao-Shi Fu and J. W. Reagan, 'Pathology of Cervical Intraepithelial Neoplasia', in Coppleson (ed.), *Gynecologic Oncology*, p. 398.

151 CCR, p. 34.

152 *Ibid.*, pp. 66, 96.

153 Green, Transcripts of Public Hearings, Day 11, 24 August 1987, p. 680, BAGC A638/3a 101B, ANZA.

154 *Ibid.*, p. 729.

155 Posner, 'What's in a Smear?', p. 173.

3. Management of Patients with Carcinoma in Situ

1 Matheson, *Fate Cries Enough*, p. 168. Matheson's book won the New Zealand Society of Authors (PEN) award for a first book of prose: *NZH*, 25 August 1990.

2 J. F. Gwynne, 'Sandra Coney and the National Women's Hospital Affair', Correspondence, *NZMJ*, vol. 103, 1990, p. 378.

3 W. A. McIndoe, M. R. McLean, R. W. Jones and P. R. Mullins, 'The Invasive Potential of Carcinoma in Situ of the Cervix', *O&G*, vol. 64, 1984, pp. 451–8.

4 In his evidence, Green referred to a much larger follow-up study in America, involving 4517 CIS cases and reported in 1968: G. H. Green, Evidence to Cartwright Inquiry, 20 July 1987, p. 98, UOAA. Per Kolstad also did a larger follow-up study: see Kolstad and Klem, 'Long-term Follow-up'.

5 Coney and Bunkle, 'An Unfortunate Experiment', pp. 58–59.

6 CCR, p. 95.

7 *Ibid.*, p. 54.

8 *Ibid.*, pp. 57, 150.

9 Jackie McAuliffe, Fertility Action, to McIndoe, 9 April 1985 and reply 7 May 1985, Coney's Appendices to Submission to Inquiry, 13a and 13b, BAGC 18489 A638 44a, ANZA.

10 Peter Mullins to Kevin Ryan, 15 June 1990, in author's possession.

11 Charlotte Paul and Linda Holloway, 'No New Evidence on the Cervical Cancer Study', *NZMJ*, vol. 103, 1990, pp. 581–3.

12 McIndoe *et al.,* 'The Invasive Potential'.

13 Margaret R. E. McCredie, Katrina J. Sharples, Charlotte Paul, Judith Branyai, Gabriele Medley, Ronald W. Jones and David C. G. Skegg, 'Natural History of Cervical Neoplasia and Risk of Invasive Cancer in Women with Cervical Intraepithelial Neoplasia 3: A Retrospective Cohort Study', *The Lancet Oncology*, vol. 9, 5, 2008, pp. 425–34.

14 Green, Transcripts of Public Hearings, Day 21, 7 September 1987, pp. 1621–2, BAGC A638/5a 101D, ANZA.

15 Coney, *The Unfortunate Experiment*, p. 17.

16 CCR, pp. 57, 96.

17 *Ibid.*, pp. 54, 57, 150.

18 *Dominion*, 29 June 1987.

19 Julie Roberts, 'I Was Lucky . . . I Lived', *NZWW*, 8 August 1988, p. 24.

20 CCR, pp. 150, 206.

21 *Ibid.*, pp. 38, 54, 95; Coney, *The Unfortunate Experiment*, p. 11.

22 Paul, 'Letter from New Zealand', p. 535.

23 CCR, p. 72.

24 *Ibid.,* p. 38.

25 *Ibid.,* p. 83.

26 *Ibid.,* p. 106.

27 NWHMC minutes, 5 May 1958, BAGC A638 38a, ANZA.

28 Aleck W. Bourne, *Recent Advances in Obstetrics and Gynaecology*, 9th edn, Churchill, London, 1958, p. 313.

29 Coppleson and Reid, *Preclinical Carcinoma of the Cervix Uteri*, p. 190.

30 *Ibid.,* p. 191. See also Coppleson, 'Cervical Intraepithelial Neoplasia', p. 427.

31 McLaren, *The Prevention of Cervical Cancer*, p. 58.

32 James S. Krieger and Lawrence J. McCormack, 'Graded Treatment for in Situ Carcinoma of the Uterine Cervix', *AJOG*, vol. 101, 2, 1968,

pp. 171–9, and 'Discussion', pp. 179–81.

33   Bruce H. Thompson, J. Donald Woodruff,
      Hugh J. Davis, Conrad G. Julian and Fred G.
      Silva II, 'Cytopathology, Histopathology, and
      Colposcopy in the Management of Cervical
      Neoplasia', *AJOG*, vol. 114, 2, 1972, pp. 329–33,
      and 'Discussion', pp. 333–8, Woodruff,
      'Closing Remarks', p. 337.

34   Leading article, 'Uncertainties of Cervical
      Cytology', *BMJ*, vol. 4, 1973, pp. 501–2.

35   Malcolm Coppleson, 'Colposcopy', in John
      Stallworthy and Gordon Bourne (eds), *Recent
      Advances in Obstetrics and Gynaecology,* 12th
      edn, Churchill Livingstone, Edinburgh, 1977,
      p. 180.

36   John Stallworthy and Gordon Bourne
      (eds), *Recent Advances in Obstetrics and
      Gynaecology,*13th edn, Churchill Livingstone,
      Edinburgh, 1979, pp. 268–9.

37   Richart, Fu and Reagan, 'Pathology of
      Cervical Intraepithelial Neoplasia', p. 398.

38   Ralph M. Richart, 'Current Concepts in
      Obstetrics and Gynecology: The Patient with
      an Abnormal Pap Smear – Screening Tech-
      niques and Management', *The New England
      Journal of Medicine*, vol. 302, 6, 1980, pp. 332–4.

39   Richart, Transcripts of Public Hearings, Day
      14, 27 August 1987, p. 935, BAGC A638/3a 101B,
      ANZA.

40   Richart, 'Current Concepts in Obstetrics
      and Gynecology', p. 332; Richart, Fu and
      Reagan, 'Pathology of Cervical Intraepithelial
      Neoplasia', p. 399.

41   Sir Norman Jeffcoate, *Principles of
      Gynaecology*, 4th edn, Butterworths, London
      and Boston, 1975 (1980 reprint), p. 400.

42   Goran Larsson, *Conization for Preinvasive
      and Early Invasive Carcinoma of the Uterine
      Cervix*, Acta Obstetricia et Gynecologica
      Scandinavica, Supplement vol. 114, Lund, 1983,
      pp. 9–10.

43   Richart, Transcripts of Public Hearings, Day
      14, 27 August 1987, p. 946, BAGC A638/3a
      101B, ANZA.

44   Green, 'Invasive Potentiality of Cervical
      Carcinoma in Situ', p. 165.

45   R. R. Margulis, R. W. Dustin, H. C. Walser and
      J. E. Ladd, 'Carcinoma in Situ of the Cervix
      with Vaginal Vault Extension', *O&G*, vol. 19,
      1962, pp. 569–74.

46   G. H. Green, 'Cervical Cytology and
      Carcinoma in Situ', *Journal of Obstetrics and
      Gynaecology of the British Commonwealth*,
      vol. 72, 1, 1965, pp. 13–22.

47   CCR, p. 27.

48   NWHMC minutes, 21 July 1958, BAGC A638
      37a, ANZA.

49   NWHMC minutes, 19 October 1964, BAGC
      A638 39b, ANZA.

50   Coppleson and Reid, *Preclinical Carcinoma of*

the Cervix Uteri, p. 137.

51   Dennis Bonham, Report on Study Leave
      – October–December 1969, 17 May 1970, p. 10,
      University of Auckland Council Reports, 1970,
      vol. 2, UOAA.

52   Coppleson, 'Colposcopy', p. 156.

53   *Ibid.*, p. 178.

54   *Ibid.*, pp. 177, 181. In his 1970 book, Coppleson
      also wrote that 'much present-day manage-
      ment of cervical lesions is imbalanced,
      illogical and too radical' because the clini-
      cian abdicated his 'traditional captaincy
      to the pathologist or exfoliative cytologist'.
      Malcolm Coppleson, Ellis Pixley and Bevan
      Reid, *Colposcopy: A Scientific and Practical
      Approach to the Cervix in Health and Disease*,
      Charles C. Thomas Publishers, Springfield,
      Illinois, 1970, p. 6.

55   CCR, p. 55.

56   Evidence of Dr Joseph Jordan, Appendix,
      Case History no. 60/69, p. 3, Case History no.
      60/64, p. 4, UOAA.

57   Jordan, Case History no. 60/64, p. 4, UOAA.

58   Green, 'Invasive Potentiality of Cervical
      Carcinoma in Situ', p. 163.

59   Vernon E. Hollyock and William Chanen,
      'The Use of the Colposcope in the Selection
      of Patients for Cervical Cone Biopsy', *AJOG*,
      vol. 114, 2, 1972, pp. 185–9; their references
      include J. R. Boyd, *AJOG*, vol. 75, 1958, p. 983;
      G. D. Byrne, *Australian and New Zealand
      Journal of Obstetrics and Gynaecology*, vol.
      6, 1966, p. 266; H. Arthure, J. Tomlinson
      and G. Organe *et al.*, *Report on Confidential
      Enquiries into Maternal Deaths in England and
      Wales, 1964–66*, (Great Britain Department of
      Health and Social Security, Reports on Public
      Health and Medical Subjects, no. 119) HMSO,
      London, 1969, pp. 80–81.

60   Thompson *et al.*, 'Cytopathology,
      Histopathology, and Colposcopy', pp. 329–33.

61   Donald R. Tredway, Duane E. Townsend,
      David N. Hovland and Richard T. Upton,
      'Colposcopy and Cryosurgery in Cervical
      Intraepithelial Neoplasia', *AJOG*, vol. 114, 8,
      1972, p. 1020.

62   Leading article, 'Outcome of Pregnancy after
      Cone Biopsy', *BMJ*, vol. 280, 1980, pp. 1393–4;
      Lisa Saffron, 'Cervical Cancer – The Politics
      of Prevention', *Spare Rib*, vol. 129, April 1983,
      reprinted in Sue O'Sullivan, *Women's Health:
      A Spare Rib Reader*, Pandora, London, 1987,
      p. 47; McCormick, 'Cervical Smears: A
      Questionable Practice?'.

63   Jeffcoate, *Principles of Gynaecology*, 1980,
      p. 492.

64   Samy Rifai, 'Gynecological Issues: Colposcopy
      Referees the Pap Smear', *The Female Patient*,
      vol. 5, 4 (April) 1980, n.p., reproduced in
      the New Zealand Women's Health Network

Newsletter, 25, August 1981, Joan Donley Archives, UOAA.

65   Coppleson, 'Cervical Intraepithelial Neoplasia', p. 420.

66   Peggy Foster, *Women and the Health Care Industry: An Unhealthy Relationship?*, Open University Press, Buckingham, 1995, p. 118.

67   G. H. Green, 'Pregnancy Following Cervical Carcinoma in Situ: A Review of 60 Cases', *Journal of Obstetrics and Gynaecology of the British Commonwealth*, vol. 73, 1966, pp. 897–902.

68   Kolstad and Klem, 'Long-term Follow-up', pp. 125, 128.

69   McIndoe *et al.*, 'Invasive Potential', p. 255.

70   Jocelyn Chamberlain, 'Failures of the Cervical Cytology Screening Programme', *BMJ*, vol. 289, 1984, p. 854.

71   McCredie *et al.*, 'Natural History of Cervical Neoplasia', p. 425.

72   Raffle and Gray, *Screening: Evidence and Practice*, pp. 20, 25; see also *Jeffcoate's Principles of Gynaecology*, revised by V. R. Tindall, Butterworths, London, 1987, p. 484, which makes a similar point.

73   Koss, *et al.*, 'Some Histological Aspects of Behaviour of Epidermoid Carcinoma In Situ'.

74   *NZH*, 6 May 2004.

75   Editorial, 'Adverse Pregnancy Outcomes after Treatment for Cervical Intraepithelial Neoplasia', *BMJ*, vol. 337, 2008, pp. 769–70; M. Arbyn, M. Kyrgiou, C. Simeons, A. O. Raifu, G. Koliopoulos, P. Martin-Hirsch, W. Prendiville and E. Paraskevaidis, 'Perinatal Mortality and Other Severe Adverse Pregnancy Outcomes Associated with Treatment of Cervical Intraepithelial Neoplasia: Meta-analysis', *BMJ*, vol. 337, 2008, pp. 798–802; S. Albrechtsen, S. Rasmussen, S. Thoreson, L. M. Irgens and O. E. Iversen, 'Pregnancy Outcome in Women Before and After Cervical Conisation: Population Based Cohort Study', *BMJ*, vol. 337, 2008, pp. 803–5.

76   Andrew Mackintosh, Transcripts of Public Hearings, Day 54, 24 November 1987, p. 4768, BAGC A638/12a 101K, ANZA.

77   Murray Jamieson, Transcripts of Public Hearings, Day 54, 24 November 1987, Background Material to Counsel Assisting Closing Address, BAGC 18492 A638 130a, ANZA.

78   Richart, Transcripts of Public Hearings, Day 14, 27 August 1987, p. 946, BAGC A638/3a 101B, ANZA; also cited in Coney, *The Unfortunate Experiment*, p. 169.

79   Coney, *The Unfortunate Experiment*, p. 172.

80   Jeffcoate, *Principles of Gynaecology,* 1980, p. 492.

81   Graham, Sotto and Paloucek, *Carcinoma of the Cervix*, pp. 92–93.

82   On the Schiller test, see Walter Schiller,

'Early Diagnosis of Carcinoma of the Cervix', *Surgery, Gynecology and Obstetrics*, vol. 56, 1933, pp. 212–22.

83   Leopold Koss, 'Dysplasia: A Real Concept or a Misnomer?', *O&G*, vol. 51, 1978, pp. 374, 375.

84   CCR, p. 24.

85   Stallworthy and Bourne, *Recent Advances*, 13th edn, 1979, p. 269; Kolstad and Klem, 'Long-term Follow-up', p. 125.

86   *Ibid.,* pp. 127, 128.

87   J. Elizabeth Macgregor and Sue Teper, 'Uterine Cervical Cytology and Young Women', *The Lancet*, vol. 311, 8072, 1978, pp. 1029–31.

88   Cited in Coney, *The Unfortunate Experiment,* p. 169.

89   Ralph M. Richart and Bruce A. Barron, 'A Follow-up Study of Patients with Cervical Dysplasia', *AJOG*, vol. 105, 1969, p. 391.

90   J. H. Robertson, Bertha E. Woodend, E. H. Crozier and June Hutchinson, 'Risk of Cervical Cancer Associated with Mild Dyskaryosis', *BMJ*, vol. 297, 1988, pp. 18–21.

91   Paul, 'Letter from New Zealand'; Slater, 'New Zealand Cervical Cancer Study: Could It Happen Again?', Correspondence.

92   Östör, 'Review: Natural History of Cervical Intraepithelial Neoplasia', pp. 186–92.

93   Green, 'Invasive Potentiality of Cervical Carcinoma in Situ', p. 158.

94   *Ibid.,* p. 163.

95   Green, Transcripts of Public Hearings, Day 11, 24 August 1987, p. 682, BAGC A638/3a 101B.

96   Jack Cuzick, Christine Clavel, Karl-Ulrich Petry, Chris J. L. M. Meijer, Heike Hoyer, Samuel Ratnam, Anne Szarewski, Philippe Birembaut, Shalini Kulasingam, Peter Sasiene and Thomas Iftner, 'Early Detection and Diagnosis: Overview of the European and North American Studies on HPV Testing in Primary Cervical Cancer Screening', *International Journal of Cancer*, vol. 119, 5, February 2006, pp. 1095–101; also reported in 'Sexual Health Matters: Pap Smear Has Had its Day', *New Zealand Doctor,* 26 July 2006, p. 8.

97   Kolstad and Klem, 'Long-term Follow-up', p. 128.

98   G. H. Green, 'Cervical Carcinoma in Situ: An Atypical View', *Australian and New Zealand Journal of Obstetrics and Gynaecology*, vol. 10, 1970, pp. 41, 43. McIndoe referred to 25 cases only which had colposcopically directed punch or wedge biopsy. McIndoe *et al.*, 'The Invasive Potential', p. 248.

99   Kolstad, 'Diagnosis and Management of Precancerous Lesions of the Cervix Uteri', p. 559.

100 Green and Donovan, 'The Natural History of Cervical Carcinoma in Situ', p. 2.

101 G. H. Green, 'Invasive Potential of Cervical Carcinoma in Situ', *International Journal of*

Gynaecology and Obstetrics, vol. 7, 4, 1969, pp. 157–71.

102 Green, 'The Progression of Pre-invasive Lesions'.

103 CCR, p. 60.

104 Green, 'The Progression of Pre-invasive Lesions'.

105 CCR, p. 51.

106 Green, Transcripts of Public Hearings, Day 21, 7 September 1987, p. 1592, BAGC A638/5a 101D, ANZA.

107 Green, Transcripts of Public Hearings, Day 13, 26 August 1987, p. 860, BAGC A638/3a 101B, ANZA.

108 See also Koss *et al.,* 'Some Histological Aspects of Behaviour of Epidermoid Carcinoma In Situ', pp. 1160–78, 1182, 1188, 1204, 1211, cited in McCredie *et al.,* 'Natural History of Cervical Neoplasia'.

109 Green, Transcripts of Public Hearings, Day 11, 24 August 1987, p. 685, BAGC A638/3a 101B, ANZA.

110 Coney, Transcripts of Public Hearings, Day 6, 11 August 1987, p. 322, BAGC A638/2a 101A, ANZA.

111 Jeffcoate, *Principles of Gynaecology,* 1980, p. 492; see also C. T. Griffiths, J. H. Austin and P. A. Younge, 'Punch Biopsy of the Cervix', *AJOG,* vol. 88, 1964, p. 695.

112 CCR, p. 51.

113 *Ibid.,* p. 116.

114 *Ibid.,* p. 119.

115 Green, Transcripts of Public Hearings, Day 12, 25 August 1987, p. 757, BAGC A638/3a 101B, ANZA.

116 Patients (Mrs S. 70W/144), Transcripts of Public Hearings, Day 10, 20 August 1987, pp. 649–51, BAGC A638/3a 101B, ANZA.

117 Green, Transcripts of Public Hearings, Day 12, 25 August 1987, p. 757, BAGC A638/3a 101B, ANZA.

118 David Cole to Vice-Chancellor C. J. Maiden, 22 December 1977, Box 245, UOAA.

119 Patients (Mrs W. 57W/71), Transcripts of Public Hearings, Day 9, 19 August 1987, p. 607, BAGC A638/3a 101B, ANZA.

120 Green, Transcripts of Public Hearings, Day 13, 26 August 1987, p. 922, BAGC A638/3a 101B, ANZA.

121 Richart, Transcripts of Public Hearings, Day 14, 27 August 1987, p. 957, BAGC A638/3a 101B, ANZA.

122 Green (re. Patient 60W/69), Transcripts of Public Hearings, Day 16, 31 August 1987, p. 1246, BAGC A638/4a 101C, ANZA.

123 CCR, p. 106.

124 Green, Transcripts of Public Hearings, Day 17, 1 September 1987, p. 1225, BAGC A638/4a 101C ANZA. In his evidence, Green provided three international journal references express-ing this concern, Evidence to Inquiry, p. 63, UOAA.

125 Green, Transcripts of Public Hearings, Day 12, 25 August 1987, pp. 758–61, BAGC A638/3a 101B, ANZA.

126 Coney, *The Unfortunate Experiment,* p. 85.

127 Green, Transcripts of Public Hearings, Day 18, 2 September 1987, pp. 1281–2, BAGC A638/4a 101C, ANZA.

128 Green, Transcripts of Public Hearings, Day 21, 7 September 1987, p. 1645, BAGC A638/5a 101D, ANZA.

129 Green (re. patient Mrs M. 61W/107), Transcripts of Public Hearings, Day 12, 25 August 1987, pp. 758, 761, BAGC A638/3a 101B, ANZA.

130 Green (re. patient Mrs P. 58W/63), Transcripts of Public Hearings, Day 12, 25 August 1987, p. 761, BAGC A638/3a 101B, ANZA.

131 Green, Transcripts of Public Hearings, Day 21, 7 September 1987, p. 1645, BAGC A638/5a 101D, ANZA.

132 Green, Transcripts of Public Hearings, Day 17, 1 September 1987, p. 1255, BAGC A638/4a 101C, ANZA.

133 Green, Transcripts of Public Hearings, Day 18, 2 September 1987, p. 1340, BAGC A638/4a 101C, ANZA.

134 Coney and Bunkle, 'An Unfortunate Experiment', p. 62; Pamela Hyde, 'At Your Cervix Madam: A Socio-historical Study of Cervical Cancer from the Late Nineteenth Century to the Late Twentieth Century', PhD, Victoria University of Wellington, 1997, p. 99.

135 Coney and Bunkle, 'An Unfortunate Experiment', p. 49.

136 *Ibid.,* p. 49.

137 Green, evidence to Inquiry, p. 77, UOAA.

138 *Ibid.,* p. 78.

139 Green, 'Cervical Cytology and Carcinoma in Situ', p. 17.

140 Coney, Transcripts of Public Hearings, Day 7, 12 August 1987, p. 377, BAGC A638/2a 101A, ANZA; G. H. Green, 'Tubal Ligation', *NZMJ,* vol. 57, 1958, p. 477.

141 Faye Hercock, 'Professional Politics and Family Planning Clinics', in L. Bryder (ed.), *A Healthy Country: Essays on the Social History of Medicine in New Zealand,* Bridget Williams Books, Wellington, 1991, pp. 181–97.

142 Green, 'Tubal Ligation', p. 477. North America in particular had a history of forcibly steri-lising those seen to be 'feeble-minded' or otherwise unsuited for motherhood: Georgina Feldberg, 'On the Cutting Edge: Science and Obstetrical Practice in a Women's Hospital, 1945–60', in Georgina Feldberg, Molly Ladd-Taylor, Alison Li and Kathryn McPherson (eds), *Women, Health, and Nation: Canada and the United States since 1945,* McGill-

Queen's University Press, Montreal and Kingston, 2003, p. 134.

143 Green, Transcripts of Public Hearings, Day 12, 25 August 1987, p. 761, BAGC A638/3a 101B, ANZA.

144 Patients (Mrs R. 68W/40), Transcripts of Public Hearings, Day 9, 19 August 1987, p. 574, BAGC A638/3a 101B, ANZA.

145 Mrs Elsie Marion Barnes, Transcripts of Public Hearings, Day 45, 9 November 1987, p. 3846, BAGC A638/10a 101I, ANZA.

146 *Ibid.*, p. 3852.

147 *Ibid.*, p. 3853.

148 *East and Bays Courier*, 16 July 2008.

149 *NZWW*, 8 August 1988, p. 20

150 Coppleson, 'Cervical Intraepithelial Neoplasia', p. 456.

151 See, for example, Matheson, *Fate Cries Enough*, p.140.

152 McIndoe *et al.*, 'The Invasive Potential', p. 255.

153 A. E. Raffle, B. Alden and E. F. D. Mackenzie, 'Detection Rates for Abnormal Cervical Smears: What Are We Screening For?', *The Lancet*, vol. 345, 1995, pp. 1469–73.

154 *Ibid.*

155 *Ibid.* Raffle, Alden and Mackenzie cited McIndoe *et al.*'s 1984 article, 'The Invasive Potential'.

156 Lerner, *Breast Cancer Wars*, pp. 126–7.

157 Coney, *The Unfortunate Experiment*, p. 261.

4   The Therapeutic Relationship and Patient Consent

1   *Dominion Sunday Times*, 26 March 1989.

2   CCR, p. 206.

3   Te Ohu Whakatupu (Maori Women's Secretariat) Ministry of Women's Affairs Submission to Inquiry, August 1987, p. 14, BAGC A638 34a, ANZA.

4   Green, 'The Progression of Pre-invasive Lesions of the Cervix to Invasion', p. 285.

5   Coppleson, 'Colposcopy', p. 178.

6   Coney, Submission to Inquiry, p. 14, UOAA.

7   Green, Transcripts of Public Hearings, Day 12, 25 August 1987, p. 757, BAGC A638/3a 101B, ANZA.

8   Posner, 'What's in a Smear?', p. 174.

9   Richard Seddon, Transcripts of Public Hearings, Day 4, 6 August 1987, p. 134, BAGC A638/2a 101A, ANZA.

10 Dennis Bonham, Submission to Inquiry, pp. 18–19, UOAA.

11 Elsie Barnes, Transcripts of Public Hearings, Day 45, 9 November 1987, p. 3846, BAGC A638/10a 101I, ANZA.

12 *Sunday Star*, 8 July 1990.

13 Patients (Mrs L 69W/132), Transcripts of Public Hearings, Day 9, 19 August 1987, p. 559, BAGC A638/3a 101B, ANZA.

14 Patients (Mrs G 64W/83), Transcripts of Public Hearings, Day 9, 19 August 1987, pp. 618–25, BAGC A638/3a 101B, ANZA.

15 Patients (Mrs L 69W/132), Transcripts of Public Hearings, Day 9, 19 August 1987, p. 559, BAGC A638/3a 101B, ANZA.

16 M. M. to Michael Bassett, 6 June 1987, Mrs A. E. B. to Bassett, 9 June 1987, 'Hospital Boards – Ministerial Inquiry 1986–8', ABQU 632 W4452 397 56/101 62694, Archives New Zealand Wellington (ANZW).

17 Gabrielle Collison, Correspondence, BAGC 18492 A638 162a, letter no. 83, ANZA.

18 Green, Transcripts of Public Hearings, Day 16, 31 August 1987, p. 1254, BAGC A638/4a 101C, ANZA.

19 Green, Transcripts of Public Hearings, Day 13, 26 August 1987, p. 903, BAGC A638/3a 101B, ANZA.

20 *Ibid.*, p. 919, BAGC A638/3a 101B, ANZA.

21 Barron H. Lerner, 'What Do You Know? Cancer, History and Medical Practice', in Jacalyn Duffin, *Clio in the Clinic: History in Medical Practice*, Oxford University Press, Oxford, 2005, p. 302.

22 Lowell Goddard's final submission to Inquiry, cited in Matheson, *Fate Cries Enough*, p. 169.

23 Coney, *The Unfortunate Experiment*, p. 107. 'One of the women said she "did hear the words carcinoma in situ", but "it didn't mean anything to me"': Coney, *The Unfortunate Experiment*, p. 127.

24 McLaren, *The Prevention of Cervical Cancer*, pp. 60, 61.

25 Coppleson and Reid, *Preclinical Carcinoma of the Cervix Uteri*, p. 198.

26 *Ibid.*, p. 185.

27 Coppleson, 'Cervical Intraepithelial Neoplasia', p. 427.

28 McKelvey, 'Carcinoma in Situ of the Cervix', p. 825.

29 Krieger and McCormack, 'Graded Treatment for in Situ Carcinoma', 'Discussion', p. 182.

30 Matheson, *Fate Cries Enough*, p. 33.

31 Jeffcoate, *Principles of Gynaecology*, 1980, p. 493.

32 Green, 'Cervical Cytology and Carcinoma in Situ', p. 19.

33 Green, 'Invasive Potentiality', p. 160.

34 Matheson, *Fate Cries Enough*, p. 30.

35 *Ibid.*, p. 31; see also chapter 2 for more discussion.

36 Sandra Coney and Phillida Bunkle, 'Interview Notes with Professor Herbert Green, 27 November – 1986', Cartwright Inquiry Exhibit 3a, p. 6, UOAA. *Jeffcoate's Principles of Gynaecology*, 1987, p. 484 made the same point.

37 Coney and Bunkle, 'Interview Notes with Professor Herbert Green', p. 7, UOAA.

38 Patients' notes, Exhibit to Cartwright Inquiry,

UOAA; unedited version of the *Metro* article in Phillida Bunkle, *Second Opinion: The Politics of Women's Health in New Zealand*, Oxford University Press, Auckland, 1988, p. 185.

39 Matheson, *Fate Cries Enough*, p. 159.

40 *Ibid.*, p. 122.

41 Green, Transcripts of Public Hearings, Day 18, 2 September 1987, p. 1310, BAGC A638/4a 101C, ANZA.

42 Green, Transcripts of Public Hearings, Day 12, 25 August 1987, p. 758, BAGC A638/3a 101B, ANZA.

43 Judith Medlicott, Review of *Fate Cries Enough*, in *Dominion Sunday Times*, 3 December 1989.

44 CCR, p. 46; Coney and Bunkle, 'An Unfortunate Experiment', p. 56. From case notes reproduced in Matheson, *Fate Cries Enough*, pp. 54, 57, 59, 68: Wedge biopsy 24 March 1970, Cone biopsy 16 February 1971, Wedge biopsy 23 November 1971 and Ring biopsy 27 January 1976.

45 Matheson, *Fate Cries Enough*, p. 130.

46 Green, Evidence to Inquiry, pp. 99, 100, UOAA.

47 Richart, Transcripts of Public Hearings, Day 14, 27 August 1987, pp. 946, 956, BAGC A638/3a 101B, ANZA; Richart, Transcripts of Public Hearings, Day 15, 28 August 1987, p. 1047, BAGC A638/4a 101C, ANZA.

48 *Ibid.*, p. 1047.

49 Green, Transcripts of Public Hearings, Day 16, 31 August 1987, p. 1115, BAGC A638/4a 101C, ANZA.

50 Green, Transcripts of Public Hearings, Day 18, 2 September 1987, pp. 1299–301, BAGC A638/4a 101C, ANZA.

51 Cited in Coney and Bunkle, 'An Unfortunate Experiment', p. 56.

52 Jeffcoate wrote that microinvasion was 'not necessarily ominous' and 'would be better described as . . . micropenetration to avoid confusion of thought which the term invasion is likely to induce': Jeffcoate, *Principles of Gynaecology*, 1980, p. 397.

53 Matheson, *Fate Cries Enough*, p. 177.

54 *Ibid.*, p. 210.

55 Kolstad and Klem, 'Long-term Follow-up', pp. 125, 128.

56 Green, Transcripts of Public Hearings, Day 18, 2 September 1987, p. 1284, BAGC A638/4a 101C, ANZA.

57 Coppleson, 'Cervical Intraepithelial Neoplasia', p. 420.

58 Jan Corbett, *Sunday Star*, 5 August 1990; a 2005 publication stated that 'two negative cytological tests over one year', along with a negative HPV DNA result, a test which was not available at that time, meant that a woman could be safely returned to routine screening: World Health Organization International

Agency for Research on Cancer, *IARC Handbooks on Cancer Prevention, Volume 10, Cervix Cancer Screening*, IARC Press, Lyon, 2005, p. 19.

59 *Sunday Star*, 5 August 1990.

60 Matheson, *Fate Cries Enough*, pp. 65, 165.

61 *Ibid.*, p. 165.

62 CCR, p. 96.

63 Matheson, *Fate Cries Enough*, p. 173.

64 Tina Posner and Martin Vessey, *Prevention of Cervical Cancer: The Patient's View*, King Edward's Hospital Fund for London, London, 1988.

65 *Ibid.*, pp. 9, 44, 55, 97.

66 *Ibid.*, p. 45.

67 *Ibid.*, p. 59.

68 *Ibid.*, p. 96.

69 W. M. Peters and J. B. Kershaw, 'Cervical Cytology Screening', Correspondence, *BMJ*, vol. 296, 1988, p. 1670, refers to V. Nathoo, 'Investigation of Non-responders at a Cervical Screening Clinic', *BMJ*, vol. 296, 1988, pp. 1041–2.

70 Coney, *The Unfortunate Experiment*, pp. 33, 135.

71 Bonham to Bunkle, 15 November 1985, Coney's Appendices to Evidence, 13d, BAGC 18489 A638 44a, ANZA.

72 Coney, *The Unfortunate Experiment*, p. 22.

73 Discussing a symposium at the Auckland Medical School in 1986, at one point she describes it as the Cervical Cancer symposium, and at another the CIS symposium; Coney, *The Unfortunate Experiment*, p. 32.

74 The Health Alternatives for Women (THAW), Submission to Inquiry, p. 7, BAGC A638 36a, ANZA.

75 Coney, Submission to Inquiry, p. 9, also cited in the *NZH*, 18 June 1987.

76 Alison Margaret Wilson, 'Primetime Television Coverage of the Cartwright Inquiry, 1987–1988', MA, University of Auckland, 1993, pp. 13–14.

77 Peters and Kershaw, Correspondence, *BMJ*; see also chapter 4.

78 Lois Taylor, Letter to Editor, *New Zealand Listener*, 8 October 1988.

79 Arnold and Sprumont, 'The "Nuremberg Code"', p. 91.

80 This was repeated in the text of the CCR, p. 134, though the appendix included the 1983 Declaration which was quite different.

81 Baker, 'Transcultural Medical Ethics and Human Rights', pp. 322–4; Arnold and Sprumont, 'The "Nuremberg Code"', pp. 91, 127; Winslade and Krause, 'The Nuremberg Code Turns Fifty', pp. 147–8, 156.

82 Green, 'Cervical Cytology and Carcinoma in Situ', p. 19.

83 Sir Austin Bradford Hill, 'Medical Ethics

and Controlled Trials', *BMJ*, vol. 1, 1963, pp. 1043–9.

84   Baker, 'Transcultural Medical Ethics and Human Rights', p. 321.

85   Coney and Bunkle, 'An Unfortunate Experiment', p. 52.

86   Gonzalo Herranz, 'The Inclusion of the Ten Principles of Nuremberg in Professional Codes of Ethics: An International Comparison', in Tröhler and Reiter-Theil (eds), *Ethics Codes in Medicine*, p. 127; Rothman, *Strangers at the Bedside*, p. 62.

87   Winslade and Krause, 'The Nuremberg Code Turns Fifty', p. 147.

88   Baker, 'Transcultural Medical Ethics and Human Rights', p. 324.

89   Annual Report Medical Protection Society No. 93, 1985, cited by W. J. Ramsay, letter to Cartwright, 12 August 1987, BAGCA638 35a, ANZA.

90   Richard Doll, 'Clinical Trials: Retrospect and Prospect', *Statistics in Medicine*, vol. 1, 1982, pp. 337–44.

91   Green, Transcripts of Public Hearings, Day 21, 7 September 1987, p. 1590, BAGC A638/5a 101D, ANZA.

92   Rothman, *Strangers at the Bedside*, p. 91.

93   NWHMC minutes, 18 June 1972, BAGC A638 21a, ANZA; Annual Report to the Auckland Medical Research Foundation, July 1973, Box 257, UOAA; on the Hospital Ethical Committee, see also chapter 8.

94   Lerner, *Breast Cancer Wars*, p. 88.

95   J. V. Hodge, Brief for Committee of Inquiry, pp. 9–10, 13, 26, UOAA.

96   'News: Medical Research Council: Scientific Secretary', *NZMJ*, vol. 68, 1968, p. 183.

97   J. V. Hodge, Brief for Committee of Inquiry, pp. 9–10, 13, 26, UOAA.

98   NWHMC minutes, 20 February 1969, BAGC A638 40b, ANZA.

99   NWHMC minutes, 24 May 1973, BAGC A638 21a, ANZA.

100   *Ibid.*; NWHMC minutes, 20 September 1973, BAGC A638 21a; NWHMC minutes, 18 October 1973, BAGC A638 21a, ANZA.

101   NWHMC minutes, 19 June 1975, BAGC A638 41a, ANZA.

102   NWHMC minutes, 20 June 1976, BAGC A638 41a, ANZA.

103   Coney, *The Unfortunate Experiment*, p. 140.

104   NWHMC minutes, 21 April 1977, BAGC A638 41b, ANZA.

105   Dr Joseph (Joe) Jordan, Submission to Cervical Cancer Inquiry, p. 12, UOAA.

106   *Ibid.*, p. 13.

107   Patients (Mrs B – 82W/330), Transcripts of Public Hearings, Day 9, 19 August 1987, p. 369, BAGC A638/3a 101D, ANZA.

108   CCR, p. 168.

109   *Auckland Star*, 1 September 1987.

110   Peter Keating and Alberto Cambrosio, 'Risk on Trial: The Interaction of Innovation and Risk in Cancer Clinical Trials', in Schlick and Tröhler (eds), *The Risks of Medical Innovation*, p. 237.

111   Green, Transcripts of Public Hearings, Day 16, 31 August 1987, p. 1131, BAGC A638/4a 101C, ANZA.

112   *Ibid.*, p. 1154.

113   Ornella Moscucci, 'The "Ineffable Freemasonry of Sex": Feminist Surgeons and the Establishment of Radiotherapy in Early Twentieth-century Britain', in David Cantor (ed.), *Cancer in the Twentieth Century*, The Johns Hopkins University Press, Baltimore, 2008, pp. 139–63.

114   Sir John Stallworthy, 'Clinical Invasive Carcinoma of Cervix: Combined Radiotherapy and Radical Hysterectomy as Primary Treatment', in Coppleson (ed.), *Gynecologic Oncology*, pp. 508, 512.

115   J. H. Nelson Jr, 'Clinical Invasive Carcinoma of Cervix: Place of Radical Hysterectomy as Primary Treatment', in Coppleson (ed.), *Gynecologic Oncology*, p. 505.

116   Coney, *The Unfortunate Experiment*, p. 202.

117   Linda Kaye, 'What Happened to the Cartwright Women? The Legal Proceedings', in Sandra Coney (ed.), *Unfinished Business: What Happened to the Cartwright Report?*, Women's Health Action, Auckland, 1993, p. 81.

118   Green, Transcripts of Public Hearings, Day 16, 31 August 1987, p. 1131, BAGC A638/4a 101C, ANZA.

119   Harrison cross-examining Green, Transcripts of Public Hearings, Day 11, 24 August 1987, p. 664, BAGC A638/3a 101B, ANZA.

120   Phillida Bunkle, 'Side-stepping Cartwright: The Cartwright Recommendations Five Years On', in Coney (ed.), *Unfinished Business*, p. 53.

5   A Profession Divided

1   CCR, p. 60.

2   *Ibid.*, p. 198.

3   *Ibid.*

4   *Ibid.*, pp. 150–1.

5   Graham, Sotto and Paloucek, *Carcinoma of the Cervix*, pp. 74–75.

6   Leading article, 'Diagnostic Problems in Cervical Cancer', *BMJ*, vol. 1, 1974 , pp. 471–2.

7   Stallworthy and Bourne (eds), *Recent Advances in Obstetrics and Gynaecology*, 11th edn, 1966, p. 356.

8   Koss, 'Dysplasia: A Real Concept or a Misnomer?', p. 377.

9   Alexander Sedlis, Sanford Sall, Yoshi Tsukada, Robert Park, Charles Mangan, Hugh Shingleton and John A. Blessing,

'Microinvasive Carcinoma of the Uterine Cervix: A Clinical-pathologic Study', *AJOG*, vol. 133, 1, 1979, pp. 64–74. Also referred to by Richart, Fu and Reagan in 'Pathology of Cervical Intraepithelial Neoplasia', in Coppleson (ed.), *Gynecologic Oncology*, p. 401; and Malcolm Coppleson, 'Preclinical Invasive Carcinoma of Cervix (Microinvasive and Occult Invasive Carcinoma): Clinical Features and Management', *ibid.*, p. 452, where he commented on the subjectivity of the diagnosis.

10   Coney, *The Unfortunate Experiment,* p. 261.

11   Lerner, *Breast Cancer Wars,* pp. 11, 126–7, 198, 246.

12   Jeffcoate, *Principles of Gynaecology*, 1980, pp. 398–9.

13   Coppleson, 'Cervical Intraepithelial Neoplasia', pp. 409–11. As recently as 2005, a textbook noted, 'The grading of CIN lesions is prone to high rates of intra-observer and inter-observer variability', 'CIN2 and CIN3 lesions associated with extensive gland involvement may be confused with microinvasive squamous-cell carcinoma, resulting in overall error': World Health Organization International Agency for Research on Cancer: *IARC Handbooks on Cancer Prevention, Volume 10*, pp. 13, 14, 60.

14   Charlotte Paul, 'Education and Debate: Internal and External Morality of Medicine: Lessons from New Zealand', *BMJ*, vol. 320, 2000, pp. 499–503.

15   *Ibid.*

16   McIndoe and McLean, Memo, 10 October 1973; G. H. Green, Memo to Medical Superintendent (through Bonham) NWH, 7 November 1973, CI Exhibit, 3b, UOAA.

17   Green, Submission to Inquiry, p. 7, UOAA.

18   'Obituary: Malcolm Robert McLean', *NZMJ*, vol. 106, 1993, p. 236.

19   'Readers Respond', *Metro*, August 1990, pp. 124–5.

20   See obituary by Dorothy Rosenthal (http://www/cytojournal. com/content/pdf/1742-6413-2-2.pdf).

21   Jones and Fitzgerald, 'The Development of Cervical Cytology and Colposcopy in New Zealand'.

22   Kolstad and Klem, 'Long-term Follow-up', pp. 127, 128. See also Coppleson, 'Preclinical Invasive Carcinoma of Cervix', p. 461, where he wrote of errors in diagnosis of microinvasive cancer at initial treatment: 'Human error in clinical and laboratory assessment in large series is not unexpected.'

23   Transcripts of Public Hearings, Day 51, 19 November 1987, pp. 4411–2, BAGC A638/11a, 101J, ANZA.

24   Transcripts of Public Hearings, Day 54, 24 November 1987, p. 4792, BAGC A638/12a, 101K, ANZA.

25   Green, 'Cervical Cytology and Carcinoma in Situ', p. 15; 'The Progression of Pre-invasive Lesions of the Cervix to Invasion', p. 285.

26   Rennie cross-examining Green, Transcripts of Public Hearings, Day 19, 3 September 1987, p. 1346, BAGC A638/4a, 101C, ANZA.

27   G. H. Green, Memo to Medical Superintendent (through Bonham) NWH, 7 November 1973, Cartwright Inquiry Exhibit, 3b, UOAA, cited in CCR, p. 80.

28   CCR, p. 78.

29   Coppleson, 'Cervical Intraepithelial Neoplasia', p. 412.

30   Andrew Mackintosh, interviewed by Jenny Carlyon, 11 April 2006.

31   Green, Memo to Medical Superintendent NWH, 7 November 1973, UOAA, cited in CCR, p. 80.

32   Sir Graham Liggins, interviewed by Jenny Carlyon, 12 May 2004.

33   McLean, Memo, 10 May 1974, in response to Green's Memo of 7 November 1973, Coney's Appendices, Appendix 1, BAGC 18489 Acc A638 44a, ANZA.

34   McLean, Memo, 18 October 1973 exhibit 5a to medical superintendent, Coney's Appendices, Appendix 1, BAGC 18489 Acc A638 44a, ANZA.

35   Green, Transcripts of Public Hearings, Day 11, 24 August 1987, p. 697, BAGC A638/3a 101B, ANZA.

36   Koss, 'Dysplasia: A Real Concept or a Misnomer?', p. 374. In 1960 he wrote: 'If invasion has been ruled out, in situ carcinoma ceases to be a surgical emergency. It can be treated electively and without urgency': 'Exfoliative Cytology of the Uterine Cervix and Vagina', *CA: A Cancer Journal for Clinicians*, vol. 10, 1960, pp. 182–93.

37   Coppleson, 'Cervical Intraepithelial Neoplasia', p. 421.

38   Green, Memo to Medical Superintendent NWH, 7 November 1973, pp. 3–6, UOAA.

39   *Ibid.*

40   *Ibid.*, p. 9.

41   Read out at Inquiry, Green, Transcripts of Public Hearings, Day 11, 24 August 1987, p. 697, BAGC A638/3a 101B, ANZA.

42   Editorial, 'Diagnostic Problems in Cervical Cancer', *BMJ*, vol. 1, 1974, pp. 471–2.

43   Richart, Transcripts of Public Hearings, Day 14, 27 August 1987, p. 962, BAGC A638/3a 101B, ANZA; Richart, Fu and Reagan, 'Pathology of Cervical Intraepithelial Neoplasia', pp. 398, 400.

44   McIndoe, Memo to Green, 15 September 1969, BAGC 18491 A638 97a 410B, ANZA; CCR, p. 73.

45   CCR, pp. 72–74; W. A. McIndoe and G. H.

Green, 'Vaginal Carcinoma in Situ Following Hysterectomy', *Acta Cytologica,* vol. 13, 3, 1969, pp. 158–62.

46 McIndoe to James Maclean, President of World Congress, 15 March 1972, ABQU 632 W 4452 119 13-72-61867, ANZW.

47 McIndoe to Duncan, 17 June 1971, BAGC 18491 A638 97a 410B, ANZA.

48 McIndoe, Memo to Warren, 14 December 1973, UOAA.

49 *Ibid.,* cited in CCR, p. 73.

50 Green, Transcripts of Public Hearings, Day 19, 3 September 1987, pp. 1389–90, BAGC A638/4a 101C, ANZA.

51 Coney, *The Unfortunate Experiment,* p. 175.

52 Green to Warren, 25 June 1974, BAGC 18491 A638 97a 410B, ANZA.

53 Coney, *The Unfortunate Experiment,* pp. 44–45, 194.

54 John France, interviewed by Jenny Carlyon, 9 December 2004.

55 Coney, *The Unfortunate Experiment,* p. 24.

56 Green, 'The Progression of Pre-invasive Lesions of the Cervix to Invasion'.

57 M. Ueki and G. H. Green, 'Cervical Carcinoma *in situ* after Incomplete Conisation', *Asia-Oceania Journal of Obstetrics and Gynaecology,* vol. 14, 2, 1988, pp. 147–53.

58 McIndoe *et al.,* 'The Invasive Potential'; CCR, p. 248.

59 Coney, *The Unfortunate Experiment,* p. 24.

60 Sadamu Noda to Cartwright, 10 July 1987, Submissions A–N, BAGC A638 34a, ANZA.

61 McIndoe, Memo to Warren, 14 Dec. 1973, UOAA.

62 Coney, *The Unfortunate Experiment,* p. 165; Matheson, *Fate Cries Enough,* p. 180.

63 Green, Transcripts of Public Hearings, Day 21, 7 September 1987, p. 1667, BAGC A638/5a 101D, ANZA.

64 Coney and Bunkle, 'Interview Notes with Professor Herbert Green', p. 10.

65 Ralph M. Richart and Thomas C. Wright, 'Controversies in the Management of Low-Grade Cervical Intraepithelial Neoplasia', *Cancer,* Supplement, vol. 71, 4, 1993, p. 1415.

66 CCR, p. 81.

67 See, for example, Warren, Transcripts of Public Hearings, Day 37, 23 October 1987, pp. 3102, 3109, 3142, BAGC A638/8a 101G, ANZA.

68 CCR, p. 108. W. A. McIndoe and S. E. Williams, 'The Value of Cytology', Correspondence, *NZMJ,* vol. 76, 1972, p. 129.

69 W. A. McIndoe, 'A Cervical Cytology Screening Programme in the Thames Area', *NZMJ,* vol. 63, 1964, p. 6; McIndoe, 'A Cervical Cytology Screening Programme in the Thames Area: Second and Third Years of Study'; McIndoe also published

with Green in the late 1960s: McIndoe and Green, 'Vaginal Carcinoma In Situ Following Hysterectomy'.

70 W. A. McIndoe, 'Cytology or Colposcopy', *Australian and New Zealand Journal of Obstetrics and Gynaecology,* vol. 8, 3, 1968, pp. 117–18.

71 Memo by McIndoe, 13 July 1972, UOAA.

72 NWHMC minutes, 29 April 1971, BAGC A638 40b; NWHMC minutes, 20 June 1971, BAGC A638 40b, ANZA.

73 NWHMC minutes, 17 June 1971, BAGC A638 40b, ANZA.

74 Jones and Fitzgerald, 'The Development of Cervical Cytology and Colposcopy in New Zealand'. For an obituary of Grieve, see 'Obituary: Bruce Walton Grieve', *NZMJ,* vol. 106, 1993, p. 532.

75 Green, Transcripts of Public Hearings, Day 19, 3 September 1987, p. 1416, BAGC A638/4aa 101c, ANZA.

76 NWHMC minutes, 23 June 1974, BAGC A638 21a, ANZA.

77 NWHMC minutes, 16 October 1975, BAGC A638 41a, ANZA.

78 CCR, p. 88.

79 Roger Cooter, 'The Ethical Body', in Roger Cooter and John Pickstone (eds), *Companion to Medicine in the Twentieth Century,* Routledge, London, 2000, pp. 456–7.

80 Cochrane and Holland, 'Validation of Screening Procedures'.

81 NWHMC minutes, 20 September 1973, BAGC A638 21a, ANZA.

82 NWHMC minutes, 18 October 1973, BAGC A638 21a, ANZA.

83 NWHMC minutes, 2 December 1976, BAGC A638 41b, ANZA.

84 NWHMC minutes, 13 October 1977, BAGC A638 41b, ANZA.

85 Read out at Inquiry, Green, Transcripts of Public Hearings, Day 11, 24 August 1987, pp. 701–2, BAGC A638/3a 101B, ANZA.

86 Coney, Transcripts of Public Hearings, Day 6, 11 August 1987, p. 409, BAGC 638/2a, 101A, ANZA.

87 Cited in Coney, *The Unfortunate Experiment,* p. 67.

88 NWHMC minutes, 13 July 1978, BAGC A638 21b, ANZA.

89 'Medicolegal: Medical Council Charges Professor Bonham', *NZMJ,* vol. 103, 1990, pp. 547-9.

90 Submission to Inquiry on behalf of the Auckland Hospital Board, pp. 5–8, BAGC A638/ 34a, ANZA.

91 Green, Transcripts of Public Hearings, Day 21, 7 September 1987, p. 1667, BAGC 638/5a 101D, ANZA.

92 Submission to Inquiry on behalf of the

Auckland Hospital Board, p. 6, BAGC A638/
34a, ANZA.

93 McIndoe *et al.*, 'The Invasive Potential'; CCR,
    p. 248.

94 Submission to Inquiry on behalf of the
    Auckland Hospital Board, p. 8, BAGC A638/
    34a, ANZA.

95 Coppleson, 'Colposcopy', p. 178.

96 Larsson, *Conization for Preinvasive and Early
    Invasive Carcinoma of the Uterine Cervix*,
    pp. 9–10.

97 Green, 'The Progression of Pre-invasive
    Lesions of the Cervix to Invasion', p. 283.

98 Murray Jamieson, 'Cervical Neoplasia: A
    Report for Meeting of the Gynaecology
    Cancer Advisory Group, 18 August 1982',
    pp. 5, 10, in Medical Superintendent's Office:
    Gynaecological Cancer Advisory Group July
    1981–; National Women's Hospital records (in
    author's possession). Jamieson recommended
    a conservative treatment for CIN using cryo-
    surgery and possibly laser therapy (the 1978
    protocol had not mentioned these treatments
    which were still being debated): Jamieson,
    p. 11. By the time of the Inquiry, local destruc-
    tive methods of removal of the lesions were
    also being used, such as laser treatment,
    cryotherapy (freezing) and diathermy exci-
    sion: Murray Jamieson, Transcripts of Public
    Hearings, Day 58, 1 December 1987, BAGC
    A638/12a 101K, p. 5123, ANZA.

99 G. H. Green, 'Screening for Cervical Cancer',
    Correspondence, *NZMJ*, vol. 98, 1985, p. 968.
    Of those 29 in Group 2 who were said to
    develop invasive cancer, fourteen were
    diagnosed as 'occult invasive (FIGO Stage
    1b occult)' which was a histopathological
    diagnosis and the clinical significance of this
    was still subject to debate: see Coppleson,
    'Cervical Intraepithelial Neoplasia', p. 452.

100 Coney, *The Unfortunate Experiment*, p. 47.

101 I. D. Ronayne, Personal communication, 1
    November 2006. For obituaries of McIndoe
    and McLean see 'Obituary: William Arthur
    McIndoe', *NZMJ*, vol. 1987, p. 37, and
    'Obituary: Malcolm Robert McLean', *NZMJ*,
    vol. 106, 1993, p. 236.

102 Professor J. Gremley Evans, Personal commu-
    nication, May 2006.

103 Letter from Peter B. Herdson, Professor and
    Chairman, Department of Pathology and
    Laboratory Medicine, Professor Emeritus,
    University of Auckland, Past-President Royal
    College of Pathologists of Australasia, to
    David Collins, 11 November 1990 (records
    held by Nancie Bonham).

104 Joseph Jordan, Submission to Inquiry, p. 23,
    UOAA.

105 Doll, 'Concluding Remarks', p. 403.

106 G. H. Green, Memo to Medical

Superintendent (through Bonham) NWH, 7
November 1973, Cartwright Inquiry Exhibit,
3b, UOAA, cited in CCR, p. 80.

107 Sir William Liley to Green, 6 May 1975
    (records held by Nancie Bonham).

108 Green, 'Screening for Cervical Cancer',
    Correspondence.

109 *Auckland University News*, November 1981,
    p. 31.

110 Green, Transcripts of Public Hearings, Day
    21, 7 September 1987, pp.1595, 1602, BAGC
    A638/5a 101D, ANZA; this was also Murray
    Jamieson's memory of Green's lectures:
    Murray Jamieson, Transcripts of Public
    Hearings, Day 58, 1 December 1987, p. 5181,
    BAGC A638/12a 101K, ANZA.

111 Barbara Smith, interviewed by Jenny Carlyon,
    16 September 2004.

112 Tony Baird, interviewed by Jenny Carlyon, 15
    September 2004.

113 Julie Roberts, 'You Saved My Mum . . . Thank
    You', *NZWW*, 8 August 1988, p. 20.

114 McIndoe cited in Jones and Fitzgerald, 'The
    Development of Cervical Cytology and
    Colposcopy in New Zealand'.

6   Population-based Cervical Screening

1   Coney, *The Unfortunate Experiment*, p. 194.

2   D. C. G. Skegg, Charlotte Paul, R. J.
    Seddon, N. W. Fitzgerald, P. M. Barham
    and C. J. Clements, 'Cancer Screening:
    Recommendations for Routine Cervical
    Screening', *NZMJ*, vol. 98, 1985, pp. 636–9.

3   D. C. G. Skegg, 'Cervical Screening',
    Correspondence, *NZMJ*, vol. 99, 1986, p. 27.

4   *Ibid.*, p. 26.

5   Green, 'Screening for Cervical Cancer',
    Correspondence, p. 968; Jean L. Marx, 'The
    Annual Pap Smear: An Idea Whose Time Has
    Gone?', *Science*, vol. 205, 4402, 1979, pp. 177–8;
    and Editorial, 'Flap about Pap', *Time*, vol. 112,
    20, (13 November) 1978, p. 77. See also Gina
    Kolata, 'Is the War on Cancer Being Won?'
    Science, vol. 229, 4713, 1985, pp. 543–4.

6   Skegg, 'Cervical Screening', Correspondence,
    pp. 26–27.

7   Department of Health, *Screening for Cervical
    Cancer in New Zealand: Proceedings of a
    Meeting called by the Department of Health
    and the Cancer Society of NZ (Inc), 8 November
    1985*, Department of Health, Wellington, April
    1986. CCR exhibit 12a, UOAA.

8   CCR, pp. 198–9. The study referred to was:
    D. Bonham, G. H. Green and G. Liggins,
    'Editorial, Cervical Human Papilloma
    Virus Infection and Colposcopy', *Australian
    and New Zealand Journal of Obstetrics and
    Gynaecology*, vol. 27, 2, 1987, p. 131.

9   Alan Gray, Transcripts of Public Hearings,

Day 43, 3 November 1987, BAGC A638/9a 101H, ANZA.

10   Coney, *The Unfortunate Experiment*, p. 34; Jenny Rankine, 'Experimenting on Women', *Broadsheet*, vol. 151, August/September 1987, p. 9.

11   Ralph Richart, Transcripts of Public Hearings, Day 14, 27 August 1987, p. 942, BAGC A638/3a 101B, ANZA. Also cited in Coney, *The Unfortunate Experiment*, p. 263; Pat Rosier, 'Screening the Doctors', *Broadsheet*, October 1987, p. 7. Pat Rosier wrote monthly columns for *Broadsheet* on the proceedings of the Inquiry.

12   Rankine, 'Experimenting on Women', p. 9.

13   Posner and Vessey, *Prevention of Cervical Cancer*, p. 9.

14   Joan Austoker, *A History of the Imperial Cancer Research Fund, 1902–1986*, Oxford University Press, Oxford, 1988, p. 214, fn. 43. Her references to this dispute include G. J. Draper, 'Screening for Cervical Cancer: Revised Policy', *British Journal of Family Planning*, vol. 8, 1982, pp. 95–100; G. A. Cook and G. J. Draper, 'Trends in Cervical Cancer and Carcinoma *in Situ* in Great Britain', *British Journal of Cancer*, vol. 50, 1984, pp. 367–75; James Le Fanu, 'Case for Treatment', *New Statesman*, 16 March 1984, pp. 10–11; 'Commentary from Westminster, Screening for Cancer of the Cervix', *The Lancet*, vol. 2, 1984, p. 1483; Jocelyn Chamberlain, 'Failures of the Cervical Cytology Screening Programme', *BMJ*, vol. 289, 2, 1984, pp. 853–4; Thomson Prentice, 'Cervical Cancer Screening Provisions a Shambles, Shadow Minister Claims', *The Times*, 18 March 1985; Ann McPherson, 'Cervical Screening', *Journal of the Royal College of General Practitioners*, vol. 35, 1985, pp. 219–22.

15   M. F. G. Murphy, M. J. Campbell and P. O. Goldblatt, 'Twenty Years' Screening for Cancer of the Uterine Cervix in Great Britain, 1964–84; Further Evidence of its Ineffectiveness', *Journal of Epidemiology and Community Health*, vol. 42, 1, 1988, pp. 49–53.

16   Comment on transcript of 1985 meeting to implement Skegg recommendations, UOAA.

17   G. H. Green, 'Cervical Cytology', Correspondence, *NZMJ*, vol. 78, 1972, pp. 449–50.

18   G. H. Green, 'The Early Diagnosis of Gynaecological Malignancy', 1967 lecture, Bonham records (held by Nancie Bonham).

19   Green, 'The Progression of Pre-invasive Lesions of the Cervix to Invasion', p. 279. (Sydney Farber founded Children's Cancer Research Foundation in 1947, see www.dana-farber.org.)

20   G. H. Green, 'Rising Cervical Cancer Mortality in Young New Zealand Women', *NZMJ*, vol. 89, 1979, p. 90.

21   See Hyde, 'At Your Cervix Madam', pp. 97, 100.

22   Marx, 'The Annual Pap Smear', pp. 177–8.

23   Raffle and Gray, *Screening: Evidence and Practice*, p. 10.

24   L. M. Franks, 'Cytology Service and the Ministry', Letter to the Editor, *BMJ*, vol. 1, 1967, p. 498; Knox, 'Cervical Cytology'; Cochrane, *Effectiveness and Efficiency*, pp. 26–27; see also Cochrane and Holland, 'Validation of Screening Procedures'; H. S. Ahluwalia and Richard Doll, 'Mortality from Cancer of the Cervix Uteri in British Columbia and other Parts of Canada', *British Journal of Preventive and Social Medicine*, vol. 22, 1968, p. 164; L. J. Kinlen and R. Doll, 'Trends in Mortality from Cancer of the Uterus in Canada and in England and Wales', *British Journal of Preventive and Social Medicine*, vol. 27, 1973, pp. 146–9.

25   Purvis L. Martin, 'How Preventable is Invasive Cervical Cancer? A Community Study of Preventable Factors', and Robert C. Goodlin, 'Discussion', *AJOG*, vol. 113, 1972, p. 548.

26   Editorial, 'Uncertainties of Cervical Cytology', *BMJ*, vol. 4 (5891), 1973, pp. 501–2.

27   J. R. Douglas, Correspondence, *Medical Journal of Australia*, 1972, p. 1376; cited in *NZH*, 29 June 1972; also cited by Green, Transcripts of Public Hearings, Day 11, 24 August 1987, p. 689, BAGC A638/3a 101B, ANZA.

28   Jeffcoate, *Principles of Gynaecology*, 1980, p. 492.

29   Richard Doll, 'Occasional Review: Prospects for Prevention', *BMJ*, vol. 286, 1983, p. 451.

30   Editorial, 'Cancer of the Cervix: Death by Incompetence', *The Lancet*, vol. 2, (17 August) 1985, pp. 363–4. (According to Raffle and Gray, this was written by Professor George Knox: Raffle and Gray, *Screening: Evidence and Practice*, p. 22.)

31   Ibid.

32   *The Times*, 27 June 1994.

33   James McCormick and Robin Fox, 'Death of Petr Skrabanek (Obituary)', *The Lancet*, vol. 344, 8914, (2 July) 1994, pp. 52–53.

34   Coney, *The Unfortunate Experiment*, p. 264.

35   E. Lâârâ, N. E. Day and M. Hakama, 'Trends in Mortality from Cervical Cancer in the Nordic Countries: Association with Organised Screening Programmes', *The Lancet*, vol. 329, 8544, 1987, pp. 1247–9.

36   Ibid.

37   Petr Skrabanek, 'Cervical Cancer Screening', Correspondence, *The Lancet*, vol. 2, 1987, pp. 1432–3. A lively debate occurred in the pages of *The Lancet* in 1987 between Skrabanek and N. E. Day and M. Hakama. Skrabanek also set out his views in Petr

Skrabanek and James McCormick, *Follies and Fallacies in Medicine*, The Tarragon Press, Glasgow, 1989. Raffle and Gray later critiqued the so-called 'Iceland effect' for its misleading statistics: see Raffle and Gray, *Screening: Evidence and Practice*, p. 105.

38   He later showed himself to be a critique of modern medicine: James Le Fanu, *The Rise and Fall of Modern Medicine*, Abacus, London, 1999, new edn 2000, reprinted 2001, 2004.

39   *Sunday Telegraph*, 5 June 1988; *NZH*, 8 August 1988.

40   *Sunday Telegraph*, 5 June 1988.

41   Alwyn Smith, Letter to the Editor, *BMJ*, vol. 296, 1670, p. 1988.

42   *Auckland Star*, 14 September 1986.

43   Murray Jamieson to Linda Waasdorp, 27 June 1986, Cartwright Inquiry Exhibit 12c, UOAA.

44   R. R. Love and A. E. Camilli, 'The Value of Screening', *Cancer*, vol. 48, 2, 1981, pp. 493–4, cited in Posner and Vessey, *Prevention of Cervical Cancer*, p. 12.

45   Jeffcoate, *Principles of Gynaecology*, 1980, p. 492; this was repeated in *Jeffcoate's Principles of Gynaecology*, 1987, p. 489.

46   Eugene D. Robin, 'Failure of the Cervical Cytology Screening Programme', Letter to the Editor, *BMJ*, vol. 290, 1985, p. 75.

47   G. H. Green, '"Break-through" Bleeding and Cervical Cancer', Letter to the Editor, *BMJ*, vol. 1, 1965, p. 997.

48   Liam Wright to Skegg, 1 March 1985, ABQU 632 W4452 119 13–72–61867, ANZW.

49   *Auckland Star*, 14 September 1986; Murray Jamieson, Transcripts of Public Hearings, Day 58, 1 December 1987, pp. 5124–8, BAGC A638/12a 101K, ANZA.

50   McCormick, 'Cervical Smears', p. 209.

51   *Ibid*.

52   *NZWW*, 23 July 1984.

53   *Sunday Star*, 14 September 1986.

54   Department of Health, *Screening for Cervical Cancer in New Zealand*, p. 36.

55   *Ibid*., pp. 47–48. Skegg pointed out that Green had first noted the trend in younger women; nevertheless the death rates were still much higher among older women.

56   *Ibid*., p. 4.

57   Petr Skrabanek and Murray Jamieson, 'Eaten by Worms: A Comment on Cervical Screening', Correspondence, *NZMJ*, vol. 98, 1985, p. 654.

58   Foster, *Women and the Health Care Industry*, p. 116.

59   Te Ohu Whakatupu (Maori Women's Secretariat) Ministry of Women's Affairs, Submission to Inquiry, August 1987, p. 22, BAGC A638 34a, ANZA. See also Sarah Calvert, 'Cervices at Risk', *Broadsheet*, vol. 123, October 1984, p. 37.

60   Coney and Bunkle, 'Interview notes with Professor Herbert Green', p. 10; Green, Transcripts of Public Hearings, Day 18, 2 September 1987, p. 1322, BAGC A638/4a 101C, ANZA.

61   Green, 'Rising Cervical Cancer Mortality in Young New Zealand Women', p. 91.

62   Jean Robinson, 'Cervical Cancer: Doctors Hide the Truth', *Spare Rib*, vol. 154, 1985, reprinted in O'Sullivan (ed.), *Women's Health*, p. 51.

63   Lara V. Marks, *Sexual Chemistry: A History of the Contraceptive Pill*, Yale University Press, New Haven and London, 2001, p. 164.

64   *Ibid*., p. 167.

65   *Ibid*., pp. 167–8.

66   Green, '"Break-through" Bleeding and Cervical Cancer', Letter to the Editor, p. 997.

67   Martin Vessey and Richard Doll, 'Evaluation of Existing Techniques: Is "the Pill" Safe Enough to Continue Using?', *Proceedings of the Royal Society of London*, vol. 195, 1976, pp. 69–80; Marks, *Sexual Chemistry*, p. 172; Lara Marks, 'Assessing the Risk and Safety of the Pill: Maternal Mortality and the Pill', in Schlich and Tröhler (eds), *The Risks of Medical Innovation*, p. 195. These studies continued: a 2003 study concluded that long duration of hormonal contraceptives is associated with an increased risk of cervical cancer: J. S. Smith, J. Green, A. B. de Gonzalez, P. Appleby, J. Peto, M. Plummer, S. Franceschi and V. Beral, 'Cervical Cancer and Use of Hormonal Contraceptives: A Systematic Review', *The Lancet*, vol. 361, 2003, pp. 1159–67.

68   McCormick, 'Cervical Smears', p. 209; also reported in *NZH*, 14 August 1989.

69   *NZH*, 14 August 1989; D. C. G. Skegg, 'Leading Article: How Not to Organise a Cervical Screening Programme', *NZMJ*, vol. 102, 1989, pp. 527–8.

70   Angela Raffle to Linda Bryder, email communication, 2 April 2008.

71   Angela E. Raffle, 'New Tests in Cervical Screening', *The Lancet*, vol. 351, 1998, p. 297.

72   Celia Lampe from Policy Unit, Ministry of Women's Affairs, Transcripts of Public Hearings, Day 60, 3 December 1987, BAGC A638/13a 101L, ANZA.

73   L. T. to Ann Hercus [signed 'Yours in sisterhood'], 14 March 1985 ABQU 632 W4452 119 13–72–61867, ANZW.

74   Germaine Greer, *The Whole Woman*, A. A. Knopf, New York, 1999, p. 123.

75   For example, Raffle, 'New Tests in Cervical Screening', p. 297.

76   Walt Bogdanich, 'Lax Laboratories: The Pap Test Misses Much Cervical Cancer Through Labs' Errors – Cut-rate "Pap Mills" Process Slides Using Screeners with Incentives to Rush

– Misplaced Sense of Security', *Wall Street Journal*, 2 November 1987, p. 3.

77 Green, 'Screening for Cervical Cancer', Correspondence, p. 968.

78 Green, Transcripts of Public Hearings, Day 11, 24 August 1987, p. 688, BAGC A6383a 101B, ANZA.

79 Green to Mrs Kill, Assistant Director, Division of Health Promotion, 12 June 1986, Exhibit Cervical Cancer Inquiry, 12d, UOAA.

80 Coney and Bunkle, 'Interview Notes with Professor Herbert Green', UOAA, p. 6.

81 Fertility Action, Submission to Inquiry, p. 75. A. Kennedy from the Department of Histopathology, Northern General Hospital, Sheffield, England, similarly believed, 'Preoccupation with this condition seems to have deprived us of our sense of proportion with respect to other malignant diseases in women': Letter to the Editor, *The Lancet*, vol. 2, 1989, p. 392; a similar point was made in *Jeffcoate's Principles of Gynaecology*, 1987, p. 486.

82 World Health Organization International Agency for Research on Cancer, *IARC Handbooks on Cancer Prevention, Volume 10*, p. 214.

83 Raffle and Gray, *Screening: Evidence and Practice*, pp. 20, 25.

84 Alan Gray, Submission to Inquiry, 19 October 1987, BAGC A638 34a, ANZA.

85 Ministry of Women's Affairs, Submission to Inquiry, p. 29, BAGC A638 34a, ANZA.

86 Comment on transcript of 1985 meeting to implement Skegg recommendations, UOAA.

87 Seddon, Transcripts of Public Hearings, Day 4, 6 August 1987, p. 93, BAGC A638/2a 101A, ANZA.

88 Colin David Mantell, Statement of Evidence to Inquiry, pp. 7–8, UOAA.

89 CCR, p. 206.

90 Closing Submissions of Counsel for University of Auckland, p. 40, UOAA.

91 Dr Bruce Ronald Phillips, 9 September 1987, Submissions from Public, BAGC Acc A638 35a N-S, ANZA.

92 D. H. C. Davidson, 30 July 1987, Submissions from Public, BAGC A638 34a, A-N, ANZA.

93 At Postgraduate Diploma courses, for example, Green usually gave one lecture out of about fifteen: Nancie Bonham records.

94 Green, Transcripts of Public Hearings, Day 21, 7 September 1987, pp. 1595–603, BAGC A638/5a 101D, ANZA.

95 Murray Jamieson, Transcripts of Public Hearings, Day 58, 1 December 1987, p. 5181, BAGC A638/12a 101K, ANZA.

96 Charlotte Paul, 'New Zealand's Cervical Cytology History: Implications for the Control of Cervical Cancer', *NZMJ*, vol. 117,

2004, pp. 1170–4.

97 Green to Mrs Kill, Assistant Director, Division of Health Promotion, 12 June 1986, Cartwright Exhibit 12 d, UOAA.

98 Green, evidence to Inquiry, p. 34, UOAA.

99 Report by Bonham on overseas trip, July 1983, Medical Superintendent's Office: Gynaecological Cancer Advisory Group, July 1981, UOAA.

100 R. W. Jones, M. L. Yeong, A. W. Stewart, G. C. Hitchcock and W. E. Dervan, 'Cervical Cytology in the Auckland Region', *NZMJ*, vol. 101, 1988, pp. 132–5; Dennis G. Bonham, 'Cervical Cytology in Auckland', Correspondence, *NZMJ*, vol. 101, 1988, p. 248.

101 Sue Neal, Submission to Inquiry, 22 June 1987, p. 2, Submissions from Public, BAGC A638 35a, ANZA.

102 Skegg *et al.*, 'Cancer Screening: Recommendations', pp. 636–9.

103 Ministry of Women's Affairs, Submission to Inquiry, Part 2, Prepared by Te Ohu Whakatupu (Maori Women's Secretariat), August 1987, p. 24, BAGC A638 34a, ANZA.

104 Reported in *NZH*, 26 January 1988.

105 B. Kyle, Memo, 23 June 1975, NWHMC minutes, 17 July 1975, BAGC A638 41a, ANZA.

106 Paul M. McNeill, *The Ethics and Politics of Human Experimentation*, Cambridge University Press, Cambridge, 1993, p. 76.

107 'Obituary: Alastair Macfarlane', *NZMJ*, vol. 99, 1986, p. 690.

108 David Cole, Unpublished Memoirs, ch. 14: 'At the Top of the Heap: The Medical Council of NZ', p. 52.

109 Coney, *The Unfortunate Experiment*, p. 178.

110 Cartwright to Noda, 31 July 1987, Submissions A-N, BAGC A638 34a, ANZA.

111 *NZH*, 27 January 1988.

112 *Ibid.*; the 'lambs to the slaughter' comment was repeated in Matheson, *Fate Cries Enough*, p. 161.

113 *NZH*, 27 January 1988; *Auckland Star*, 27 January 1988.

114 Fertility Action, Transcripts of Public Hearings, Day 63, 9 December 1987, p. 5561, BAGC A638/13a 101L, ANZA; Fertility Action Submission to Cervical Cancer Inquiry, August 1987, p. 74, UOAA; *NZH*, 10 December 1987.

7   Four Women Take on the Might of the Medical Profession

1 Leading Article, 'Screening for Cervical Cancer', *BMJ*, vol. 2, 1976, pp. 659–60.

2 Diana Wichtel, 'Delivering with Style', *New Zealand Listener*, 21 September 1985, p. 14.

3 Michael Bassett, interviewed by Jenny Carlyon, 23 May 2006; see also Michael

Bassett, *Working with David: Inside the Lange Cabinet*, Hodder Moa, Auckland, 2008, p. 261.

4   Sandra Coney, *Standing in the Sunshine: A History of New Zealand Women Since They Won the Vote,* Penguin Books, Auckland, 1993, p. 142.

5   Mary Holmes, *The Representation of Feminists as Political Actors: Challenging Liberal Democracy*, VDM Verlag Dr. Muller, Saarbrucken, 2008.

6   Boston Women's Health Book Collective, *Our Bodies Ourselves: A Book by and for Women*, Simon and Schuster, New York; first published in 1971, this went through many subsequent editions: 1973, 1976, 1984, 1985, 1995, 1998, 2005.

7   Rothman, *Strangers at the Bedside,* pp. 143–4; see also Elizabeth Siegel Watkins, 'Changing Rationale for Long-term Hormone Replacement Therapy in America, 1960–2000', *Health and History*, vol. 4, 1, 2001, p. 24.

8   Christine Dann, *Up From Under: Women and Liberation in New Zealand 1970–1985,* Allen and Unwin/Port Nicolson Press, Wellington, 1985, p. 81; Bunkle, *Second Opinion*, pp. 15–16; see also Georgina Feldberg, Molly Ladd-Taylor, Alison Li and Kathryn McPherson, 'Comparative Perspectives on Canadian and American Women's Health Care since 1945', in Feldberg, *et al.* (eds), *Women, Health, and Nation: Canada and the United States since 1945*, pp. 26–32.

9   Sandra Coney, 'Health Organisations', in Anne Else (ed.), *Women Together: A History of Women's Organisations in New Zealand: Ngā Rōpū Wāhine o te Motu*, Historical Branch, Department of Internal Affairs and Daphne Brasell Associates Press, Wellington, 1993, p. 249; and Sandra Coney, 'Fertility Action 1984– ', *ibid.*, pp. 284–6.

10 Sandra Morgen, *Into Our Own Hands: The Women's Health Movement in the United States, 1969–1990*, Rutgers University Press, New Brunswick, 2002, p. 22.

11 Sheryl Burt Ruzek, *The Women's Health Movement: Feminist Alternatives to Medical Control*, Praegers Publishers, New York, 1978, p. 53.

12 Dann, *Up From Under*, p. 82.

13 Roberts, 'We Had to Speak Out About What We Knew'.

14 *NZWW*, 23 July 1990, p. 19.

15 *Ibid.*

16 Phillida Bunkle, 'Personal Crisis/Global Crisis', in Maud Cahill and Christine Dann (eds), *Changing our Lives: Women Working in the Women's Liberation Movement, 1970–1990*, Bridget Williams Books Ltd, Wellington, 1991, pp. 171–8. Stallworthy's talk was no doubt based on research published in *The Lancet*

in which he stated, 'There has been almost a conspiracy of silence in declaring its risks': J. Stallworthy, A. S. Moolgaoker and J. J. Walsh, 'Legal Abortion: Critical Assessment of its Risks', *The Lancet*, vol. 2, 1971, pp. 1245–9.

17 Bunkle, *Second Opinion*, pp. viii–ix.

18 Coney, 'Fertility Action', p. 284; Phillida Bunkle, 'Women, Health and Politics: Divisions and Connections', in Rosemary du Plessis and Lynne Alice (eds), *Feminist Thought in Aotearoa New Zealand: Connections and Differences*, Oxford University Press, Oxford, 1988, pp. 238–44; Bunkle, *Second Opinion*, pp. 102–29.

19 *Dominion*, 6 October 1970, cited in Bunkle, *Second Opinion*, p. 9.

20 G. H. Green, 'The Foetus Began to Cry . . . Abortion' (Part 1), *New Zealand Nursing Journal*, vol. 63, 7, 1970, pp. 11-12; (Part 2), *NZNJ*, vol. 63, 9, 1970, pp. 6-7; (Part 3), vol. 63, 9, 1970, pp. 11–12.

21 Paul Patten, interviewed by Jenny Carlyon, 11 April 2006.

22 Cited by Jan Corbett, 'Second Thoughts on the Unfortunate Experiment at National Women's', *Metro,* July 1990, p. 58.

23 Cited by Sandra Coney, 'From Here to Maternity', *Broadsheet*, vol. 123, October 1984, p. 5.

24 Mary Dobbie, *The Trouble with Women: The Story of Parents Centre New Zealand,* Cape Catley Ltd, Queen Charlotte Sound, 1990, pp. 83, 84, 104.

25 NWHMC minutes, 20 May 1964, BAGC A638 39b, 17 August 1964, BAGC A638 39b, ANZA.

26 Helen Smyth, *Rocking the Cradle: Contraception, Sex and Politics in New Zealand,* Steele Roberts Ltd, Wellington, 2000, p. 127.

27 Submission to the Royal Commission on Contraception, Sterilisation and Abortion, 24 May 1976, NZFPA publications, COM26 3J, ANZW.

28 Smyth, *Rocking the Cradle*, p. 243, fn. 8.

29 Abortion doctors published in *Broadsheet*, vol. 68, April 1979, p. 20.

30 *NZH*, 9 July 1980.

31 Transcript of interview with Professor Dennis Bonham, 9 December 1986, Cartwright Inquiry Exhibit 3c, UOAA; Dennis Bonham, Submission to Cartwright Inquiry, p. 27, UOAA.

32 NWHMC minutes, 23 July 1981, BAGC A638 21c, ANZA. At that time, Graham Liggins was Senior Lecturer in the University of Auckland School of Obstetrics and Gynaecology.

33 Lesley McCowan, interviewed by Jenny Carlyon, 19 December 2005; Hilary Liddell, interviewed by Jenny Carlyon, 1 April 2005; Cynthia Farquhar, interviewed by Jenny

Carlyon, 14 April 2006; Lynda Batcheler, interviewed by Jenny Carlyon, 31 March 2006; see also Lesley McCowan, Hilary Liddell, Cindy Farquhar, Lynda Batchelor [sic], Correspondence, *NZMJ*, vol. 103, 1990, p. 543.

34 John France also believed that Bonham was at the forefront of women's health issues: John France, interviewed by Jenny Carlyon, 9 December 2004; John Werry, transcript, 'Bonham Memorial', 6 May 2005 (in author's possession).

35 McLeod, 'The Importance of Being Sandra Coney', p. 60; Nancie Bonham, personal notes, 2004 (in author's possession).

36 Coney, *The Unfortunate Experiment*, pp. 20–21.

37 Coney, Evidence to Inquiry, cited in *NZH*, 11 December 1987; *Dominion*, 11 December 1987.

38 Fertility Action, Transcripts of Public Hearings, Day 64, 10 December 1987, pp. 5571-2, BAGC A638/13a 101L, ANZA.

39 Neville R. Butler and Dennis G. Bonham, *Perinatal Mortality: The First Report of the 1958 British Perinatal Mortality Survey under the Auspices of the National Birthday Trust Fund*, E. and S. Livingstone Ltd, Edinburgh, 1963; Joan Donley, *Save the Midwife*, New Women's Press, Auckland, 1986. See also Ann Oakley, *The Captured Womb: A History of the Medical Care of Pregnant Women*, Blackwell, Oxford and New York, 1984, pp. 205–6.

40 Coney, *The Unfortunate Experiment*, p. 19.

41 A. Susan Williams, *Women and Childbirth in the Twentieth Century: A History of the National Birthday Trust Fund 1928–93*, Sutton, Gloucestershire, 1997, pp. 196, 197, 199.

42 *Ibid.*; see also Geoffrey Chamberlain, *Special Delivery: The Life of the Celebrated British Obstetrician William Nixon*, Royal College of Obstetricians and Gynaecologists, London, 2004, p. 88.

43 Maternity Services Committee of the Board of Health, *Maternity Services in New Zealand*, Report Series No. 26, Government Printer, Wellington, 1976, see especially pp. 38–40.

44 Women's Health Network, *Newsletter*, April 1985, Joan Donley Archives, UOAA. In 1962 there were 27 private maternity hospitals in New Zealand; by 1983 there were only five: Hay, *Caring Commodity*, p. 181.

45 Roger A. Rosenblatt, Judith Reinken and Phil Shoemack, 'Is Obstetrics Safe in Small Hospitals? Evidence from New Zealand's Regionalised Perinatal System', *The Lancet*, vol. 326, 8452, 1985, pp. 429–32.

46 Maternity Action included the following groups: Auckland Childbirth Education Association, Auckland East Parents' Centre, Auckland Home Birth Association, Auckland Parents' Centre Central, Auckland Women's Health Collective, Caesarean Support Group Inc, Helensville Hospital Community Committee, Manukau Parents' Centre, New Mothers' Support Groups Inc, Obstetric Watch, Papakura Parents' Centre, Save the Midwives Association, West Auckland Parents' Centre and West Auckland Women's Centre. Judy Larkin, Submission to Inquiry by Spokesperson, Maternity Action, 20 June 1987, BAGC A638 34a. See also Joan Donley, 'Having the Baby at Home', *Broadsheet*, vol. 132, September 1985, pp. 17, 47; and Dann, *Up From Under*, p. 84.

47 *NZH*, 19 October 1976; Donley, 'Having the Baby at Home', p. 17.

48 Auckland Hospital Board minutes, 1976, Joan Donley Archives, UOAA.

49 Bunkle, *Second Opinion*, p. xiv.

50 *Auckland Star*, 4 October 1978; see also Joan Donley, *Herstory of N.Z. Homebirth Association*, Domiciliary Midwives Society of New Zealand, Wellington, 1992, p. 4.

51 NWHMC minutes, 19 June 1980, BAGC A638 21B, ANZA.

52 Submissions to Maternity Services Committee, 1980, ABQU 632 W4550 5 29 21 1979–81, ANZW.

53 Dobbie, *Trouble with Women*, p. 127.

54 Ruzek, *Women's Health Movement*, p. 117.

55 Sandra Coney, 'Alienated Labour – Foetal Monitoring', *Broadsheet*, May 1979, pp. 16–17, 38–39. See also Adrienne Rich, *Of Woman Born: Motherhood as Experience and Institution*, Norton, New York, 1976.

56 Submission to Maternity Services Committee, 3 July 1980, ABQU 632 W4550 5 29–21 1979–81, ANZW.

57 Joan Mackay to Nancie Bonham, 11 October 2003: enclosed letter D. G. Bonham to B. J. McKay, 20 October 1983, Nancie Bonham, private collection.

58 Coney, *The Unfortunate Experiment*, p. 30.

59 Helen Clark to Joan Donley, 26 March 1985, Joan Donley Archives, UOAA.

60 Dann, *Up From Under*, p. 83.

61 Women's Health Network, *Newsletter*, 'The New Women's Board of Health', 1985, sent to Helen Clark, Joan Donley Archives, UOAA.

62 *Ibid.*

63 Women's Health Network, *Newsletter*, April 1985, Joan Donley Archives, UOAA.

64 *Ibid.*

65 Letter from a member of the Network to Women's Health Network, August 1985, Joan Donley Archives, UOAA.

66 Christine Bird, 'Using Women's Health Groups', *Broadsheet*, vol. 137, March 1986, pp. 19–21.

67 Phillida Bunkle, 'Dalkon Shield Disaster', *Broadsheet*, vol. 122, September 1984, p. 22;

Bunkle, *Second Opinion*, pp. 116–18.

68 Bunkle, 'Dalkon Shield Disaster', p. 36.

69 Coney, *The Unfortunate Experiment,* p. 20.

70 Dann, *Up From Under,* p. 86.

71 Janet Hadley, 'The Case against Depo-Provera', in O'Sullivan (ed.), *Women's Health*, p. 170.

72 Coney, *The Unfortunate Experiment,* p. 17.

73 NWHMC minutes, 20 June 1971, BAGC A638 40b, ANZA.

74 'Health Survey: New Zealand Contraception and Health Study: Design and Preliminary Report (Chairman Sir Graham Liggins)', *NZMJ*, 23 April 1986, pp. 283–6, reported in Phillida Bunkle, 'Calling the Shots? The International Politics of Depo-Provera', in Rita Arditti, Renate Duelli Klein and Shelley Minden (eds), *Test-Tube Women: What Future for Motherhood?*, Pandora Press, London, 1984, p. 171–2.

75 Jill Rakusen, 'Healthy Women: Depo Provera', *Broadsheet*, vol. 39, May 1976, pp. 32–33.

76 Sue Neal, 'Only Women Bleed', *Broadsheet*, vol. 75, December 1979, pp. 12–15.

77 *Auckland Star*, 12 March 1980.

78 David Young, 'Contraceptive Controversy', *NZ Listener*, 8 March 1980, pp. 30–33.

79 Ruth Bonita, 'Contraceptive Research: For Whose Protection?', *Broadsheet*, vol. 76, January/February 1980, pp. 6–8; also cited in Bunkle, *Second Opinion*, p. 66.

80 P. J. Scott, 'Memorandum to the Director, Medical Research Council of New Zealand from the Standing Committee on Therapeutic Trials, re. The New Zealand Contraception and Health Study Protocol', 1984.

81 'Contraceptive Controversy', p. 33.

82 Lyn Potter, 'Hysterectomy and Hospital Hassles', *Broadsheet*, vol. 120, June 1984, pp. 28–33.

83 Tony Hegh (ed.), *Frank Answers from Thursday: Girls, Women Ask About their Bodies and Sexual Needs*, Wilson and Horton, Auckland, 1974.

84 'Curing the Hysterectomy Hangups', October 1981, reprinted from *Thursday* magazine; cited in Douglas Jenkin, 'A Question of Trust', *NZ Listener*, 14 November 1987, p. 18.

85 Jenkin, 'A Question of Trust', p. 18.

86 Sandra Coney, 'Do You Really Have to Have a Hysterectomy?', *NZWW*, 13 April 1987, pp. 64–67. Potter and Coney later wrote a book on hysterectomy: Sandra Coney and Lyn Potter, *Hysterectomy,* Heinemann Reid, Auckland, 1990.

87 Green, Transcripts of Public Hearings, Day 21, 7 September 1987, p. 1658, BAGC A638/5a 101D, ANZA.

88 Calvert, 'Cervices at Risk', p. 37.

89 Marion Kleist, 'Pap Smears', *Broadsheet*, vol. 125, December 1984, pp. 3–4.

90 Robinson, 'Cervical Cancer: Doctors Hide the Truth', p. 50.

91 Saffron, 'Cervical Cancer – The Politics of Prevention', pp. 47–48.

92 Ruth Henderson, for THAW to Cartwright enclosed Women's Health Information Centre leaflet, London, 10 December 1987, BAGC A638 36a, ANZA.

93 Robinson, 'Cervical Cancer: Doctors Hide the Truth', pp. 50–51. See also Joy Bickley, 'Safety Screen or Smoke-screen', *NZNJ*, vol. 81, 8, August 1987, p. 11.

94 Posner, 'What's in a Smear?'.

95 *Ibid.*, p. 176.

96 Coney, *The Unfortunate Experiment*, p. 11.

97 Posner, 'What's in a Smear?', p. 174.

98 Ellen Lewlin and Virginia Olesen (eds), *Women, Health and Healing: Towards a New Perspective,* Tavistock, New York, 1985, pp. 8–9.

99 Coney, *The Unfortunate Experiment*, p. 80.

100 Erich Geiringer, 'Trial in Error', *NZ Listener*, 26 November 1988, p. 46; Erich Geiringer, 'The Triumph of Victimocracy', *Metro*, 113, November 1990, pp. 134–8.

101 Fertility Action, 1989–90, ABQU 632 W4452 1750 358/1/11, ANZW.

102 Wichtel, 'Delivering with Style', p. 14.

103 Pat Rosier, 'Broadcast: A Feminist Victory', *Broadsheet,* vol. 161, September 1988, pp. 6–7.

104 *Sunday Star*, 6 August 1989.

8 The Cervical Cancer Inquiry and the 'full story'

1 Michael Bassett, interviewed by Jenny Carlyon, 23 May 2006; see also Bassett, *Working with David*, p. 261.

2 D. G. Bolitho, 'Some Financial and Medico-political Aspects of the New Zealand Medical Profession's Reaction to the Introduction of Social Security', *New Zealand Journal of History*, vol. 18, 1, 1984, pp. 34–49.

3 Derek A. Dow, *Safeguarding the Public Health: A History of the New Zealand Department of Health,* Victoria University Press, Wellington, 1995, pp. 214–16.

4 Hay, *The Caring Commodity,* p.161, cites the *New Zealand Medical Association Annual Report for 1984*, p. 4.

5 Joan Donley to Michael Bassett, 15 February 1985, Joan Donley Archives, UOAA.

6 S. M. for the Auckland Women's Health Collective to Bassett, 18 June 1987, 'Hospital Boards – Ministerial Inquiry 1986–88', ABQU 632 W4452 397 56/101 62694, ANZW.

7 J. R. to Bassett, 9 June 1987, 'Hospital Boards – Ministerial Inquiry 1986–88', ABQU 632 W4452 397 56/101 62694, ANZW.

8 E. J. to Bassett, 8 June 1987, 'Hospital Boards

9   – Ministerial Inquiry 1986–88', ABQU 632 W4452 397 56/101 62694, ANZW.

9   H. W. to Bassett, 4 June 1987, 'Hospital Boards – Ministerial Inquiry 1986–88', ABQU 632 W4452 397 56/101 62694, ANZW.

10   Secretary, Cancer Society to Bassett, 8 June 1987, 'Hospital Boards – Ministerial Inquiry 1986–88', ABQU 632 W4452 397 56/101 62694, ANZW.

11   David Skegg to George Salmond, 4 June 1987, 'Hospital Boards – Ministerial Inquiry 1986–88', ABQU 632 W4452 397 56/101 62694, ANZW.

12   Rankine, 'Experimenting on Women', p. 10. The address given, according to Corbett ('Second Thoughts', p. 60), was Coney's.

13   Coney, *The Unfortunate Experiment*, p. 75.

14   Corbett, 'Second Thoughts', p. 60.

15   *Dominion*, 14 June 1987.

16   Corbett, 'Second Thoughts', p. 60, cited United Nations Economic and Social Commission for Asia and Pacific Countries Monograph Series No. 12, 1985, p. 191.

17   Coney, *The Unfortunate Experiment*, p. 74.

18   Sandra Coney to Bassett, 16 June 1987; Bassett to Coney, 19 June 1987; C. J. Thompson, Deputy Solicitor-General to Attorney General, 2 July 1987; P. J. Trapsi, Chief Judge to Geoffrey Palmer, Minister of Justice; R. A. Patston, Acting Manager, Corporate Management Support, to Sandra Coney, 7 July 1987; ABQU 632 W4452 397 56/101 62694, ANZW.

19   *Fertility Action Newsletter*, vol. 11, February/March 1988, Joan Donley Archives, UOAA.

20   Bunkle, 'Side-stepping Cartwright', p. 50.

21   CCR, p. 4.

22   Green, Transcripts of Public Hearings, Day 12, 25 August 1987, pp. 808–10, BAGC A638/3a 101B, ANZA.

23   Green, Transcripts of Public Hearings, Day 16, 31 August 1987, p. 1148, BAGC A638/4a 101D, ANZA.

24   *Ibid.*, p. 1193, BAGC A638/4a 101D, ANZA.

25   Green, Transcripts of Public Hearings, Day 17, 1 September 1987, p. 1219, BAGC A638/4a 101C, ANZA.

26   Green, Transcripts of Public Hearings, Day 18, 2 September 1987, p. 1337, BAGC A638/4a 101C, ANZA.

27   *Ibid.*, p. 1340.

28   John France, interviewed by Jenny Carlyon, 9 December 2004.

29   Cole, Unpublished Memoirs, p. 52.

30   See also chapter 3.

31   *Sunday Star,* 20 September 1987; *Auckland Star,* 28 September 1987; Pat Rosier, 'The Speculum Bites Back', *Broadsheet*, vol. 153, November 1987, p. 5. Coney and Bunkle had included a statement about hospitals using anaesthetised women in training in the original article, which *Metro* cut as it was not relevant to the Green story, see Bunkle, *Second Opinion*, p. 184.

32   J. Bibby, N. Boyd, C. W. E. Redman and D. M. Luesley, 'Consent for Vaginal Examinations by Students on Anaesthetised Patients', Correspondence, *The Lancet,* vol. 332, 8620 (12 November) 1988, p. 1150; L. Rogers, 'Anaesthetised Women Suffer Unauthorised Medical Probes', *Sunday Times,* 21 May 1995.

33   *Sunday Star,* 11 October 1987; *Dominion Sunday Times,* 11 October 1987.

34   National Women's Hospital Medical Committee minutes, 19 June 1980, p. 2747, BAGC A638 21B, ANZA.

35   *NZH,* 19 September 1987; *Auckland Star,* 18 September 1987; Rosier, 'The Speculum Bites Back'; Coney, *The Unfortunate Experiment,* p. 204.

36   Rosier, 'The Speculum Bites Back'; Cahill and Dann (eds), *Changing Our Lives*, p. 152.

37   Jenkin, 'A Question of Trust', pp. 17–18.

38   *Ibid.*

39   Coney, *The Unfortunate Experiment,* p. 203.

40   *Ibid.*, p. 204.

41   *Ibid.*

42   H. M. to Collison, 1 October 1987, Collison's Correspondence, BAGC 18492, A638 162a, ANZA.

43   D. R. to Collison, 21 September 1987, BAGC 18492, A638 162a, ANZA.

44   M. W. to Collison, 1 October 1987, BAGC 18492, A638 162a, ANZA.

45   M. K. to Collison, 18 December 1987, BAGC 18492, A638 162a, ANZA.

46   V. S. to Collison, 24 September 1987, BAGC 18492, A638 162a, ANZA.

47   *Auckland Star,* 11 November 1987.

48   NWHMC minutes, 21 October 1963, BAGC A638 22b, ANZA.

49   'Final Summary of Information Concerning Vaginal Swabs of Newborn Babies, Research on Foetal Uteri and Clinical Photography of Cervices of Anaesthetised Women: For Public Distribution', Sallyann Thompson to Valerie Smith by request, 30 September 1988 (in author's possession).

50   *Auckland Star,* 12 November 1987.

51   *NZH,* 12 November 1987.

52   Gabrielle Collison, 'Position Paper', July 1988, p. 49 (in author's possession).

53   *The Auckland Sun,* 20 January 1988; Bonham, Transcripts of Public Hearings, Day 65, 19 January 1988, p. 5604, BAGC A638/13a 101L, ANZA.

54   *NZH,* 9 August 1988.

55   CCR, p. 141.

56   *Auckland Star,* 10 December 1987; Fertility Action Submission, p. 70, UOAA.

57 *NZH*, 18 June 1982.

58 *NZH*, 10 December 1987; other doctors were also mentioned in the CCR, pp. 172, 220–1.

59 Mrs M. R. to Collison, 12 November 1987, BAGC 18491 A638 97a 410B, ANZA.

60 Coney, *The Unfortunate Experiment*, pp. 220–1.

61 *Ibid.*, pp. 144–5.

62 *NZH*, 1 December 1987.

63 *Ibid.*; Pat Rosier, 'Listen to the Women', *Broadsheet*, vol. 155, January/February 1988, p. 6.

64 Mrs C. M. Purdue, Submission to the Inquiry, 25 June 1987, BAGC A638 35a, ANZA.

65 J. M. Snow, 'National Women's', Letter to the Editor, *NZH*, 9 October 1987.

66 *NZH*, 7 December 1987. Other letters of support from ex-patients include: Olga Carr and Family, Letter to the Editor, *Sunday Star*, 8 August 1990; Betty Woolliams (Waiau Pa), 'Readers Respond', *Metro*, August 1990, pp. 125–6; B. Willett, 'Readers Respond', *Metro*, August 1990, p. 126.

67 S. U. A. to Collison, 1 October 1987, BAGC 18492 A638 162a, ANZA.

68 P. W. to Nurse Ashton, 1 October 1987, BAGC 18492 A638 162a, ANZA.

69 CCR, p. 172.

70 *Ibid.*, pp. 167, 171.

71 Wilson, 'Primetime Television Coverage', p. 46.

72 CCR, pp. 166, 168.

73 These letters are held in the Inquiry files, ANZ Auckland, and Health Department files, ANZ Wellington. Some further examples of letters to Collison include S. C. to Collison, 20 November 1987; J. D. to Collison, 20 November 1987; B. G. to Collison, 26 November 1987; M. K. to Collison, 18 December 1987; A. H. to Collison, 12 November 1987; P. L. M. to Collison, 28 September 1987; F. P. to Collison, 5 October 1987; J. I. W. to Collison, BAGC 18492 A638 162a, ANZA.

74 Submission, Sierra Moon, for the Auckland Women's Health Collective, 24 June 1987, BAGC A638 34a, ANZA.

75 Submission, Wellington Women's Health Collective, 17 September 1987, BAGC A638 35a, ANZA.

76 Submission, New Zealand Women's Health Network, pp. 2, 5, BAGC A638 35a, ANZA.

77 *Ministry of Women's Affairs Newsletter*, vol. 7, December 1987–January 1988.

78 Miriama Evans, Ministry of Women's Affairs, Transcripts of Public Hearings, Day 60, 3 December 1987, pp. 5305–9, BAGC A638/13a 101L, ANZA.

79 Ruth Norman to Cartwright, 15 March 1988, BAGC A638 34a, ANZA.

80 Cited by Bunkle, 'Side-stepping Cartwright', p. 53.

81 CCR, p. 116.

82 Ministry of Women's Affairs, Transcripts of Public Hearings, Day 60, 3 December 1987, BAGC A638/13a 101L, ANZA.

83 Rosier, 'Listen to the Women', p. 8.

84 Ministry of Women's Affairs, questioning by Sandra Coney, Transcripts of Public Hearings, Day 60, 3 December 1987, BAGC A638/13a 101L, ANZA.

85 Wilson, 'Primetime Television Coverage', pp. 30–33, 37.

86 Coney, *The Unfortunate Experiment*, p. 154.

87 *Ibid.*, p. 149.

88 Interview by Sandra Coney and Phillida Bunkle with Dennis Bonham, 9 December 1986, p. 7, Exhibit 3c, Cartwright Inquiry, UOAA.

89 *NZH*, 16 June 1987; Sandra Coney, Submission to Inquiry, August 1987, p. 8, UOAA.

90 *NZH*, 19 June 1987; Coney, Submission to Inquiry, p. 8, UOAA.

91 Lerner, *Breast Cancer Wars*, pp. 200, 201; see also chapter 2.

92 Coney, Submission to Inquiry, p. 9, UOAA.

93 Bunkle, 'Side-stepping Cartwright', p. 54.

94 Reported by Rosier, 'Listen to the Women', p. 8, UOAA.

95 Fertility Action, Submission to Inquiry, pp. 10–11, UOAA.

96 Fertility Action, Submission to Inquiry: 'Drug Company Funded Research – Upjohn's New Zealand Contraception and Health Study', pp. 31–32. Foster, *Women and the Health Care Industry*, p. 21. In fact New Zealand was not alone in allowing use of this form of contraceptive at that time: in 1984 manufacturers of Depo-Provera were granted a long-term British licence.

97 Fertility Action, Submission to Inquiry, pp. 32–33; see chapter 5.

98 Fertility Action, Submission to Inquiry, pp. 73–74. Read by Coney, Transcripts of Public Hearings, Day 63, 9 December 1987, BAGC A638/13a 101L, ANZA.

99 M. B., Women's Electoral Lobby, West Auckland, 25 July 1987, 'Hospital Boards – Ministerial Inquiry 1986–88', ABQU 632 W4452 397 56/101 62694, ANZW.

100 *NZH*, 21 November 1987.

101 *NZH*, 18 November 1987.

102 *Ibid.*

103 *NZH*, 21 November 1987.

104 *Auckland Star*, 23 November 1987.

105 Pamela Hayward, interviewed by Jenny Carlyon, 2 September 2004.

106 Barbara Smith, interviewed by Jenny Carlyon, 16 September 2004.

107 CCR, p. 172.

108 Judy Yarwood, Letter to the Editor, *NZH*, 28 November 1987.

109 Chris Rothman, 'Cervical Cancer Enquiry', Letter to the Editor, *NZNJ*, vol. 81, 12, December 1987, p. 4.

110 Both were nurse educators, and Bickley's qualifications included university degrees in social sciences and sociology: Transcripts of Public Hearings, Day 52, 20 November 1987, pp. 4459–61, BAGC A638/11A 101J, ANZA.

111 'News and Events: NZNA Speaks to Cervical Cancer Inquiry', *NZNJ*, vol. 81, 12, December 1987, p. 7.

112 'News and Events: NZNA at the Cervical Cancer Inquiry', *NZNJ*, vol. 81, 2, February 1988, p. 4.

113 Joy Bickley, 'Watchdogs or Wimps? Nurses' Response to the Cartwright Report', in Coney (ed.), *Unfinished Business*, pp. 132–3.

114 New Zealand Nurses' Industrial Union of Workers Submission to the Inquiry, December 1987, BAGC A638 35a, ANZA.

115 'Conference/AGM 1987, Motions passed at the AGM', *NZNJ*, vol. 81, 10, October 1987, p. 19.

116 *NZH*, 23 September 1987.

117 Julie Nicholls, 'Patients' Rights', Letter to the Editor, *NZNJ*, vol. 81, 11, November 1987, pp. 3–4.

118 S. Pemberton, Letter to the Editor, *NZ Listener*, 8 October 1988.

119 *NZH*, 16 August 1988.

120 Coney, *The Unfortunate Experiment*, p. 9.

121 Rosier, 'Broadcast: A Feminist Victory'.

122 *Ibid.*; 'The Women's Book Festival', reviews by Pat Rosier, *Broadsheet*, vol. 161, September 1988, pp. 12–13. The article by Bunkle did not include the original opening statement however, which *Metro* lawyer Rhys Harrison had advised might be defamatory: McLeod, 'The Importance of Being Sandra Coney', p. 65.

123 *NZ Listener*, 24 September 1988.

124 *Dominion Sunday Times*, 4 September 1988.

125 Michael King, 'Books: Glimpses of Reality', *Metro*, October 1988, pp. 218–19.

126 *Sunday Star*, 14 August 1988; Linda Kaye came from a broadly feminist perspective, as indicated from a letter she published in *Broadsheet* on artificial insemination by donor which she believed was 'consistent with patriarchy and the suppression of women's sexuality, and political control (by way of increasing medicalisation) of women's reproductive lives': Letter to the Editor, *Broadsheet*, April 1984, p. 15.

127 Klim McPherson, 'Life After Cartwright', Correspondence, *NZMJ*, vol. 102, 1989, p. 169. *NZH*, 19 April 1989.

128 Coney, *The Unfortunate Experiment*, p. 273.

129 *Ibid.*, p. 266.

130 *Ibid.*, p. 218.

131 Cole, Unpublished Memoirs, p. 52.

132 *Ibid.*

133 Coney, *The Unfortunate Experiment*, p. 74.

134 *Ibid.*, p. 36.

135 *Ibid.*, p. 161.

136 *Ibid.*, p. 166.

137 *Ibid.*, p. 246.

138 Coney, Submission to Inquiry, 'The Breadth of the Problem', pp. 1–6, UOAA.

139 Fertility Action, Submission to Inquiry, p. 75, UOAA.

140 Coney, *The Unfortunate Experiment*, p. 161.

141 See CCR, Appendix 1, p. 219.

142 *Jeffcoate's Principles of Gynaecology*, p. 404. The same author also points out that if colposcopy were available, then a punch biopsy was preferable, given the hazards associated with cone biopsy: *ibid.*, p. 489.

143 Coney, *The Unfortunate Experiment*, p. 186. Pat Rosier also wrote, 'Bonham was pretty slippery under cross-examination': Rosier, 'The Speculum Bites Back', p. 5.

144 Coney, *The Unfortunate Experiment*, p. 232.

145 *Ibid.*, p. 264.

146 *Ibid.*, p. 205.

147 Sandra Coney, *Out of the Frying Pan: Inflammatory Writings 1972–1989*, Penguin Books, Auckland, 1990, pp. 67–68; *Broadsheet*, 6 April 1973.

148 Coney, *The Unfortunate Experiment*, pp. 148, 149.

149 *Ibid.*, p. 149.

150 *Ibid.*, p. 238; CCR, p. 72.

151 Coney, *The Unfortunate Experiment*, p. 236.

152 Green, Transcripts of Public Hearings (cross-examined by Rennie), Day 19, 3 September 1987, pp. 1398–9, BAGC A638/4a 101C, ANZA.

153 McIndoe, Memo to Medical Superintendent, NWH, 14 December 1973, UOAA.

154 Green, Transcripts of Public Hearings, Day 13, 26 August 1987, pp. 841–2, BAGC A638/3a 101B, ANZA.

155 Coney, *The Unfortunate Experiment*, p. 239.

156 *Ibid.*, p. 165.

157 *Ibid.*, p. 24.

158 *Ibid.*, p. 40.

159 *Ibid.*, pp. 191, 194. The interlude referred to appeared to have been a private joke between McLean, Harrison and Rennie. During questioning by Collins, McLean suddenly said, 'Mr Harrison is making fun of me Your Honour.' Commission replied, 'I am sure he meant no harm.' Collins attempted to get back to the questioning; the transcript read: '. . . Dr McLean to get back to NWH, a topic I am sure you will be relieved to get back to. (much laughter!). Commission – Do you want an adjournment? Rennie – I am sorry Ma'am. (still much laughter!).' Collins continued with

the cross-examination. Transcripts of Public Hearings, Day 51, 19 November 1987, pp. 4437–8, BAGC A638/11a, 101J, ANZA.

160 Richard Seddon, interviewed by Jenny Carlyon, 2 September 2005. Transcripts of Public Hearings, Day 4, 6 August 1987, p. 145, BAGC A638/2a 101A, ANZA, refers to Seddon's comment about fathers being present at deliveries, and Hugh Rennie's promise to Cartwright not to pursue this line.

161 Coney, *The Unfortunate Experiment,* p. 234.

162 *Ibid.,* pp. 234–5.

163 *Ibid.,* p. 231.

164 *Ibid.,* p. 253.

165 *Ibid.,* p. 246.

166 *Ibid.,* p. 180 (Dr Noda who supported Green formed an exception to this, see chapter 6).

167 Coney, *The Unfortunate Experiment,* p. 168.

168 *Ibid.,* p. 271.

169 Roberts, 'You Saved My Mum'.

170 Ronald W. Jones and Malcolm R. McLean, 'Carcinoma In Situ of the Vulva: A Review of 31 Treated and Five Untreated Cases', *O&G,* vol. 68, 4, 1986, pp. 499–503.

171 Evidence of Dr Joseph Jordan, Appendix, Case History 60/64, p. 4, UOAA.

172 Green, Transcripts of Public Hearings, Day 11, 24 August 1987, pp. 723–5, BAGC A368/3a, 101B, ANZA.

173 Coney, *The Unfortunate Experiment,* p. 76.

174 Green, Transcripts of Public Hearings, Day 11, 24 August 1987, p. 720, BAGC A638/3a 101B, ANZA.

175 Coney, *The Unfortunate Experiment,* pp. 197–8.

176 Sandra Coney, 'Sandra Coney and the National Women's Hospital Affair', Correspondence, *NZMJ,* vol. 103, 1990, pp. 355.

177 Coney, *The Unfortunate Experiment,* p. 101.

178 Kolstad and Klem, 'Long-term Follow-up', pp. 125, 128.

179 Macgregor and Teper, 'Uterine Cervical Cytology and Young Women'.

180 Coney, *The Unfortunate Experiment,* pp. 59–60.

181 Evidence of Dr Joseph Jordan, Appendix, Case History 60/64, September 1987, p. 4, UOAA.

182 Green, Transcripts of Public Hearings, Day 17, 1 September 1987, pp. 1224–6, BAGC A638/4a 101C, ANZA.

183 Kolstad and Klem, 'Long-term Follow-up', pp. 127, 128.

184 Coney, 'Do You Really Have to Have a Hysterectomy?'; see also chapter 7.

185 Coney, *The Unfortunate Experiment,* pp. 124–9. Angela Raffle cited McIndoe's paper that also stated that reverting to negative cytology is no guarantee that cancer will not develop

later: Raffle, Alden and Mackenzie, 'Detection Rates'.

186 Coney, *The Unfortunate Experiment,* p. 124.

187 *Ibid.,* p. 266.

9   Media Wars: The Report's Reception

1   Dann, *Up From Under,* p. 86.

2   Roberts, 'We Had to Speak Out About What We Knew'.

3   *NZH,* 16 August 1988.

4   *Auckland Star,* 6 September 1988.

5   Sandra Coney, 'The End of the Experiment', *NZ Listener,* 10 September 1988, pp. 22–24.

6   See CCR, p. 172 (this related to the 1960s baby swabs).

7   Coney, 'The End of the Experiment'.

8   Mantell to Maiden, 10 October 1988, Box 432, UOAA.

9   Mrs J. M. to Maiden, 5 October 1988, Box 432, UOAA.

10 Lynda Williams to Dean Medical School, 1 September 1988, Box 432, UOAA.

11 Coney, 'The End of the Experiment'.

12 NWHMC minutes, 19 August 1963, BAGC A638 22b, ANZA.

13 *Sunday Star,* 25 September 1988.

14 *Sunday Star,* 2 October 1988.

15 Murray Jamieson to N. Holford, 23 September 1988, Box 432, UOAA.

16 G. C. Liggins, 'The George Addlington Syme Oration: Winds of Change', *Australia and New Zealand Journal of Surgery,* vol. 61, 1991, p. 169.

17 *Auckland Star,* 7 October 1988.

18 *Auckland Star,* 14 March 1989.

19 *Dominion Sunday Times,* 26 March 1989.

20 *Dominion Sunday Times,* 12 March 1989.

21 Graeme Overton, interviewed by Jenny Carlyon, 8 November 2004.

22 *Auckland Star,* 14 March 1989.

23 *Dominion Sunday Times,* 19 March 1989.

24 J. Little, 'Cartwright Report "Based on Scam"', *Dominion Sunday Times,* 12 March 1989; *NZH,* 14 March 1989.

25 *NZWW,* 8 August 1988, p. 16.

26 Paul and Holloway, 'No New Evidence'; McIndoe *et al.,* 'The Invasive Potential', reprinted in CCR, Appendix 7, p. 255.

27 Skrabanek, 'Cervical Cancer Study'.

28 *Auckland Star,* 14 March 1989.

29 *Dominion Sunday Times,* 19 March 1989.

30 *Dominion Sunday Times,* 26 March 1989.

31 Karen Chapman, 'A Question of Ethics: An Update on the Implementation of the Cartwright Report', *NZNJ,* vol. 82, 2, March 1989, p. 25.

32 Sandra Coney 'Doctors in Charge', *NZ Listener,* 20 May 1989, pp. 16–18.

33 *Sunday Star,* 24 June 1990.

34 Kolstad and Klem, 'Long-term Follow-up', pp. 125, 128.

35 Skrabanek, 'Cervical Cancer Study'; McIndoe *et al.*, 'The Invasive Potential', reprinted in CCR, Appendix 7, pp. 247–57.

36 Coney, *The Unfortunate Experiment*, p. 271.

37 Coney, 'The End of the Experiment', p. 22.

38 Corbett, 'Second Thoughts', p. 70.

39 Lynda Williams, 'Looking Back at the 1987 Cervical Cancer Inquiry – and Response', Correspondence, *NZMJ*, vol. 117, 2004 (URL: http://www.nzma.org.nz.ezproxy.auckland.ac.nz/journal/117-1202/1084).

40 Tony Baird, 'Cervical Cancer Inquiry', Correspondence, *NZMJ*, vol. 117, 2004 (URL: http://www.nzma.org.nz.ezproxy.auckland.ac.nz/journal/117-1204/1136). During the Inquiry, two National Women's Hospital doctors, Murray Jamieson and Andrew Mackintosh, reviewed all 3037 cases of carcinoma in situ of the cervix registered at the hospital from 1955 to 1986. This included 1222 from 1955 to 1976 and 1815 from 1977 to 1986. Thirty-two patients were identified as having invasive genital tract disease following persistent CIS of the cervix and eight of those patients had died of genital tract cancer by the end of 1986: Gabrielle Collison, 'National Women's Hospital Auckland, New Zealand: A Position Paper', July 1988, p. 46 (in author's possession).

41 Bruce Faris, 'Sandra Coney and the National Women's Hospital Affair', Correspondence, *NZMJ*, vol. 103, 1990, p. 354; Coney, 'Sandra Coney and the National Women's Hospital Affair', Correspondence, *NZMJ*, vol. 103, 1990, pp. 354–5.

42 John Holdem, Helen Holdem, Briar Green, Rick Pearson and Nicola Green, Letter to the Editor, *NZ Listener*, 16 April 1988.

43 John and Helen Holdem, Letter to the Editor, *NZH*, 30 September 1988; Jordan, 'Minor Degrees of Cervical Intraepithelial Neoplasia'.

44 Valerie Smith, Submission to Inquiry, February 1988, BAGC A638 36a, ANZA.

45 Valerie Smith, 'Cartwright Inquiry or Inquisition', n.d., c. 1990, Box 432, UOAA.

46 *Dominion Sunday Times*, 19 March 1989.

47 Corbett, 'Second Thoughts', p. 61.

48 Valerie Smith, Letter to the Editor, *Metro*, October 1990, pp. 16–17.

49 *Auckland Star*, 1 August 1990.

50 *Auckland Star*, 8 August 1990.

51 Debbie Smith, Letter to the Editor, *Auckland Star*, 10 July 1990.

52 Corbett, 'Second Thoughts', p. 55.

53 Sandra Coney, 'Unfinished Business: The Cartwright Report Five Years On', in Coney (ed.), *Unfinished Business*, p. 43.

54 *Sunday Star*, 15 July 1990.

55 Warwick Roger, Letter to the Editor, *Sunday Star*, 11 November 1990. The two overseas experts were Laverty and Richart who disagreed with some of McLean's classifications: Warwick Roger, 'My Town: Intellectual Thuggery', *Metro*, September 1990, pp. 8–13. Helen Clark's statement was reported in *Sunday Star*, 23 September 1990.

56 Joan Donley to Australian Consolidated Press Ltd, 24 July 1990, Joan Donley Archives, UOAA; also referred to in Jan Corbett, 'Have You Been Burned at the Stake Yet?', *Metro*, October 1990, p. 158.

57 McLeod, 'The Importance of Being Sandra Coney', p. 55.

58 Corbett, 'Have You Been Burned at the Stake Yet?'.

59 *NZH*, 23 October 1990.

60 *NZH*, 27 November 1990. Harrison's position as convener was reported in *NZH*, 24 October 1990. See also Warwick Roger, Letter to the Editor, *Sunday Star*, 11 November 1990.

61 *NZH*, 6 August 1988. David Caygill had taken over as Health Minister from Michael Bassett on 28 August 1987.

62 David Caygill, address to NZ Congress of Obstetricians and Gynaecologists, Wellington, 2 March 1988, Joan Donley Archives, UOAA.

63 Coney, *The Unfortunate Experiment*, p. 231.

64 Sandra Coney, 'Central Issue is a Return to Trust', *Dominion Sunday Times*, 26 March 1989; Matheson, *Fate Cries Enough*, pp. 82–84.

65 R. W. Jones, 'Viewpoint: Reflections on Carcinoma In Situ', *NZMJ*, vol. 104, 1991, pp. 339–41.

66 Green, 'Cervical Carcinoma in Situ: An Atypical Viewpoint'.

67 Ron Jones, interviewed by Jenny Carlyon, 2 December 2004.

68 New Zealand Royal College of Obstetricians and Gynaecologists (NZRCOG), *Newsletter*, 2, 1988, Joan Donley Archives, UOAA.

69 *Ibid.*

70 M. A. H. Baird to R. J. Seddon, 13 December 1988, Joan Donley Archives, UOAA.

71 NZRCOG, *Newsletter*, 2, 1988.

72 Green's comments on 8 November 1985 conference on screening, 12 June 1986, CCR Exhibit 12d, UOAA.

73 Ian St George, Letter to the Editor, *New Zealand Family Physician*, vol. 54, Spring 1988, n.p., Joan Donley's scrapbook, Joan Donley Archives.

74 Drs Jenny Simpson, Rosy Fenwicke, Sue Pullon, Tim Cookson, Christopher Ryan, Nina Sawicki, Winifred Kennedy, Juliette Broadmore, Margaret Sparrow, Julie Kimber and Nalayini Pasupati, 'Letter to the Editor', *NZ Listener*, 22 October 1988 (nine were

women and eight qualified in medicine at Otago Medical School after 1970, five of those in 1980–81).

75   John Neutze, Letter to the Editor, *NZ Listener*, 8 April 1989; see also John Neutze, 'Ethics after Cartwright: What is the Real Issue', Correspondence, *NZMJ*, vol. 104, 1991, pp. 146, 236.

76   Peter Cairney, 'National Women's Hospital Inquiry', Correspondence, *NZMJ*, vol. 102, 1989, p. 261.

77   W. J. Pryor, 'Doubts about the Cartwright Findings', Correspondence, *NZMJ*, vol. 103, 1990, p. 355.

78   Corbett, 'Have You Been Burned at the Stake Yet?', pp. 162–5.

79   Michael Mackay, 'Readers Respond', *Metro*, August 1990, p. 126; see also Michael Mackay, 'The Cartwright Report', Correspondence, *NZMJ*, vol. 104, 1991, p. 125.

80   Cited in Matheson, *Fate Cries Enough*, pp. 227–8.

81   'Medicolegal: Medical Council Charges Professor Bonham'.

82   See chapter 8.

83   McNeill, *The Ethics and Politics of Human Experimentation*, p. 76.

84   *Auckland Star*, 7 October 1988.

85   Ibid.

86   *NZMJ*, vol. 103, 1990, p. 109; L. H. Varlow, 'NZMA Complaint', Correspondence, *NZMJ*, vol. 103, 1990, pp. 191–2.

87   NZMA *Newsletter*, January 1990, p. 9, Joan Donley Archives, UOAA.

88   Ibid.

89   Doug Baird, Letter to the Editor, *NZ General Practice*, 22 June 1990, n.p., Joan Donley's scrapbook, Joan Donley Archives.

90   *New Zealand General Practice*, 6 May 1991.

91   Stanley C. Simmons to Katherine O'Regan, 26 March 1991, RCOG Archives, A4/26/37, London, UK.

92   Keith Sinclair, *A History of the University of Auckland*, Oxford University Press, Auckland, 1983, p. 275.

93   CCR, p.180.

94   Algar Warren to Secretary, AHB, Joint Relations Committee minutes, 18 April 1961, p. 27, Box 157, UOAA.

95   Algar Warren to Chief Executive, AHB, 7 October 1976, p. 4, Box 245, UOAA.

96   W. J. Pryor, Central Ethical Committee, 6 May 1977, Box 245, UOAA.

97   *NZH*, 6 August 1988; *Central Leader*, 16 August 1988.

98   Secretary, Auckland Women's Health Council to the Chancellor, University of Auckland, 8 February 1989, Joan Donley Archives, UOAA. A patient interviewed by Jenny Carlyon found Bonham far from insensitive: Trish Hegerty,

interviewed by Jenny Carlyon, 18 November 2005.

99   *Sunday Star*, 14 October 1990.

100 *New Zealand Doctor*, 5 November 1990.

101 *NZH*, 8 December 1990; *Sunday Star*, 9 December 1990.

102 'Medicolegal: Medical Council Charges Professor Bonham'; *Sunday Star*, 1 July 1990; *Sunday Star*, 15 July 1990; *NZH*, 15 October 1990, 'Bonham Verdict: Disgraceful Conduct', the article stated, 'The council's interim findings, with explanation are as follows . . .'.

103 Kolstad, 'Diagnosis and Management of Precancerous Lesions of the Cervix Uteri', p. 559; Koss, 'Exfoliative Cytology Koss', pp. 182–93; Koss, 'Dysplasia: A Real Concept or a Misnomer?', p. 374.

104 Lyndsey Swan, Editorial, 'Second Thoughts', *NZ General Practice*, 10 July 1990.

105 *Sunday Star*, 9 December 1990.

106 Natalie S. Gilmore, Letter to the Editor, *Sunday Star*, 30 September 1990.

107 Lesley McCowan, Hilary Liddell, Cindy Farquhar and Lynda Batchelor [*sic*], Correspondence, *NZMJ*, vol. 103, 1990, p. 543.

108 J. A. Malloch, 'The Medical Council and Professor Bonham', Correspondence, *NZMJ*, vol. 103, 1990, p. 567.

109 John Scott, 'Obituary, Professor Dennis Geoffrey Bonham', The University of Auckland News, vol. 35, 5, June 2005, pp. 20–21.

110 'An Unfortunate Experiment', *NZH*, 14 May 2005. The article began in large typeface: 'Professor's role in cervical cancer treatment at National Women's led to charges of disgraceful conduct', and proceeded to provide an obituary.

111 *NZH*, 29 July 1995.

112 Cole, Unpublished memoirs, ch. 14, p. 53. The letter referred to was Skrabanek and Jamieson, "Eaten By Worms: A Comment on Cervical Screening", Correspondence, see also Chapter 6.

113 C. S. Harison, 'Ethics after Cartwright', Correspondence, *NZMJ*, vol. 104, 1991, p. 235 (Harison qualified in medicine in 1954).

114 'A Chronology of Events related to the Cartwright Report 1987–1993', in Coney (ed.), *Unfinished Business*, p. 11.

115 Kaye, 'What Happened to the Cartwright Women?'.

116 D. A. Purdie, 'The National Women's Hospital Settlement', Correspondence, *NZMJ*, vol. 105, 1992, p. 270.

117 *NZH*, 9 May 1992.

118 *Sunday Star*, 10 May 1992.

119 Purdie, 'The National Women's Hospital Settlement', Correspondence.

120 Bickley, 'Watchdogs or Wimps?', pp. 127–8.

'News and Events: Health Workers on Cervical Cancer Enquiry', *NZNJ*, vol. 81, 9, September 1988, p. 5.

121 'News and Events: NZNA Welcomes Cervical Cancer Report', *NZNJ*, vol. 81, 7, August 1988, p. 5.

122 Joy Bickley, 'What the Cervical Cancer Inquiry Report Means for Nurses', *NZNJ*, vol. 81, 8, September 1988, pp. 14–15.

123 Bronwyn West (Wellington), 'Nurses Must Act', Letter to the Editor, *NZNJ*, vol. 81, 8, September 1988, p. 3.

124 *Auckland Star*, 6 September 1988.

125 *Auckland Star*, 14 September 1988.

126 Trevor Billing, Letter to the Editor, *Auckland Star*, 20 September 1988.

127 Valerie Edwards and Mary Brayshaw, Letter to the Editor, *Auckland Star*, 20 September 1988.

128 Valerie Fleming, 'Angst over Cancer Inquiry Comments', Letter to the Editor, *NZNJ*, vol. 81, 10, October 1988, p. 3.

129 *NZNJ*, vol. 81, 12, December/January 1989, p. 3.

130 'News and Events: Metro Article "Scurrilous"', *NZNJ*, vol. 83, 7, August 1990, p. 7.

131 Editorial, *Dominion*, 10 April 1989, cited in Coney, *Out of the Frying Pan*, p. 239.

132 Richard Clime, Letter to the Editor, *Metro*, November 1990, pp. 15, 17–18; on Dr Kildare, see Joseph Turow, *Playing Doctor: Television, Storytelling, and Medical Power*, Oxford University Press, New York, 1989.

133 Editorial, *Auckland Star*, 14 March 1989.

134 Rosier, 'Broadcast: A Feminist Victory', p. 6.

10 New World, Better World? Implementing Cartwright

1 'News: Inquiry Faults National Women's Hospital', *NZMJ*, vol. 101, 1988, p. 558.

2 CCR, pp. 212–18.

3 *NZJ*, 7 August 1988; Sandra Coney, 'The First Post-Cartwright Year: A Case Study in Institutional Resistance', in Coney, *Out of the Frying Pan*, p. 240; Auckland Women's Health Council, 'Recommendations Coming from the Plenary Session: The Cartwright Report – One Year On, 5 August 1989', Box 432, UOAA.

4 Coney, 'The First Post-Cartwright Year', p. 240.

5 Coney, 'Health Organisations', p. 252.

6 *NZH*, 24 August 1988.

7 Lisa Sabbage, 'A Spur to Action', *Broadsheet*, October 1988, p. 8.

8 Sallyann Thompson, 'The Cartwright Inquiry', Correspondence, *NZMJ*, vol. 103, 1990, p. 436.

9 CCR, pp. 121–6, 211 (at the time of the Cartwright Report, 123 women were on the list).

10 Harrison during cross-examination of Green,

Transcripts of Public Hearings, Day 16, 31 August 1987, p. 1105, BAGC A638/41 101C, ANZA; CCR, p. 211.

11 Coney, 'Doctors in Charge', p. 17.

12 *Ibid.*, p. 18, reprinted in Coney, *Out of the Frying Pan*, pp. 203–12. Bassett later denied calling the doctors 'naughty boys'. Judith Bassett, interviewed by Jenny Carlyon, 18 May 2006.

13 Coney, *Out of the Frying Pan,* p. 212.

14 *New Zealand General Practice,* 24 April 1989.

15 John McLeod, Gay Keating and Beverley Carey, 'Implementation of the Recommendations of the Report of the Cervical Cancer Inquiry', 1 August 1989, Box 432, UOAA.

16 Matheson, *Fate Cries Enough*, p. 233.

17 CCR, pp. 212–15.

18 Harvey Carey newspaper clippings, 1950s, NWH records.

19 Ian Hutchison to Lyn Potter, Medical Superintendent's Office, 1983, NWH records.

20 CCR, p. 159.

21 Judith Bassett, interviewed by Jenny Carlyon, 18 May 2006.

22 Board of Health Maternity Services Committee 1966–68: Minutes 5 May 1966, H1 29/21 33724, ANZW.

23 Board of Management for Glasgow Maternity and Women's Hospitals, Medical Committee, 3 September 1969, p. 46, HB45 1/70, Mitchell Library, Glasgow, Scotland.

24 Memo from Chief Executive to Medical Superintendent, 27 September 1977 – copy of letter 9 September 1977 from National President, Feminists for Life (NZ) Inc with 'Maternity Patient's Bill of Rights' prepared by that organisation: NWHMC minutes, 13 October 1977, BAGC A638 41b, ANZA.

25 NWHMC minutes, 13 October 1977, BAGC A638 41b, ANZA.

26 Coney, *The Unfortunate Experiment,* pp. 234–5.

27 Matheson, *Fate Cries Enough,* p. 133

28 M. A. H. Baird, Letter to the Editor, *NZH*, 14 October 1990.

29 Submission to Inquiry, Sarah J. Calvert, PhD, for the New Zealand Women's Health Network, pp. 5–6, BAGC A638 35a, ANZA.

30 CCR, pp. 171–3, 213.

31 *NZH*, 11 August 1989.

32 *NZH*, 11 August 1989; see also Donley, *Herstory of N.Z. Homebirth Association*, pp. 38–39.

33 Auckland Women's Health Council *Newsletter*, September 1989, Joan Donley Archives, UOAA.

34 'News and Events: First Advocate', *NZNJ*, September 1989, p. 7.

35 Bunkle, 'Side-stepping Cartwright', pp. 59–60.

36 Lynda Williams, 'Dreaming the Impossible Dream: The Fate of Patient Advocacy', in

Coney (ed.), *Unfinished Business,* pp. 88–102.

37   *Ibid.*, p. 90.

38   *Ibid.*, p. 95.

39   *Ibid.*, p. 99.

40   Bunkle, 'Side-stepping Cartwright', p. 59.

41   Peter Davis, Submission to the Inquiry, p. 18, UOAA.

42   *NZH,* 11 October 1990.

43   Christine Cheyne, Mike O'Brien and Michael Belgrave, *Social Policy in Aotearoa: A Critical Introduction,* Oxford University Press, Auckland, 1997, p. 227. Commissioner for the Health and Disability Commission, *Code of Rights for Consumers of Health and Disability Services,* Wellington, Health and Disability Commission, 1995. See also, Bunkle, 'Women, Health and Politics'.

44   'News and Events: No Consumer Input – Coney', *NZNJ,* December/January 1989, p. 6.

45   David Cole to Lynda Williams, Spokesperson, Maternity Action, 22 September 1988, Box 432, UOAA.

46   The University of Auckland Report of the University Council Sub-Committee on the Cartwright Report, 9 February 1989, pp. 3, 10, Box 432, UOAA.

47   *Ibid.,* p. 13.

48   J. S. Robinson, Report to Professor D. North, Dean, Faculty of Medicine, University of Auckland on Teaching on Cervical Cancer in the Departments of Obstetrics and Gynaecology, Pathology, Community Health and General Practice, 17 July 1989, Box 432, UOAA.

49   The University of Auckland Report of the University Council Sub-Committee on the Cartwright Report, 9 February 1989, p. 17, Box 432, UOAA.

50   J. D. K. North, 'Response of the School of Medicine to the University Council Sub-Committee (Ryburn Committee) on the Cartwright Report, July 1989', p. 10, Box 432, UOAA.

51   See, for example, Ann Oakley, *From Here to Maternity: Becoming a Mother,* Penguin, Harmondsworth, 1981.

52   North, 'Response of the School of Medicine', p. 8.

53   Sandra Coney, 'The Unfortunate Experiment: Implications for the Profession', 18 October 1988, Box 432, UOAA.

54   Joan Donley to Karen Guilliland, 3 December 1989, Joan Donley Archives, UOAA.

55   'Women Do Battle With Med School', *Sunday Star,* 3 December 1989.

56   'Pioneering Work Continues', *The University of Auckland News,* vol. 24, 1, February 1994, pp. 1–3.

57   Joan Donley to 'Bonney', 31 August 1988, Joan Donley Archives, UOAA.

58   Wendy Savage, *A Savage Inquiry: Who Controls Childbirth,* Virago, London, 1986; see also Jo Murphy-Lawless, *Reading Birth and Death,* Cork University Press, Cork, 1998, p. 197, fn.

59   *Sunday Star,* 23 October 1988.

60   *Auckland Star,* 13 September 1988.

61   Kathleen Stratford, Letter to the Editor, *Auckland Star,* 20 September 1988.

62   Melva Firth, Letter to the Editor, *Auckland Star,* 27 September 1988.

63   Anna Watson, Letter to the Editor, *Auckland Star,* 26 September 1988.

64   W. Llewellyn, Letter to the Editor, *Auckland Star,* 30 September 1988.

65   *Sunday Star,* 23 October 1988.

66   Donley to Mantell, 11 November 1988, Joan Donley Archives, UOAA.

67   D. A. Lewis, for Maternity Action, Auckland, to Mantell, 17 November 1988, UOAA.

68   Karen Guilliland, Chair, National Midwives Section NZNA, 21 November 1988, Joan Donley Archives, UOAA.

69   J. C. on behalf of the Academic Women's Group to Ryburn, 19 October 1989, Box 432, UOAA.

70   Whangarei Home Birth Support Group Inc., *Newsletter* vol. 34, December/January 1989, Joan Donley Archives, UOAA. See also, *Northern Advocate,* 5 January 1989.

71   Auckland Women's Health Council *Newsletter,* March 1990, p. 3, Joan Donley Archives, UOAA.

72   *NZH,* 11 April 1990.

73   Allan Maclean, interviewed by Linda Bryder, London, August 2006.

74   *Sunday Star,* 9 February 1992.

75   *Sunday Star,* 12 April 1992.

76   *NZH,* 26 December 1992.

77   Gillian Turner, personal communication to Linda Bryder, June 2006 (in author's possession).

78   CCR, p. 212.

79   *Ibid.,* p. 213.

80   Glasgow Royal Maternity and Women's Hospital, HB45 1/73, Mitchell Library, Glasgow, Scotland.

81   Per Kolstad, Transcript of Public Hearings, Day 61, 7 December 1987, p. 5427, BAGC A638/13a 101L, ANZA.

82   CCR, pp. 176, 214.

83   Judi Strid, 'Ethical Dilemmas: A Consumer Perspective on the Performance of Ethical Committees', in Coney (ed.), *Unfinished Business,* p. 111.

84   *Ibid.,* p. 112.

85   *Ibid.,* p. 123.

86   Clinical Review, NWH, Professor Jeffrey Robinson, April 1995, p. 36 (in author's possession).

87 Sandra Coney, 'Against All Odds: The Experience of a Consumer Representative in the Establishment of the National Cervical Screening Programme', in Coney (ed.), *Unfinished Business*, p. 165.

88 Ministry of Women's Affairs, Transcripts of Public Hearings, Day 60, 3 December 1987, p. 5290, BAGC A638/13a 101L, ANZA.

89 Coney, 'Against All Odds', p. 166.

90 Cervical Screening, ABQU 632 W4452 1789 358/1/8 63072, ANZW.

91 Floss Caughey, Advisory Officer, Primary Health Care Programme to Fiona Reid, Christchurch South Health Centre, 4 March 1988, 'Primary Health Care Screening', ABQU 632 W4452 1789, ANZW.

92 New Zealand Ministry of Health, *A Brief Narrative on Maori Women and the National Cervical Screening Programme*, Ministry of Health, Wellington, 1997, pp. 15–16.

93 Meeting with Women in Nelson Inc (WIN), 12 January 1988, Primary Health Care Screening, ABQU 632 W4452 1789, ANZW. G. S. Loveridge, Chairman sub-faculty of RNZCGP to Esme Palliser, WIN, 20 June 1988, ANZW.

94 Hyde, 'At Your Cervix Madam', pp. 291–2, 295, 324.

95 Bickley, 'Safety Screen or Smoke-Screen'.

96 *NZNJ*, March 1990, p. 8.

97 *NZH*, 2 February 1991.

98 *Auckland Star*, 28 July 1989.

99 Coney, 'Against All Odds', p. 175.

100 Skegg, 'Leading Article: How Not to Organise a Cervical Screening Programme'.

101 Joan Austoker, Editorial, 'Gaining Informed Consent of Screening', *BMJ*, vol. 19, 1999, pp. 722–3. On the ethical responsibilities attached to screening programmes, see also World Health Organization International Agency for Research on Cancer, *IARC Handbooks on Cancer Prevention, Volume 10*, p. 46; and J. Austoker, 'Cancer Prevention in Primary Care: Screening for Cervical Cancer', *BMJ*, vol. 309, 1994, pp. 241–8; and David Slater, 'The Cervical Screening Muddle', *The Lancet*, vol. 351, 1998, p. 1130. The current pamphlet issued by the New Zealand Ministry of Health provides a comprehensive account: *Cervical Screening: Understanding Cervical Smear Test Results*, National Cervical Screening Programme, Ministry of Health, revised December 2007, reprinted June 2008, Code HE4598 (in author's possession).

102 Hyde, 'At Your Cervix Madam', pp. 313–14, 323.

103 Sandra Coney, *The Menopause Industry: A Guide to Medicine's 'Discovery' of the Mid-life Woman*, Penguin Books, Auckland, 1994, p. 21. On continued feminist critiques of cervical screening, see Alexandra Howson, 'Cervical Screening, Compliance and Moral Obligation', *Sociology of Health and Illness*, vol. 21, 4, 1999, pp. 401–25; and Mary Holmes, *What is Gender? Sociological Approaches*, Sage Publications, London, 2007, p. 100.

104 Cervical Screening, ABQU 632 W4452 1789 358/1/8 63072, ANZW; see also Eru W. Pomare and Gail M. de Boer, *Hauora: Maori Standards of Health: A Study of the Years 1970–1984*, Medical Research Council and Department of Health Special Report Series 78, Wellington, 1988.

105 A. M. Spence, Chief Executive of Auckland Division Cancer Society of New Zealand, to Phillida Bunkle, 16 February 1987, Exhibit 11a, Cartwright Inquiry, UOAA.

106 Ministry of Health, *A Brief Narrative*, p. 15.

107 *NZNJ*, March 1988 p. 18, referred to: Vicki Grace, PhD thesis, University of Waikato.

108 Maori Health Committee of the HRC, assisted by Lorna Dyall, 'Rangahau Hauora Maori: Maori Health Research Themes', 1998 (http://www.hrc.govt.nz/themes.htm).

109 Review of the Health, Cervical Cancer (Kaitiaki) Regulations, 1995, p. 3.

110 Ronald W. Jones, 'Cervical Cancer Prevention, Feminism, and Herb Green', *NZMJ*, vol. 120, 2007 (URL: http://www.nzma.org.nz.ezproxy.auckland.ac.nz/journal/120-1260/2694).

111 *Time*, Special Summer Issue 1990, pp. 22–23.

112 *Ibid.*

113 Ann Oakley, 'Report from Dr Ann Oakley ASB Visiting Professor 1989', 20 October 1989, Box 432, UOAA.

114 Rothman, *Strangers at the Bedside*, p. 144.

115 Brandt and Gardner, 'The Golden Age of Medicine?', p. 32.

11   The Aftermath: Public Perceptions of Unethical Practice

1   Coney, 'Unfinished Business', p. 41.

2   E. M. Symonds to George Pinker, President, RCOG, 1 October 1990, A4/26/37, RCOG Archives, London, UK.

3   John France, interviewed by Jenny Carlyon, 9 December 2004.

4   *NZH*, 8 December 1987; *Auckland Sun*, 8 December 1987, repeated in Coney, *The Unfortunate Experiment*, p. 176.

5   Per Kolstad, Transcript of Public Hearings, Day 61, 7 December 1987, pp. 5374–460, BAGC A638/13a 101L, ANZA.

6   'Death Watch', *The Guardian*, 23 August 1988.

7   Gerber and Coppleson, 'Clinical Research after Auckland'.

8   Dr Joseph Jordan, Submission to Inquiry, p. 23, UOAA.

9   Coppleson, 'Colposcopy', p. 178.

10 Gerber and Coppleson, 'Clinical Research after Auckland', p. 233.

11 CCR, p. 96.
12 Coney, *The Unfortunate Experiment,*
    pp. 103–4.
13 Paul M. McNeill, 'The Implications for
    Australia of the New Zealand *Report of
    the Cervical Cancer Inquiry*: No Room
    for Complacency', *The Medical Journal of
    Australia*, vol. 150, 1989, pp. 264–71.
14 McNeill, *The Ethics and Politics of Human
    Experimentation*, p. 17.
15 Karen Trenfield, 'Tinker, Tailor, Soldier,
    Scientist: Women and Science Today', *Hecate:
    An Interdisciplinary Journal of Women's
    Liberation*, vol. 21, 1, 1995, pp. 157–8.
16 Matheson, *Fate Cries Enough*; *Auckland Star*, 7
    September 1988.
17 See Hay, *The Caring Commodity*.
18 Pomare and de Boer, *Hauora: Maori
    Standards of Health*, pp. 28, 134.
19 Baruch A. Brody, *The Ethics of Biomedical
    Research: An International Perspective*, Oxford
    University Press, New York, 1998, p. 33.
20 World Health Organization International
    Agency for Research on Cancer, *IARC
    Handbooks on Cancer Prevention, Volume 10*,
    p. 55.
21 L. A. Reynolds and E. M. Tansey, '*Prenatal
    Corticosteroids for Reducing Morbidity and
    Mortality after Preterm Birth*', *The Transcript of
    a Witness Seminar held by the Wellcome Trust
    Centre for the History of Medicine at UCL,
    London, on 15 June 2004*, Wellcome Witnesses
    to Twentieth Century Medicine vol. 25, Well-
    come Trust, London, 2005, pp. 15–16, fn. 33.
22 *NZH*, 10 May 2008.
23 McCredie *et al.*, 'Natural History of Cervical
    Neoplasia'.
24 G. H. Green, 'The Significance of Cervical
    Carcinoma In Situ', *American Journal of
    Obstetrics and Gynecology*, vol. 94, 7, 1966,
    pp. 1009–22.
25 Coney and Bunkle, 'An Unfortunate
    Experiment', p. 16, repeated in Coney, *The
    Unfortunate Experiment*, p. 54, and *NZH*, 6
    August 1988.
26 CCR, p. 96.
27 *Ibid.*, p. 156.
28 Folke Pettersson (ed.), *Annual Report on
    the Results of Treatment in Gynecological
    Cancer*, 12th vol., International Federation of
    Gynecology and Obstetrics, Stockholm, 1988,
    pp. 56–74.
29 Jackie McAuliffe, Fertility Action, to McIndoe,

9 April 1985 and reply, 7 May 1985, Coney's
    Appendices to Submission to Inquiry, 13a and
    13b, BAGC 18489 A638 44a, ANZA.
30 Paul, 'Letter from New Zealand', p. 534.
31 Anil Sharma, 'Pregnancy Care in NZ: A Brief
    History', *New Zealand Doctor,* 12 September
    2007, p. 29.
32 Amanda Cameron, 'How Cartwright Changed
    It All . . .', *New Zealand Doctor*, 10 September
    2008, pp. 12–13.

12 Conclusion: An 'Unfortunate Experiment'?

1   *NZH*, 10 May 2008.
2   McCredie *et al.*, 'Natural History of Cervical
    Neoplasia': Green's management of CIS was
    described as 'an unethical clinical study' in the
    first sentence of the report: p. 425.
3   Stafl and Wilkinson, 'Cervical and Vaginal
    Intra-epithelial Neoplasia', p. 258.
4   Richart, Fu and Reagan, 'Pathology of
    Cervical Intraepithelial Neoplasia', p. 398.
5   B. Willett, Letter to the Editor, *Metro,* August
    1990, p. 126.
6   Coney, 'Sandra Coney and the National
    Women's Hospital Affair', Correspondence.
7   Wendy Mitchinson, 'Agency, Diversity, and
    Constraints: Women and the Physicians,
    Canada, 1850–1950', in The Feminist Health
    Care Ethics Research Network, Susan Sherwin
    (co-ordinator), *The Politics of Women's Health:
    Exploring Agency and Autonomy*, Temple
    University Press, Philadelphia, 1998, pp. 122,
    127, 146 note 3.
8   Thomas Schlick, 'Risk and Medical
    Innovation: a Historical Perspective', in
    Schlick and Tröhler (eds), *The Risks of Medical
    Innovation*, p. 7.
9   W. J. Ramsay to Cartwright, 12 August 1987,
    BAGC A638 35a, ANZA.
10 Coney, 'Unfinished Business', p. 35.
11 Watkins, 'Changing Rationale for Long-term
    Hormone Replacement Therapy', p. 33.
12 Coney, *Out of the Frying Pan*, p. 240;
    Dominion, 10 April 1989.
13 Watkins, 'Changing Rationale', pp. 24, 30.
14 *Northern Advocate*, 26 September 1988.
15 *NZH*, 10 May 2008.
16 Janet McCalman, *Sex and Suffering: Women's
    Health and a Women's Hospital: The Royal
    Women's Hospital Melbourne 1856–1996*,
    Melbourne University Press, Melbourne, 1998,
    p. 315.

Bibliography

ARCHIVES

Archives New Zealand, Auckland Regional Office (ANZA)

Series BAGC A638 – Commission of Inquiry into Cervical Cancer, National Women's Hospital. This incorporates transcripts of the public hearings, final submissions, correspondence, background material for the Inquiry and records pertaining to National Women's Hospital, including the minutes of the Senior Medical Staff, Hospital Medical Committee (1954–87) and Ethical Committee; Tumour Panel Notes; and Dr Collison's correspondence.
Series YCBZ – Auckland Hospital Board, Green Lane and Cornwall Park Hospital Committee minutes 1946–52.

Archives New Zealand, Wellington (ANZW)

Series ABQU 632 W4452 – Auckland Hospital Board – National Women's Hospital 1965–89. This includes ministerial papers and correspondence relating to the Inquiry, files on the Cervical Screening Programme and records relating to Fertility Action.
Series ABQU 632 W4550 – Board of Health Maternity Services Committee.
Series H1 29/21 alt no 33724 – Board of Health Maternity Services Committee.
Series COM26 3J – New Zealand Family Planning Association publications.

Archives, Office of the Vice-Chancellor, University of Auckland (UOAA)

Boxes 157, 245, 257, 314 and 432 relate to the Postgraduate School of Obstetrics and Gynaecology; Box 432 includes reports on the Cervical Cancer Inquiry by the Ryburn Committee, by Professor Jeffrey Robinson and by the Auckland Women's Health Council.
University of Auckland Council Reports, 1970, vol. 2 – Dennis Bonham, Report on Study Leave, October–December 1969, 17 May 1970.
Records relating to the Cervical Cancer Inquiry, 1988, including submissions to the Inquiry, along with various exhibits and correspondence.
Misc. – Refresher Course in Obstetrics, February–March 1967, Postgraduate School of Obstetrics and Gynaecology; MRCNZ, Referee Report on Grant Application, Form MRC/4, 1981; MRC folder, Bonham, Epidemiology and Data Processing in Obstetrics, Neonatology and Gynaecology, 82/137.

University of Auckland Library, Special Collections

Joan Donley Papers: This collection includes an extensive set of newspaper clippings; newsletters for the New Zealand Women's Health Network, the Auckland Women's Health Council, Fertility Action, the Ministry of Women's Affairs and Whangarei Home Birth Support Group Inc.; along with other miscellaneous papers, such as Auckland Hospital Board minutes, David Caygill's address to the New Zealand Congress of Obstetricians and Gynaecologists (2 March 1988), and newsletters of the New Zealand Royal College of Obstetricians and Gynaecologists and the New Zealand Medical Association.

Other Archives

Glasgow Maternity and Women's Hospitals Records, HB45 series, Mitchell Library, Glasgow, Scotland.
Royal College of Obstetrics and Gynaecology Archive, London, England.
Wellcome Library, London, England: SA/NWF/F13, 56 – Cervical Cancer Prevention Campaign, 1965.

INTERVIEWS

Dr Jenny Carlyon conducted over 90 interviews as part of a general history of National Women's Hospital, of which the principal investigator was Linda Bryder. The interviews cited here are the ones directly referred to in the text. They include Tony Baird, 15 September 2004; Judith Bassett, 18 May 2006; Michael Bassett, 23 May 2006; Lynda Batcheler, 31 March 2006; Cynthia Farquhar, 14 April 2006; John France, 9

December 2004; Pamela Hayward, 2 September 2004; Trish Hegerty, 18 November 2005; Ron Jones, 2
December 2004; Hilary Liddell, 1 April 2005; Sir Graham (Mont) Liggins, 12 May 2004; Lesley McCowan,
19 December 2005; Paul Patten, 11 April 2006; Richard Seddon, 2 September 2005; Barbara Smith, 16
September 2004.

Koss, Leopold G., interviewed by Stephen G. Silverberg, M.D., Vancouver March 2004, *International Journal of Gynecological Pathology* (URL: http://ovidsp.tx.ovid.com.ezproxy.auckland.ac.nz/spb/ovidweb.cgi).

Maclean, Allan, interviewed by Linda Bryder, London, August 2006.

MISCELLANEOUS REPORTS AND PERSONAL COMMUNICATIONS (IN AUTHOR'S POSSESSION)

Bonham, Nancie, private papers and notes.

Cole, David, Unpublished Memoirs.

Collison, Gabrielle, 'National Women's Hospital Auckland, New Zealand: A Position Paper', July 1988.

Evans, Professor J. Gremley, personal communication, May 2006.

Jamieson, Murray, 'Cervical Neoplasia: A Report for Meeting of the Gynaecology Cancer Advisory Group, 18 August 1982', National Women's Hospital records.

Mullins, Peter R., letter to Kevin Ryan, Barrister and Solicitor, 5 June 1990.

Raffle, Angela, personal communication, 2 April 2008.

Robinson, Jeffrey, Clinical Review of National Women's Hospital, April 1995.

Ronayne, Ian, personal communication, 1 November 2006.

Scott, P. J., 'Memorandum to the Director, Medical Research Council of New Zealand from the Standing Committee on Therapeutic Trials, re. The New Zealand Contraception and Health Study Protocol', 1984.

Turner, Gillian, personal communication, June 2006.

OFFICIAL PUBLICATIONS

Commissioner for the Health and Disability Commission, *Code of Rights for Consumers of Health and Disability Services,* Health and Disability Commission, Wellington, 1995.

Department of Health, *Screening for Cervical Cancer in New Zealand: Proceedings of a Meeting called by the Department of Health and the Cancer Society of NZ (Inc), 8 November 1985,* Department of Health, Wellington, April 1986.

Maternity Services Committee of the Board of Health, *Maternity Services in New Zealand*, Board of Health Report Series No. 26, Government Printer, Wellington, 1976.

Maori Health Committee of the Health Research Council, 'Rangahau Hauora Maori: Maori Health Research Themes', 1998 (http://www.hrc.govt.nz/themes.htm).

New Zealand Ministry of Health, *A Brief Narrative on Maori Women and the National Cervical Screening Programme*, Ministry of Health, Wellington, 1997.

New Zealand Official Year-book, 1951–52, 57th issue, compiled in the Census and Statistics Department New Zealand, Government Printer, Wellington, 1952.

Pomare, Eru W. and Gail M. de Boer, *Hauora: Maori Standards of Health: A Study of the Years 1970–1984*, Medical Research Council and Department of Health Special Report Series 78, Wellington, 1988.

The Report of the Committee of Inquiry into Allegations Concerning the Treatment of Cervical Cancer at National Women's Hospital and into Other Related Matters, Government Printing Office, Auckland, 1988.

NEWSPAPERS AND MAGAZINES

Auckland Star; Auckland Sun; Auckland University News; Central Leader; Dominion; Dominion Sunday Times; East and Bays Courier; Guardian; Journal of General Practice; Metro; New Zealand Doctor; New Zealand Herald; New Zealand Listener; New Zealand Nursing Journal: Kai Tiaki; New Zealand Woman's Weekly; North and South; Northern Advocate; Sunday Star; Sunday Telegraph; The Times; Time.

BOOKS AND ARTICLES

Ahluwalia, H.S. and Richard Doll, 'Mortality from Cancer of the Cervix Uteri in British Columbia and other Parts of Canada', *British Journal of Preventive and Social Medicine*, vol. 22, 1968, pp. 161–4.

Albrechtsen, S., S. Rasmussen, S. Thoreson, L. M. Irgens and O. E. Iversen, 'Pregnancy Outcome in Women Before and After Cervical Conisation: Population Based Cohort Study', *British Medical Journal (BMJ)*, vol. 337, 2008, pp. 803–5.

Anderson, George H., David A. Boyes, John L. Benedet, Jean C. Le Riche, Jasenka P. Matisic, Kenneth C. Suen, Ann J. Worth, Amelia Millner and Owen M. Bennett, 'Organisation and Results of the Cervical Cytology Screening Programme in British Columbia, 1955–85', *BMJ*, vol. 296, 1988, pp. 975–8.

Anon, 'Cancer of the Cervix: Death by Incompetence', *The Lancet*, vol. 2, 1985, pp. 363–4.

Arbyn, M., M. Kyrgiou, C. Simeons, A. O. Raifu, G. Koliopoulos, P. Martin-Hirsch, W. Prendiville and E. Paraskevaidis, 'Perinatal Mortality and Other Severe Adverse Pregnancy Outcomes Associated with Treatment of Cervical Intraepithelial Neoplasia: Meta-analysis', *BMJ*, vol. 337, 2008, pp. 798–802.

Arditti, Rita, Renate Duelli Klein and Shelley Minden (eds), *Test-Tube Women: What Future for Motherhood?*, Pandora Press, London, 1984.

Arnold, Pascal and Dominique Sprumont, 'The "Nuremberg Code": Rules of Public International Law', in Ulrich Tröhler and Stella Reiter-Theil in cooperation with Eckhard Herych (eds), *Ethics Codes in Medicine: Foundations and Achievements of Codification since 1947*, Ashgate, Aldershot, 1998, pp. 84–96.

Ashley, David J. B., 'Evidence for the Existence of Two Forms of Cervical Carcinoma', *Journal of Obstetrics and Gynaecology of the British Commonwealth*, vol. 73, 1966, pp. 382–9.

Auckland Hospital Board, Obstetrical and Gynaecological Unit, Cornwall Hospital, Auckland, *First Clinical Report, for the Year Ended 31st March 1949*, prepared by the registrars F. L. Clark and G. H. Green, 1950.

Auckland Hospital Board, Obstetrical and Gynaecological Unit, Cornwall Hospital, Auckland, *Second Clinical Report, for the Year Ended 31st March 1950*, prepared by G. H. Green, 1951.

Austoker, Joan, *A History of the Imperial Cancer Research Fund, 1902–1986*, Oxford University Press, Oxford, 1988.

Austoker, Joan, 'Cancer Prevention in Primary Care: Screening for Cervical Cancer', *BMJ*, vol. 309, 1994, pp. 241–8.

Austoker, Joan, 'Editorial: Gaining Informed Consent of Screening', *BMJ*, vol. 19, 1999, pp. 722–3.

Baeyertz, J. D., 'Results of a Cervical Smear Campaign in Wanganui', *New Zealand Medical Journal (NZMJ)*, vol. 64, 1965, pp. 618–25.

Baker, Robert, 'Transcultural Medical Ethics and Human Rights', in Tröhler and Reiter-Theil in cooperation with Herych (eds), *Ethics Codes in Medicine*, pp. 312–31.

Bassett, Michael, *Working with David: Inside the Lange Cabinet*, Hodder Moa, Auckland, 2008.

Bickley, Joy, 'Safety Screen or Smoke-screen?', *New Zealand Nursing Journal: Kai Tiaki (NZNJ)*, vol. 81, 8, August 1987, pp. 10–12.

Bickley, Joy, 'What the Cervical Cancer Inquiry Report Means for Nurses', *NZNJ*, vol. 81, 9, September 1988, pp. 14–15.

Bickley, Joy, 'Watchdogs or Wimps? Nurses' Response to the Cartwright Report', in Sandra Coney (ed.), *Unfinished Business: What Happened to the Cartwright Report?*, Women's Health Action, Auckland, 1993, pp. 125–36.

Bird, Christine, 'Using Women's Health Groups', *Broadsheet*, vol. 137, March 1986, pp. 19–21.

Bogdanich, Walt, 'Lax Laboratories: The Pap Test Misses Much Cervical Cancer Through Labs' Errors: Cut-rate "Pap Mills" Process Slides Using Screeners With Incentives to Rush: Misplaced Sense of Security', *Wall Street Journal*, 2 November 1987, pp. 1–6.

Bolitho, D. G., 'Some Financial and Medico-political Aspects of the New Zealand Medical Profession's Reaction to the Introduction of Social Security', *New Zealand Journal of History*, vol. 18, 1, 1984, pp. 34–49.

Bonham, Dennis G., G. H. Green and G. Liggins, 'Editorial, Cervical Human Papilloma Virus Infection and Colposcopy', *Australian and New Zealand Journal of Obstetrics and Gynaecology*, vol. 27, 2, 1987, p. 131.

Bonita, Ruth, 'Contraceptive Research: For Whose Protection?', *Broadsheet*, vol. 76, January/February 1980, pp. 6–8.

Boston Women's Health Book Collective, *Our Bodies Ourselves: A Book by and for Women*, Simon and Schuster, New York, 1971.

Bourne, Aleck W., *Recent Advances in Obstetrics and Gynaecology*, 9th edn, Churchill, London, 1958.

Boyes, D. A., H. K. Fidler and D. R. Lock, 'Significance of in Situ Carcinoma of the Uterine Cervix', *BMJ*, vol. 1, 1962, p. 203.

Brandt, Allan M. and Martha Gardner, 'The Golden Age of Medicine?', in Roger Cooter and John Pickstone (eds), *Companion to Medicine in the Twentieth Century,* Routledge, London, 2003, pp. 21–37.

Brody, Baruch A., *The Ethics of Biomedical Research: An International Perspective*, Oxford University Press, New York, 1998.

Bryder, Linda, *A Voice for Mothers: The Plunket Society and Infant Welfare 1907–2000*, Auckland University Press, Auckland, 2003.

Bryder, Linda, 'Liley, Albert William', in W. F. Bynum and Helen Bynum (eds), *Dictionary of Medical Biography,* Volume 3, Greenwood Publishing Group, Westport CT, 2007, p. 793.

Bryder, Linda, 'Debates about Cervical Screening: An Historical Overview', *Journal of Epidemiology and Community Health*, vol. 62, 2008, pp. 284–7.

Bunkle, Phillida, 'Calling the Shots? The International Politics of Depo-Provera', in Rita Arditti, Renate Duelli Klein and Shelley Minden (eds), *Test-Tube Women: What Future for Motherhood?*, Pandora Press, London, 1984, pp. 165–87.

Bunkle, Phillida, 'Dalkon Shield Disaster', *Broadsheet*, vol. 122, September 1984, pp. 18–22, 36.

Bunkle, Phillida, *Second Opinion: The Politics of Women's Health in New Zealand*, Oxford University Press, Auckland, 1988.

Bunkle, Phillida, 'Women, Health and Politics: Divisions and Connections', in Rosemary du Plessis and Lynne Alice (eds), *Feminist Thought in Aotearoa New Zealand: Connections and Differences*, Oxford University Press, Oxford, 1988, pp. 238–44.

Bunkle, Phillida, 'Personal Crisis/Global Crisis', in Maud Cahill and Christine Dann (eds), *Changing Our Lives: Women Working in the Women's Liberation Movement, 1970–1990*, Bridget Williams Books Ltd, Wellington, 1991, pp. 169–86.

Bunkle, Phillida, 'Side-stepping Cartwright: The Cartwright Recommendations Five Years On', in Coney (ed.), *Unfinished Business*, pp. 50–64.

Butler, Neville R. and Dennis G. Bonham, *Perinatal Mortality: The First Report of the 1958 British Perinatal Mortality Survey under the Auspices of the National Birthday Trust Fund*, E. & S. Livingstone Ltd, Edinburgh, 1963.

Cahill, Maud and Christine Dann (eds), *Changing Our Lives: Women Working in the Women's Liberation Movement, 1970–1990*, Bridget Williams Books, Wellington, 1991.

Calvert, Sarah, 'Cervices at Risk', *Broadsheet*, vol. 123, October 1984, p. 37.

Cameron, Amanda, 'Sexual Health Matters: Pap Smear Has Had Its Day', *New Zealand Doctor*, 26 July 2006, p. 8.

Cameron, Amanda, 'How Cartwright Changed It All . . .', *New Zealand Doctor*, 10 September 2008, pp. 12–13.

Campbell, A. V., 'Leading Article: Ethics After Cartwright', *NZMJ*, vol. 104, 1991, pp. 36–37.

Carey, H. M. and S. E. Williams, 'Cytological Diagnosis of Pre-Clinical Carcinoma of the Cervix', *NZMJ*, vol. 57, 1958, pp. 227–35.

Casper, Monica J. and Adele E. Clarke, 'Making the Pap Smear into the "Right Tool" for the Job: Cervical Cancer Screening in the USA, circa 1940–95', *Social Studies of Science*, vol. 28, 2, 1998, pp. 255–90.

Chalmers, Iain, 'Cochrane, Archibald Leman', in Bynum and Bynum (eds), *Dictionary of Medical Biography*, Volume 2, pp. 353–5.

Chamberlain, Geoffrey, *Special Delivery: The Life of the Celebrated British Obstetrician William Nixon*, Royal College of Obstetricians and Gynaecologists, London, 2004.

Chamberlain, Jocelyn, 'Failures of the Cervical Cytology Screening Programme', *BMJ*, vol. 289, 1984, pp. 853–4.

Chang, A. R., 'Health Screening: An Analysis of Abnormal Cervical Smears at Dunedin Hospital 1963–82', *NZMJ*, vol. 98, 1985, pp. 104–7.

Chang, Alexander R., 'Medical History: Historical Aspects of Cervical Cancer in New Zealand: Concepts and Treatment up to 1941', *NZMJ*, vol. 101, 1988, pp. 514–7.

Chapman, Karen, 'A Question of Ethics: An Update on the Implementation of the Cartwright Report', *NZNJ*, vol. 82, 2, March 1989, pp. 22–25.

Cheyne, Christine, Mike O'Brien and Michael Belgrave, *Social Policy in Aotearoa: A Critical Introduction*, Oxford University Press, Auckland, 1997.

Cochrane, A. L., *Effectiveness and Efficiency: Random Reflections on Health Services*, The Nuffield Provincial Hospitals Trust, London, 1972.

Cochrane, A. L. and W. W. Holland, 'Validation of Screening Procedures', *British Medical Bulletin*, vol. 27, 1, 1971, pp. 3–8.

Coney, Sandra, 'Alienated Labour – Foetal Monitoring', *Broadsheet*, May 1979, pp. 16 17, 38–39.

Coney, Sandra, 'From Here to Maternity', *Broadsheet*, vol. 123, October 1984, pp. 4–6.

Coney, Sandra, 'Dalkon Shield News', *Broadsheet*, vol. 128, April 1985, pp. 9–10.

Coney, Sandra, 'Do You Really Have to Have a Hysterectomy?', *New Zealand Woman's Weekly*, 13 April 1987, pp. 64–67.

Coney, Sandra, 'The End of the Experiment', *New Zealand Listener*, 10 September 1988, pp. 22–24.

Coney, Sandra, *The Unfortunate Experiment: The Full Story Behind the Inquiry into Cervical Cancer Treatment*, Penguin Books, Auckland, 1988.

Coney, Sandra, 'Doctors in Charge', *New Zealand Listener*, 20 May 1989, pp. 16–18.

Coney, Sandra, *Out of the Frying Pan: Inflammatory Writings 1972–1989*, Penguin Books, Auckland, 1990.

Coney, Sandra, 'The First Post-Cartwright Year: A Case Study in Institutional Resistance', in Coney, *Out of the Frying Pan*, pp. 213–45.

Coney, Sandra (ed.), *Unfinished Business: What Happened to the Cartwright Report?*, Women's Health Action, Auckland, 1993.

Coney, Sandra, 'Against All Odds: The Experience of a Consumer Representative in the Establishment of the National Cervical Screening Programme', in Coney (ed.), *Unfinished Business*, pp. 165–78.

Coney, Sandra, 'Unfinished Business: The Cartwright Report Five Years On', in Coney (ed.), *Unfinished Business*, pp. 19–49.

Coney, Sandra, 'Fertility Action 1984–', in Anne Else (ed.), *Women Together: A History of Women's Organisations in New Zealand*, Ngā Rōpū Wāhine o te Motu, Historical Branch, Department of Internal Affairs and Daphne Brasell Associates Press, Wellington, 1993, pp. 284–6.

Coney, Sandra, 'Health Organisations', in Else (ed.), *Women Together*, pp. 241–54.

Coney, Sandra, *Standing in the Sunshine: A History of New Zealand Women Since They Won the Vote*, Penguin Books, Auckland, 1993.

Coney, Sandra, *The Menopause Industry: A Guide to Medicine's 'Discovery' of the Mid-life Woman*, Penguin Books, Auckland, 1994.

Coney, Sandra, 'Reforms in Patients' Rights and Health Service Restructuring: Obstacles to Change', *Reproductive Health Matters*, no. 6, 1995, pp. 72–83.

Coney, Sandra, 'Cartwright Ten Years On: Access to Health Care: The Over-Riding Issue', *Women's Health Watch*, 42, November 1997 (URL: http://www.womens-health.org.nz/publications/WHW/whwnov97.htm#cartten).

Coney, Sandra and Phillida Bunkle, 'An Unfortunate Experiment at National Women's', *Metro*, June 1987, pp. 47–65.

Coney, Sandra and Lyn Potter, *Hysterectomy*, Heinemann Reid, Auckland, 1990.

Cooter, Roger, 'The Ethical Body', in Cooter and Pickstone (eds), *Companion to Medicine in the Twentieth Century*, pp. 451–68.

Coppleson, L. W. and B. W. Brown, 'Control of Carcinoma of Cervix: Role of the Mathematical Model', in Malcolm Coppleson (ed.), *Gynecologic Oncology: Fundamental Principles and Clinical Practice*, Volume 1, Churchill Livingstone, New York, 1981, pp. 390–7.

Coppleson, M., 'Cervical Intraepithelial Neoplasia: Clinical Features and Management', in Coppleson (ed.), *Gynecologic Oncology*, pp. 408–33.

Coppleson, Malcolm, 'Colposcopy', in John Stallworthy and Gordon Bourne (eds), *Recent Advances in Obstetrics and Gynaecology*, 12th edn, Churchill Livingstone, Edinburgh, 1977, pp. 155–83.

Coppleson, Malcolm, 'Preclinical Invasive Carcinoma of Cervix (Microinvasive and Occult Invasive Carcinoma): Clinical Features and Management', in Coppleson (ed.), *Gynecologic Oncology*, pp. 451–64.

Coppleson, Malcolm and Bevan Reid, with the assistance of Ellis Pixley, *Preclinical Carcinoma of the Cervix Uteri: Its Nature, Origin and Management*, Pergamon Press, Oxford, 1967.

Coppleson, Malcolm, Ellis Pixley and Bevan Reid, *Colposcopy: A Scientific and Practical Approach to the Cervix in Health and Disease*, Charles C. Thomas Publishers, Springfield, Illinois, 1970.

Corbett, Jan, 'Second Thoughts on the Unfortunate Experiment at National Women's', *Metro*, July 1990, pp. 54–73.

Corbett, Jan, 'Have You Been Burned at the Stake Yet?', *Metro*, October 1990, pp. 156–65.

Cuzick, Jack, Christine Clavel, Karl-Ulrich Petry, Chris J. L. M. Meijer, Heike Hoyer, Samuel Ratnam, Anne Szarewski, Philippe Birembaut, Shalini Kulasingam, Peter Sasiene and Thomas Iftner, 'Early Detection and Diagnosis: Overview of the European and North American Studies on HPV Testing in Primary Cervical Cancer Screening', *International Journal of Cancer*, vol. 119, 5, February 2006, pp. 1095–101.

Dann, Christine, *Up From Under: Women and Liberation in New Zealand 1970–1985*, Allen and Unwin/Port Nicolson Press, Wellington, 1985.

Darby, R. E. W. and S. E. Williams, 'The Cytological Diagnosis of Carcinoma of the Cervix', *NZMJ*, vol. 64, 1965, pp. 98–102.

Davis, Kathy, *The Making of Our Bodies, Ourselves: How Feminism Travels across Borders*, Duke University Press, Durham and London, 2007.

Day, Emerson, 'The 24-hour Cancer Cure', in Walter Sanford Ross, *The Climate is Hope: How They Triumphed over Cancer*, Prentice-Hall, New Jersey, 1965; reprinted in New Zealand, A. H. & A. W. Reed, Wellington, 1967. pp. 25–33.

Dobbie, Mary, *The Trouble with Women: The Story of Parents Centre New Zealand*, Cape Catley Ltd, Queen Charlotte Sound, 1990.

Doll, Sir Richard, 'Concluding Remarks', (Given at the Medical Research Council Epidemiology Symposium, Green Lane Hospital, Auckland, 2 November 1973), *NZMJ*, vol. 80, 1974, pp. 403–4.

Doll, Sir Richard, 'Clinical Trials: Retrospect and Prospect', *Statistics in Medicine*, vol. 1, 1982, pp. 337–44.

Doll, Sir Richard, 'Occasional Review: Prospects for Prevention', *BMJ*, vol. 286, 1983, pp. 445–53.

Donley, Joan, 'Having the Baby at Home', *Broadsheet*, vol. 132, September 1985, pp. 17, 47.

Donley, Joan, *Save the Midwife*, New Women's Press, Auckland, 1986.

Donley, Joan, *Herstory of N.Z. Homebirth Association*, Domiciliary Midwives Society of New Zealand, Wellington, 1992.

Dow, Derek A., *Safeguarding the Public Health: A History of the New Zealand Department of Health*, Victoria University Press, Wellington, 1995.

du Plessis, Rosemary and Lynne Alice (eds), *Feminist Thought in Aotearoa New Zealand: Connections and Differences*, Oxford University Press, Oxford, 1988.

Duncan, G. R., 'Viewpoint: Cervical Cytology 1980: Its Place and Value', *NZMJ*, vol. 93, 1981, pp. 119–22.

Editorial, 'Adverse Pregnancy Outcomes after Treatment for Cervical Intraepithelial Neoplasia', *BMJ*, vol. 337, 2008, pp. 769–70.

Editorial, 'Cancer of the Cervix: Death by Incompetence', *The Lancet*, vol. 2, 1985, pp. 363–4.

Editorial, 'Cervical Epithelial Dysplasia', *BMJ*, vol. 1, 1975, pp. 294–5.

Editorial, 'Colposcopy', *BMJ*, vol. 282, 1981, pp. 250–1.

Editorial, 'Diagnostic Problems in Cervical Cancer', *BMJ*, vol. 1, 1974, pp. 471–2.

Editorial, 'Flap about Pap', *Time*, vol. 112, 20, (13 November) 1978, p. 77.

Editorial, 'Management of Abnormal Cervical Smears', *BMJ*, vol. 280, 1980, pp. 1239–40.

Editorial, 'Outcome of Pregnancy after Cone Biopsy', *BMJ*, vol. 280, 1980, pp. 1393–4.

Editorial, 'Preventing Cancer of Uterine Cervix', *BMJ*, vol. 1, 1962, pp. 1817–18.

Editorial, 'Screening for Cervical Cancer', *BMJ*, vol. 2, 1976, pp. 659–60.

Editorial, 'Uncertainties of Cervical Cytology', *BMJ*, vol. 4, 1973, pp. 501–2.

Feldberg, Georgina, 'On the Cutting Edge: Science and Obstetrical Practice in a Women's Hospital, 1945–60', in Georgina Feldberg, Molly Ladd-Taylor, Alison Li and Kathryn McPherson (eds), *Women, Health, and Nation: Canada and the United States since 1945*, McGill-Queen's University Press, Montreal and Kingston, 2003, pp. 123–43.

Feldberg, Georgina, Molly Ladd-Taylor, Alison Li and Kathryn McPherson, 'Comparative Perspectives on Canadian and American Women's Health Care since 1945', in Feldberg, Ladd-Taylor, Li and McPherson (eds), *Women, Health, and Nation*, pp. 26–32.

Fisher, G. J. S., 'Carcinoma of the Cervix', *NZMJ*, vol. 54, 1955, pp. 371–6.

Foster, Peggy, *Women and the Health Care Industry: An Unhealthy Relationship?*, Open University Press, Buckingham, 1995.

Geiringer, Erich, 'Trial in Error', *New Zealand Listener*, 26 November 1988, pp. 18–19, 44–46.

Geiringer, Erich, 'The Triumph of Victimocracy', *Metro*, vol. 113, November 1990, pp. 134–8.

Gerber, Paul and Malcolm Coppleson, 'Leading Article: Clinical Research after Auckland', *The Medical Journal of Australia*, vol. 150, 1989, pp. 230–3.

Graham, J. B., L. S. J. Sotto and F. P. Paloucek, *Carcinoma of the Cervix*, W. B. Saunders Co., Philadelphia and London, 1962.

Grant, Marie P. S., 'Cytology in Prevention of Cancer of Cervix', *BMJ*, vol. 1, 1963, pp. 1637–40.

Green, G. H., 'Tubal Ligation', *NZMJ*, vol. 57, 1958, pp. 470–7.

Green, G. H., *Introduction to Obstetrics: An Elementary Text for Students*, N. M. Peryer, Christchurch, 1962, 2nd edn 1964, 3rd edn 1966, 4th edn 1970, 5th edn 1975.

Green, G. H., 'Carcinoma in Situ of the Uterine Cervix: Conservative Management in 84 of 190 Cases', *Australian and New Zealand Journal of Obstetrics and Gynaecology*, vol. 2, 2, 1962, pp. 49–57.

Green, G. H., 'The Use of "Oxoid" Membrane Filters in Exfoliative Cytology', *Journal of Medical Laboratory Technology*, vol. 19, 1962, pp. 266–71.

Green, G. H., 'Cervical Carcinoma in Situ: True Cancer or Non-Invasive Lesion?', *Australian and New Zealand Journal of Obstetrics and Gynaecology*, vol. 30, 1964, pp. 165–73.

Green, G. H., 'Cervical Cytology and Carcinoma in Situ', *Journal of Obstetrics and Gynaecology of the British Commonwealth*, vol. 72, 1, 1965, pp. 13–22.

Green, G. H., 'Cervical Cone Biopsy with Octapressin', *Australian and New Zealand Journal of Obstetrics and Gynaecology*, vol. 6, 3, 1966, pp. 259–65.

Green, G. H., 'Pregnancy Following Cervical Carcinoma in Situ: A Review of 60 Cases', *Journal of Obstetrics and Gynaecology of the British Commonwealth*, vol. 73, 6, 1966, pp. 897–902.

Green, G. H., 'The Significance of Cervical Carcinoma in Situ', *Australian and New Zealand Journal of Obstetrics and Gynaecology*, vol. 6, 1, 1966, pp. 42–44.

Green, G. H., 'The Significance of Cervical Carcinoma in Situ', *American Journal of Obstetrics and Gynecology (AJOG)*, vol. 94, 7, 1966, pp. 1009–22.

Green, G. H., 'Trends in Maternal Mortality', *NZMJ*, vol. 65, 1966, pp. 80–86.

Green, G. H., 'Is Cervical Carcinoma in Situ Significant?', *International Journal of Surgery*, vol. 47, 6, 1967, pp. 511–7.

Green, G. H., 'Maori Maternal Mortality in New Zealand', *NZMJ*, vol. 66, 1967, pp. 295–9.

Green, G. H., 'Invasive Potentiality of Cervical Carcinoma in Situ', *International Journal of Gynaecology and Obstetrics*, vol. 7, 4, 1969, pp. 157-69.

Green, G. H., 'Cervical Carcinoma in Situ: An Atypical View', *Australian and New Zealand Journal of Obstetrics and Gynaecology*, vol. 10, 1970, pp. 41–48.

Green, G. H., 'Duration of Symptoms and Survival Rates for Invasive Cervical Cancer', *Australian and New Zealand Journal of Obstetrics and Gynaecology*, vol. 10, 4, 1970, pp. 238–43.

Green, G. H., 'The Foetus Began to Cry . . . Abortion. 1', *NZNJ*, vol. 63, 7, 1970, pp. 11–12.

Green, G. H., 'The Foetus Began to Cry . . . Abortion. 2', *NZNJ*, vol. 63, 9, 1970, pp. 6–7.

Green, G. H., 'The Foetus Began to Cry. . .Abortion. 3', *NZNJ*, vol. 63, 9, 1970, pp. 11–12.

Green, G. H., 'The Progression of Pre-invasive Lesions of the Cervix to Invasion', *NZMJ*, vol. 80, 1974, pp. 279–87.

Green, G. H., 'Cytology and Cervical Cancer in New Zealand', *Irish Medical Journal*, vol. 70, 12, 1977, pp. 361–3.

Green, G. H., 'Cervical Cancer and Cytology Screening in New Zealand', *British Journal of Obstetrics and Gynaecology*, vol. 85, 12, 1978, pp. 881–6.

Green, G. H., 'Rising Cervical Cancer Mortality in Young New Zealand Women', *NZMJ*, vol. 89, 1979, pp. 89–91.

Green, G. H., 'William Liley and Fetal Transfusion: A Perspective in Fetal Medicine', *Fetal Therapy*, vol. 1, 1, 1986, pp. 18–22.

Green, G. H., A. W. Liley and G. S. Liggins, 'The Place of Foetal Transfusion in Haemolytic Disease: A Report of 22 Transfusions in 16 Patients', *Australian and New Zealand Journal of Obstetrics and Gynaecology*, vol. 90, 1965, pp. 53–59.

Green G. H. and J. W. Donovan, 'The Natural History of Cervical Carcinoma in Situ', *Journal of Obstetrics and Gynaecology of the British Commonwealth*, vol. 77, 1, 1970, pp. 1–9.

Greer, Germaine, *The Whole Woman*, A. A. Knopf, New York, 1999.

Griffiths, C. T., J. H. Austin and P. A. Younge, 'Punch Biopsy of the Cervix', *AJOG*, vol. 88, 1964, pp. 695–703.

Hadley, Janet, 'The Case against Depo-Provera', in Sue O'Sullivan (ed.), *Women's Health: A Spare Rib Reader*, Pandora, London, 1987, pp. 170–4.

Harbutt, Jefcoate, 'New Concepts in the Treatment of Carcinoma of the Cervix', *NZMJ*, vol. 54, 1955, pp. 356–70.

Hay, Iain, *The Caring Commodity: The Provision of Health Care in New Zealand*, Oxford University Press, Auckland, 1989.

'Health Survey: New Zealand Contraception and Health Study: Design and Preliminary Report (Chairman Sir Graham Liggins)', *NZMJ*, 23 April 1986, pp. 283–6.

Hegh, Tony (ed.), *Frank Answers from Thursday: Girls, Women Ask About Their Bodies and Sexual Needs*, Wilson and Horton, Auckland, 1974.

Hercock, Faye, 'Professional Politics and Family Planning Clinics', in Linda Bryder (ed.), *A Healthy Country: Essays on the Social History of Medicine in New Zealand*, Bridget Williams Books, Wellington, 1991, pp. 181–97.

Herranz, Gonzalo, 'The Inclusion of the Ten Principles of Nuremberg in Professional Codes of Ethics: An International Comparison', in Tröhler and Reiter-Theil (eds), *Ethics Codes in Medicine*, pp. 127-39.

Hertig, Arthur T. and Paul A. Younge, 'A Debate: What is Cancer in Situ of the Cervix? Is it the Preinvasive Form of True Carcinoma?', *AJOG*, vol. 64, 4, 1952, pp. 807–15.

Heslop, Barbara, '"All about Research": Looking Back at the 1987 Cervical Cancer Inquiry', *NZMJ*, vol. 117, 2004, p. 1199 (URL: http://www.nzma.org.nz.ezproxy.auckland.ac.nz/journal/117-1199/1000).

Hill, Sir Austin Bradford, 'Medical Ethics and Controlled Trials', *BMJ*, vol. 1, 1963, pp. 1043–9.

Hollyock, Vernon E. and William Chanen, 'The Use of the Colposcope in the Selection of Patients for Cervical Cone Biopsy', *AJOG*, vol. 114, 2, 1972, pp. 185–9.

Holmes, Mary, *What is Gender? Sociological Approaches*, Sage Publications, London, 2007.

Holmes, Mary, *The Representation of Feminists as Political Actors: Challenging Liberal Democracy*, VDM Verlag Dr. Muller, Saarbrucken, 2008.

Howson, Alexandra, 'Cervical Screening, Compliance and Moral Obligation', *Sociology of Health and Illness*, vol. 21, 4, 1999, pp. 401–25.

Hyde, Pamela, 'Science Friction: Cervical Cancer and the Contesting of Medical Beliefs', *Sociology of Health and Illness*, vol. 22, 2, 2000, pp. 217–34.

Imperial Cancer Research Fund Coordinating Committee on Cervical Screening, 'Organisation of a Programme for Cervical Cancer Screening', *BMJ*, vol. 289, 1984, p. 8945.

Jeffcoate, Sir Norman, *Principles of Gynaecology*, 4th edn, Butterworths, London and Boston, 1975 (1980 reprint).

Jeffcoate, T. N. A., *Principles of Gynaecology*, 3rd edn, Butterworths, London, 1967.

Jeffcoate's Principles of Gynaecology, revised by V. R. Tindall, Butterworths, London, 1987.

Jenkin, Douglas, 'A Question of Trust', *New Zealand Listener*, 14 November 1987, pp. 16–18.

Jones, Howard W., Georgeanna S. Jones and William E. Ticknor, *Richard Wesley TeLinde*, Williams and Wilkins, Baltimore, 1986.

Jones, James H., *Bad Blood: The Tuskegee Syphilis Experiment: A Tragedy of Race and Medicine*, Free Press, New York, 1981.

Jones, R. W., 'Viewpoint: Reflections on Carcinoma in Situ', *NZMJ*, vol. 104, 1991, pp. 339–41.

Jones, R. W., M. L. Yeong, A. W. Stewart, G. C. Hitchcock and W. E. Dervan, 'Cervical Cytology in the Auckland Region', *NZMJ*, vol. 101, 1988, pp. 132–5.

Jones, Ronald and Norman Fitzgerald, 'The Development of Cervical Cytology and Colposcopy in New Zealand: 50 Years since the First Cytology Screening Laboratory at National Women's Hospital', *NZMJ*, vol. 117, 2004, pp. 1179–88 (URL: http://www.nzma.org.nz.ezproxy.auckland.ac.nz/journal/117-1206/1179).

Jones, Ronald W., 'Cervical Cancer Prevention, Feminism, and Herb Green', *NZMJ*, vol. 120, 2007 (URL: http://www.nzma.org.nz.ezproxy.auckland.ac.nz/journal/120-1260/2694).

Jones, Ronald W. and Malcolm R. McLean, 'Carcinoma in Situ of the Vulva: A Review of 31 Treated and Five Untreated Cases', *Obstetrics and Gynecology: Journal of the American College of Obstetricians and Gynecologists*, vol. 68, 4, 1986, pp. 499–503.

Jordan, J. A., 'Minor Degrees of Cervical Intraepithelial Neoplasia: Time to Establish a Multicentre Prospective Study to Resolve the Question', *BMJ*, vol. 297, 1988, p. 6. ??

Kaye, Linda, 'What Happened to the Cartwright Women? The Legal Proceedings', in Coney (ed.), *Unfinished Business*, pp. 79–87.

Keating, Peter and Alberto Cambrosio, 'Risk on Trial: The Interaction of Innovation and Risk in Cancer Clinical Trials', in Thomas Schlick and Ulrich Tröhler (eds), *The Risks of Medical Innovation: Risk Perception and Assessment in Historical Context*, Routledge, London, 2006, pp. 225–41.

Keith, Jocelyn, 'Bad Blood: Another Unfortunate Experiment', *NZNJ*, vol. 81, 12, December/January 1989, pp. 20–21.

King, Michael, 'Books: Glimpses of Reality', *Metro*, October 1988, pp. 216–19.

Kinlen, L. J. and R. Doll, 'Trends in Mortality from Cancer of the Uterus in Canada and in England and Wales', *British Journal of Preventive and Social Medicine*, vol. 27, 1973, pp. 146–9.

Knox, E. G., 'Cervical Cytology: A Scrutiny of the Evidence', in Gordon McLachlan (ed.), *Problems and Progress in Medical Care, Essays on Current Research, Second Series*, Nuffield Provincial Hospitals Trust, Oxford University Press, London, 1966, pp. 277–307.

Knox, E. G., 'Cervical Cancer', in Thomas McKeown (ed.), *Screening in Medical Care: Reviewing the Evidence: A Collection of Essays with a Preface by Lord Cohen of Birkenhead*, Nuffield Provincial Hospitals Trust, Oxford University Press, London, 1968, pp. 43–54.

Kolata, Gina, 'Is the War on Cancer Being Won?', *Science*, vol. 229, 4713, 1985, pp. 543–4.

Kolstad, Per, 'Diagnosis and Management of Precancerous Lesions of the Cervix Uteri', *International Journal of Gynaecology and Obstetrics*, vol. 8, 4, 2, 1970, pp. 551–60.

Kolstad, Per and Adolf Stafl, *Atlas of Colposcopy*, University Park Press Inc., Baltimore, and Universitetsforlaget, Oslo, 1972.

Kolstad, Per and Valborg Klem, 'Long-term Follow-up of 1121 Cases of Carcinoma in Situ', *Obstetrics and Gynecology: Journal of the American College of Obstetricians and Gynecologists*, vol. 48, 2, 1976, pp. 125–9.

Koss, L. G., 'Exfoliative Cytology of the Uterine Cervix and Vagina', *CA: A Cancer Journal for Clinicians*, vol. 10, 1960, pp. 182–93.

Koss, L. G., 'Concept of Genesis and Development of Carcinoma of the Cervix', *Obstetrical and Gynecological Survey*, vol. 24, 7, 2, 1969, pp. 850–9.

Koss, Leopold G., 'Dysplasia: A Real Concept or a Misnomer?', *Obstetrics and Gynecology: Journal of the American College of Obstetricians and Gynecologists*, vol. 51, 1978, pp. 374–7.

Koss, Leopold G., 'The Papanicolaou Test for Cervical Cancer Detection: A Triumph and a Tragedy', *Journal of the American Medical Association*, vol. 261, 5, 1989, pp. 737–43.

Koss, Leopold G., Fred W. Stewart, Michael J. Jordan, Frank W. Foote, Genevieve M. Baker and Emerson Day, 'Some Histological Aspects of Behavior of Epidermoid Carcinoma in Situ and Related Lesions of the Uterine Cervix: A Long-term Prospective Study', *Cancer*, vol. 16, 2, 9, 1963, pp. 1160–211.

Krieger, James S. and Lawrence J. McCormack, 'Graded Treatment for in Situ Carcinoma of the Uterine Cervix', *AJOG*, vol. 101, 2, 1968, pp. 171–9, 'Discussion', pp. 179–82.

Lâârâ, E., N. E. Day and M. Hakama, 'Trends in Mortality from Cervical Cancer in the Nordic Countries: Association with Organised Screening Programmes', *The Lancet*, vol. 329, 1987, pp. 1247–9.

Larsson, Goran, *Conization for Preinvasive and Early Invasive Carcinoma of the Uterine Cervix*, Acta Obstetricia et Gynecologica Scandinavia, Supplement 114, Lund, 1983.

Le Fanu, James, *The Rise and Fall of Modern Medicine*, Abacus, London, 1999, new edn 2000.

Leading article, 'Diagnostic Problems in Cervical Cancer', *BMJ*, vol. 1, 1974, pp. 471–2.

Leading article, 'Outcome of Pregnancy after Cone Biopsy', *BMJ*, vol. 280, 1980, pp. 1393–4.

Leading article, 'Screening for Cervical Cancer', *BMJ*, vol. 2, 1976, pp. 659–60.

Leading article, 'Uncertainties of Cervical Cytology', *BMJ*, vol. 4, 1973, pp. 501–2.

Lerner, Barron H., *The Breast Cancer Wars: Fear, Hope and the Pursuit of a Cure in Twentieth-century America*, Oxford University Press, Oxford, 2001.

Lerner, Barron H., 'What Do You Know? Cancer, History and Medical Practice', in Jacalyn Duffin (ed.), *Clio in the Clinic: History in Medical Practice*, Oxford University Press, Oxford, 2005, pp. 299–307.

Lewlin, Ellen and Virginia Olesen (eds), *Women, Health and Healing: Towards a New Perspective*, Tavistock, New York, 1985.

Liggins, G. C., 'The George Addlington Syme Oration: Winds of Change', *Australia and New Zealand Journal of Surgery*, vol. 61, 1991, pp. 169–72.

Love, R. R. and A. E. Camilli, 'The Value of Screening', *Cancer*, vol. 48, 2, 1981, pp. 493–4.

Lund, Curtis J., 'An Epitaph for Cervical Carcinoma', *Journal of the American Medical Association*, vol. 175, 2, 1961, pp. 98–99.

Macgregor, J. Elizabeth and Sue Teper, 'Uterine Cervical Cytology and Young Women', *The Lancet*, vol. 311, 8072, 1978, pp. 1029–31.

Margulis, R. R., R. W. Dustin, H. C. Walser and J. E. Ladd, 'Carcinoma in Situ of the Cervix with Vaginal Vault Extension', *Obstetrics and Gynecology: Journal of the American College of Obstetricians and Gynecologists*, vol. 19, 5, 1962, pp. 569–74.

Marks, Lara, 'Assessing the Risk and Safety of the Pill: Maternal Mortality and the Pill', in Schlick and Tröhler (eds), *The Risks of Medical Innovation*, pp. 187–203.

Marks, Lara V., *Sexual Chemistry: A History of the Contraceptive Pill*, Yale University Press, New Haven and London, 2001.

Marshall, E. J., 'Notices: The College of General Practitioners: Cervical Smear Research Survey', *NZMJ*, vol. 61, 1962, p. 473.

Marshall, E. J., 'Cervical Smear Survey', *NZMJ*, Supplement, vol. 63, 1964, pp. 18–22.

Martin, Purvis L., 'How Preventable is Invasive Cervical Cancer? A Community Study of Preventable Factors', and Robert C. Goodlin, 'Discussion', *AJOG*, vol. 113, 1972, pp. 541–8.

Marx, Jean L., 'The Annual Pap Smear: An Idea Whose Time Has Gone?', *Science*, vol. 205, 4402, 1979, pp. 177–8.

Matheson, Clare, *Fate Cries Enough*, Sceptre NZ, Auckland, 1989.

McCalman, Janet, *Sex and Suffering: Women's Health and a Women's Hospital: The Royal Women's Hospital Melbourne 1856-1996*, Melbourne University Press, Melbourne, 1998.

McCormick, James and Robin Fox, 'Death of Petr Skrabanek (Obituary)', *The Lancet*, vol. 344, 1994, pp. 52–53.

McCormick, James S., 'Cervical Smears: A Questionable Practice?', *The Lancet*, vol. 1, 1989, pp. 207–9.

McCredie, Margaret R. E., Katrina J. Sharples, Charlotte Paul, Judith Branyai, Gabriele Medley, Ronald W. Jones and David C. G. Skegg, 'Natural History of Cervical Neoplasia and Risk of Invasive Cancer in Women with Cervical Intraepithelial Neoplasia 3: A Retrospective Cohort Study', *The Lancet Oncology*, vol. 9, 5, 2008, pp. 425–34.

McIndoe, W. A., 'A Cervical Cytology Screening Programme in the Thames Area', *NZMJ*, vol. 63, 1964, p. 6.

McIndoe, W. A., 'A Cervical Cytology Screening Programme in the Thames Area: Second and Third Years of Study', *NZMJ*, vol. 65, 1966, pp. 647–51.

McIndoe, W. A., 'Cytology or Colposcopy', *Australian and New Zealand Journal of Obstetrics and Gynaecology*, vol. 8, 3, 1968, pp. 117–18.

McIndoe, W. A. and G. H. Green, 'Vaginal Carcinoma in Situ Following Hysterectomy', *Acta Cytologica*, vol. 13, 3, 1969, pp. 158–62.

McIndoe, W. A., M. R. McLean, R. W. Jones and P. R. Mullins, 'The Invasive Potential of Carcinoma in Situ of the Cervix', *Obstetrics and Gynecology: Journal of the American College of Obstetricians and Gynecologists*, vol. 64, 1984, pp. 451–8.

McKelvey, John L., 'Carcinoma in Situ of the Cervix: A General Consideration', *AJOG*, vol. 64, 4, 1952, pp. 816–25.

McKeown, Thomas and E. G. Knox, 'The Framework Required for Validation of Prescriptive Screening', in McKeown (ed.), *Screening in Medical Care*, pp. 159–73.

McLaren, Hugh C. *The Prevention of Cervical Cancer*, The English Universities Press, London, 1963.

McLeod, Rosemary, 'The Importance of Being Sandra Coney', *North and South*, July 1988, pp. 54–67.

McNeill, Paul M., 'The Implications for Australia of the New Zealand *Report of the Cervical Cancer Inquiry*: No Room for Complacency', *The Medical Journal of Australia*, vol. 150, 1989, pp. 264–71.

McNeill, Paul M., *The Ethics and Politics of Human Experimentation*, Cambridge University Press, Cambridge, 1993.

'Medicolegal: Medical Council Charges Professor Bonham', *NZMJ*, vol. 103, 1990, pp. 547–9.

Melville, R. P., 'Cancer Detection in Australia', *NZMJ*, vol. 64, 1965, pp. 196–201.

Miller, A. B., 'Control of Carcinoma In Situ by Exfoliative Cytology Screening', in Coppleson (ed.), *Gynecologic Oncology*, pp. 381–3.

Miller, A. B., J. Lindsay and G. B. Hill, 'Mortality from Cancer of the Uterus in Canada and its Relationship to Screening for Cancer of the Cervix', *International Journal of Cancer*, vol. 17, 5, 1976, pp. 602–12.

Mitchinson, Wendy, 'Agency, Diversity, and Constraints: Women and the Physicians, Canada, 1850–1950', in The Feminist Health Care Ethics Research Network, Susan Sherwin (coordinator), *The Politics of Women's Health: Exploring Agency and Autonomy*, Temple University Press, Philadelphia, 1998, pp. 122–49.

Morgen, Sandra, *Into Our Own Hands: The Women's Health Movement in the United States, 1969–1990*, Rutgers University Press, New Brunswick, 2002.

Moscucci, Ornella, *The Science of Woman: Gynaecology and Gender in England 1800–1929*, Cambridge University Press, Cambridge, 1990.

Moscucci, Ornella, 'The "Ineffable Freemasonry of Sex": Feminist Surgeons and the Establishment of Radiotherapy in Early Twentieth-century Britain', in David Cantor (ed.), *Cancer in the Twentieth Century*, The Johns Hopkins University Press, Baltimore, 2008, pp. 139–63.

Murphy, M. F. G., M. J. Campbell and P. O. Goldblatt, 'Twenty Years' Screening for Cancer of the Uterine Cervix in Great Britain, 1964–84: Further Evidence of its Ineffectiveness', *Journal of Epidemiology and Community Health*, vol. 42, 1, 1988, pp. 49–53.

Murphy-Lawless, Jo, *Reading Birth and Death*, Cork University Press, Cork, 1998.

Neal, Sue, 'Only Women Bleed', *Broadsheet*, vol. 75, December 1979, pp. 12–15.

Nelson Jr, J. H., 'Clinical Invasive Carcinoma of Cervix: Place of Radical Hysterectomy as Primary Treatment', in Coppleson (ed.), *Gynecologic Oncology*, pp. 504–7.

Oakley, Ann, *From Here to Maternity: Becoming a Mother*, Penguin, Harmondsworth, 1981.

Oakley, Ann, *The Captured Womb: A History of the Medical Care of Pregnant Women*, Blackwell, Oxford and New York, 1984.

'Obituary: Alastair Macfarlane', *NZMJ*, vol. 99, 1986, p. 690.

'Obituary: Bruce Walton Grieve', *NZMJ*, vol. 106, 1993, p. 532.

'Obituary: Malcolm Robert McLean', *NZMJ*, vol. 106, 1993, p. 236.

'Obituary: Sydney Wallace Jefcoate Harbutt', *NZMJ*, vol. 111, 1998, pp. 60–61.

'Obituary: William Arthur McIndoe', *NZMJ*, vol. 1987, p. 37.

Olson, James S. (compiler), *The History of Cancer: An Annotated Bibliography*, Greenwood Press, New York, 1989.

Östör, Andrew G., 'Review: Natural History of Cervical Intraepithelial Neoplasia: A Critical Review', *International Journal of Gynecological Pathology*, vol. 12, 2, 1993, pp. 186–92.

Papanicolaou, G. N and H. F. Traut, *Diagnosis of Uterine Cancer by Vaginal Smear*, Commonwealth Fund, Oxford University Press, Oxford, 1943.

Papanicolaou, George N. and Herbert F. Traut, 'The Diagnostic Value of Vaginal Smears in Carcinoma of the Uterus', *AJOG*, vol. 42, 1941, pp. 193–206.

Parker, Leigh, 'Sandra Coney: Sympathy is Not Enough', *New Zealand Woman's Weekly*, 23 July 1990, pp. 18–19.

Patterson, James T., *The Dread Disease: Cancer and Modern American Culture*, Harvard University Press, Cambridge, Mass., 1987.

Paul, Charlotte, 'Letter from New Zealand: The New Zealand Cervical Cancer Study: Could It Happen Again?', *BMJ*, vol. 297, 1988, pp. 533–9.

Paul, Charlotte, 'Education and Debate: Internal and External Morality of Medicine: Lessons from New
    Zealand', *BMJ*, vol. 320, 2000, pp. 499–503.
Paul, Charlotte, 'New Zealand's Cervical Cytology History: Implications for the Control of Cervical
    Cancer', *NZMJ*, vol. 117, 2004, pp. 1170–4.
Paul, Charlotte and Linda Holloway, 'No New Evidence on the Cervical Cancer Study', *NZMJ*, vol. 103,
    1990, pp. 581–3.
Paul, Charlotte, Susan Bagshaw, Ruth Bonita, Gillian Durham, N. W. Fitzgerald, R. W. Jones, Betty
    Marshall and Brian R. McAvoy, 'Cancer Screening: 1991 Cervical Screening Recommendations: A
    Working Group Report', *NZMJ*, vol. 104, 1991, pp. 291–5.
Pettersson, Folke (ed.), *Annual Report on the Results of Treatment in Gynecological Cancer*, 12th Volume,
    International Federation of Gynecology and Obstetrics, Stockholm, 1988.
Pinell, Patrice, 'Cancer', in Cooter and Pickstone (eds), *Companion to Medicine in the Twentieth Century*,
    pp. 671–87.
Popkin, D. R., 'Editorial: Cervical Cancer Screening Programs: A Gynecologist's Viewpoint', *Canadian
    Medical Association Journal*, vol. 114, 1976, pp. 982–3.
Posner, Tina, 'What's in a Smear? Cervical Screening, Medical Signs and Metaphors', *Science as Culture*,
    vol. 2, 2, 11, 1991, pp. 167–87.
Posner, Tina and Martin Vessey, *Prevention of Cervical Cancer: The Patient's View*, King Edward's Hospital
    Fund for London, London, 1988.
Potter, Lyn, 'Hysterectomy and Hospital Hassles', *Broadsheet*, vol. 120, June 1984, pp. 28–33.
Raffle, A. E., B. Alden and E. F. D. Mackenzie, 'Detection Rates for Abnormal Cervical Smears: What Are
    We Screening For?', *The Lancet*, vol. 345, 1995, pp. 1469–73.
Raffle, A. E., B. Alden, M. Quinn, P. J. Babb and M. T. Brett, 'Outcomes of Screening to Prevent Cancer:
    Analysis of Cumulative Incidence of Cervical Abnormality and Modelling of Cases and Deaths
    Prevented', *BMJ*, vol. 326, 2003, pp. 901–4.
Raffle, Angela and Muir Gray, *Screening: Evidence and Practice*, Oxford University Press, Oxford, 2007.
Rakusen, Jill, 'Healthy Women: Depo Provera', *Broadsheet*, vol. 39, May 1976, pp. 32–33.
Rankine, Jenny, 'Experimenting on Women', *Broadsheet*, vol. 151, August/September 1987, pp. 9–10.
Rawls, W. E., K. Iwamoto, E. Adam, J. L. Melnick and G. H. Green, 'Herpesvirus-type-2 Antibodies and
    Carcinoma of the Cervix', *The Lancet*, vol. 2, 1970, pp. 1142–3.
Reynolds L. A. and E. M. Tansey, '*Prenatal Corticosteroids for Reducing Morbidity and Mortality after
    Preterm Birth*', *The Transcript of a Witness Seminar held by the Wellcome Trust Centre for the History
    of Medicine at UCL, London, on 15 June 2004*, Wellcome Witnesses on Twentieth Century Medicine
    Volume 25, The Wellcome Trust, London, 2005.
Rich, Adrienne, *Of Woman Born: Motherhood as Experience and Institution*, Norton, New York, 1976.
Richart, R. M., Yao-Shi Fu and J. W. Reagan, 'Pathology of Cervical Intraepithelial Neoplasia', in Coppleson
    (ed.), *Gynecologic Oncology*, pp. 398–407.
Richart, Ralph, 'Natural History of Cervical Intraepithelial Neoplasia', *Clinical Obstetrics and Gynaecology*,
    vol. 10, 1967, pp. 748-94.
Richart, Ralph M., 'Current Concepts in Obstetrics and Gynecology: The Patient with an Abnormal Pap
    Smear – Screening Techniques and Management', *The New England Journal of Medicine*, vol. 302, 6,
    1980, pp. 332–4.
Richart, Ralph M. and John J. Sciarra, 'Treatment of Cervical Dysplasia by Outpatient
    Electrocauterization', *AJOG*, vol. 101, 2, 1968, pp. 200–5.
Richart, Ralph M. and Bruce A. Barron, 'A Follow-up Study of Patients with Cervical Dysplasia', *AJOG*, vol.
    105, 1969, pp. 386–92.
Richart, Ralph M. and Thomas C. Wright, 'Controversies in the Management of Low-Grade Cervical
    Intraepithelial Neoplasia', *Cancer*, Supplement, vol. 71, 4, 1993, pp. 1413–21.
Rifai, Samy, 'Gynecological Issues: Colposcopy Referees the Pap Smear', *The Female Patient*, vol. 5, 4, 1980,
    n.p.
Roberts, Julie, 'I Was Lucky . . . I Lived', *New Zealand Woman's Weekly*, 8 August 1988, p. 24.
Roberts, Julie, 'We Had To Speak Out about What We Knew', *New Zealand Woman's Weekly*, 8 August 1988,
    p. 16.
Roberts, Julie, 'You Saved My Mum . . . Thank You', *New Zealand Woman's Weekly*, 8 August 1988, p. 20.
Robertson, J. H., Bertha E. Woodend, E. H. Crozier and June Hutchinson, 'Risk of Cervical Cancer
    Associated with Mild Dyskaryosis', *BMJ*, vol. 297, 1988, pp. 18–21.
Robinson, Jean, 'Cervical Cancer: Doctors Hide the Truth', *Spare Rib*, vol. 154, 1985, reprinted in O'Sullivan
    (ed.), *Women's Health*, pp. 49–51.
Rosenberg, Charles E., 'Disease in History: Frames and Framers', *Millbank Quarterly*, vol. 67 (Suppl. 1),
    1989, pp. 1–15.

Rosenblatt, Roger A., Judith Reinken and Phil Shoemack, 'Is Obstetrics Safe in Small Hospitals? Evidence from New Zealand's Regionalised Perinatal System', *The Lancet*, vol. 326, 8452, 1985, pp. 429–32.

Roger, Warwick, 'My Town: Intellectual Thuggery', *Metro*, September 1990, pp. 8–13.

Rosier, Pat, 'Screening the Doctors', *Broadsheet*, October 1987, p. 7.

Rosier, Pat, 'The Speculum Bites Back', *Broadsheet*, vol. 153, November 1987, p. 5.

Rosier, Pat, 'Listen to the Women', *Broadsheet*, vol. 155, January/February 1988, pp. 5–8.

Rosier, Pat, 'Broadcast: A Feminist Victory', *Broadsheet*, vol. 161, September 1988, pp. 6–8.

Ross, Walter Sanford, *The Climate is Hope: How They Triumphed over Cancer*, Prentice-Hall, New Jersey, 1965; reprinted in New Zealand, A. H. & A. W. Reed, Wellington, 1967.

Ross, Walter Sanford, *Crusade: The Official History of the American Cancer Society*, Arbor House, New York, 1987.

Rothman, David J., *Strangers at the Bedside: A History of How Law and Bioethics Transformed Medical Decision Making*, Basic Books, New York, 1991.

Ruzek, Sheryl Burt, *The Women's Health Movement: Feminist Alternatives to Medical Control*, Praegers Publishers, New York, 1978.

Sabbage, Lisa, 'A Spur to Action', *Broadsheet*, October 1988, p. 8

Saffron, Lisa, 'Cervical Cancer – The Politics of Prevention', *Spare Rib*, vol. 129, April 1983, reprinted in O'Sullivan (ed.), *Women's Health*, pp. 42–49.

Savage, Edward W., 'Current Developments: Microinvasive Carcinoma of the Cervix', *AJOG*, vol. 113, 5, 1972, pp. 708–17.

Savage, Wendy, *A Savage Inquiry: Who Controls Childbirth?*, Virago, London, 1986.

Schiller, Walter, 'Early Diagnosis of Carcinoma of the Cervix', *Surgery, Gynecology and Obstetrics*, vol. 56, 1933, pp. 212–22.

Schlick, Thomas, 'Risk and Medical Innovation: a Historical Perspective', in Schlick and Tröhler (eds), *The Risks of Medical Innovation*, pp. 1–19.

Scott, John, 'Obituary: Professor Dennis Geoffrey Bonham', *The University of Auckland News*, vol. 35, 5, June 2005, pp. 20–21.

Sedlis, Alexander, Sanford Sall, Yoshi Tsukada, Robert Park, Charles Mangan, Hugh Shingleton and John A. Blessing, 'Microinvasive Carcinoma of the Uterine Cervix: A Clinical-pathologic Study', *AJOG*, vol. 133, 1, 1979, pp. 64–74.

'Sexual Health Matters: Pap Smear Has had its Day', *New Zealand Doctor*, 26 July 2006, p. 8.

Sharma, Anil, 'Pregnancy Care in NZ: A Brief History', *New Zealand Doctor*, 12 September 2007, p. 29.

Shelley, Jose, 'The Smear', *SHE*, June 1964, n.p.

Sinclair, Keith, *A History of the University of Auckland*, Oxford University Press, Auckland, 1983.

Skegg, D. C. G., 'Leading Article: How Not to Organise a Cervical Screening Programme', *NZMJ*, vol. 102, 1989, pp. 527–8.

Skegg, D. C. G., Charlotte Paul, R. J. Seddon, N. W. Fitzgerald, P. M. Barham and C. J. Clements, 'Cancer Screening: Recommendations for Routine Cervical Screening', *NZMJ*, vol. 98, 1985, pp. 636–9.

Skrabanek, Petr and James McCormick, *Follies and Fallacies in Medicine*, The Tarragon Press, Glasgow, 1989.

Smith, J. S., J. Green, A. B. de Gonzalez, P. Appleby, J. Peto, M. Plummer, S. Franceschi and V. Beral, 'Cervical Cancer and Use of Hormonal Contraceptives: A Systematic Review, *The Lancet*, vol. 361, 2003, pp. 1159–67.

Smyth, Helen, *Rocking the Cradle: Contraception, Sex and Politics in New Zealand*, Steele Roberts Ltd, Wellington, 2000.

Stafl, A. and E. J. Wilkinson, 'Cervical and Vaginal Intra-epithelial Neoplasia', in John Stallworthy and Gordon Bourne (eds), *Recent Advances in Obstetrics and Gynaecology*, 13th edn, Churchill Livingstone, Edinburgh, 1979, pp. 257–77.

Stafl, Adolf, Eduart G. Friedrick Jnr and Richard F. Mattingly, 'Detection of Cervical Neoplasia: Reducing the Risk of Error', *CA: A Cancer Journal for Clinicians*, vol. 24, 1974, pp. 22–30.

Stallworthy, J., A. S. Moolgaoker and J. J. Walsh, 'Legal Abortion: Critical Assessment of its Risks', *The Lancet*, vol. 2, 1971, pp. 1245–9.

Stallworthy, John and Gordon Bourne (eds), *Recent Advances in Obstetrics and Gynaecology*, 11th edn, Churchill, London, 1966.

Stallworthy, John and Gordon Bourne (eds), *Recent Advances in Obstetrics and Gynaecology*, 12th edn, Churchill Livingstone, Edinburgh, 1977.

Stallworthy, John and Gordon Bourne (eds), *Recent Advances in Obstetrics and Gynaecology*, 13th edn, Churchill Livingstone, Edinburgh, 1979.

Stallworthy, Sir John, 'Clinical Invasive Carcinoma of Cervix: Combined Radiotherapy and Radical Hysterectomy as Primary Treatment', in Coppleson (ed.), *Gynecologic Oncology*, pp. 508–16.

Straton, Judith, 'Leading Article: Progress in Cervical Screening in New Zealand', *NZMJ*, vol. 107, 1994, pp. 261–2.

Strid, Judi, 'Ethical Dilemmas: A Consumer Perspective on the Performance of Ethical Committees', in Coney (ed.), *Unfinished Business*, pp. 110–24.

Strid, Judi, 'Trust Me, I'm a Doctor: The Story of Informed Consent', in Coney (ed.), *Unfinished Business*, pp. 137–51.

Swan, Lyndsey, Editorial, 'Second Thoughts', *New Zealand General Practice*, 10 July 1990, n.p.

Szreter, Simon, 'The Importance of Social Intervention in Britain's Mortality Decline c.1850–1914: A Re-interpretation of the Role of Public Health', *Social History of Medicine*, vol. 1, 1, 1988, pp. 1–38.

The Feminist Health Care Ethics Research Network, Susan Sherwin (co-ordinator), *The Politics of Women's Health: Exploring Agency and Autonomy*, Temple University Press, Philadelphia, 1998.

Thompson, Bruce H., J. Donald Woodruff, Hugh J. Davis, Conrad G. Julian and Fred G. Silva II, 'Cytopathology, Histopathology, and Colposcopy in the Management of Cervical Neoplasia', *AJOG*, vol. 114, 2, 1972, pp. 329–37.

Tredway, Donald R., Duane E. Townsend, David N. Hovland and Richard T. Upton, 'Colposcopy and Cryosurgery in Cervical Intraepithelial Neoplasia', *AJOG*, vol. 114, 8, 1972, pp. 1020–4.

Trenfield, Karen, 'Tinker, Tailor, Soldier, Scientist: Women and Science Today', *Hecate: An Interdisciplinary Journal of Women's Liberation*, vol. 21, 1, 1995, pp. 149–72.

Turow, Joseph, *Playing Doctor: Television, Storytelling, and Medical Power*, Oxford University Press, New York, 1989.

Ueki, M. and G. H. Green, 'Cervical Carcinoma in Situ after Incomplete Conisation', *Asia-Oceania Journal of Obstetrics and Gynaecology*, vol. 14, 2, 1988, pp. 147–53.

Vessey, Martin and Richard Doll, 'Evaluation of Existing Techniques: Is "the Pill" Safe Enough to Continue Using?', *Proceedings of the Royal Society of London*, vol. 195, 1976, pp. 69–80.

Walton, R. J. (Chairman), 'Cervical Cancer Screening Programs: Report of the Task Force Appointed by the Conference of Deputy Minister of Health in December 1973 and Submitted to the Conference March 16 & 17 1976: Part I: Epidemiology and Natural History of Carcinoma of the Cervix', *Canadian Medical Association Journal*, vol. 114, 1976, pp. 1003–33.

Watkins, Elizabeth Siegel, 'Changing Rationale for Long-term Hormone Replacement Therapy in America, 1960–2000', *Health and History*, vol. 4, 1, 2002, pp. 20–36.

Wichtel, Diana, 'Delivering with Style', *New Zealand Listener*, 21 September 1985, pp. 14–15.

Williams, A. Susan, *Women and Childbirth in the Twentieth Century: A History of the National Birthday Trust Fund 1928–93*, Sutton, Gloucestershire, 1997.

Williams, Lynda, 'Dreaming the Impossible Dream: The Fate of Patient Advocacy', in Coney (ed.), *Unfinished Business*, pp. 88–102.

Wilson, Max, Jocelyn Chamberlain and A. L. Cochrane, 'Screening for Cervical Cancer', *The Lancet*, vol. 1, 1971, pp. 297–8.

Winslade, William J. and Todd L. Krause, 'The Nuremberg Code Turns Fifty', in Tröhler and Reiter-Theil in cooperation with Herych (eds), *Ethics Codes in Medicine*, pp. 140–62.

World Health Organization International Agency for Research on Cancer, *IARC Handbooks on Cancer Prevention, Volume 10, Cervix Cancer Screening*, IARC Press, Lyon, 2005.

Yamagata, S., and G. H. Green, 'Cellular Immune Status of Patients with Pelvic Cancer', *Australian and New Zealand Journal of Obstetrics and Gynaecology*, vol. 16, 3, 1976, pp. 129–42.

Yamagata, S., and G. H. Green, 'Radiation-induced Immune Changes in Patients with Cancer of the Cervix', *British Journal of Obstetrics and Gynaecology*, vol. 83, 5, 1976, pp. 400–8.

UNPUBLISHED THESES

Hyde, Pamela, 'At Your Cervix Madam: A Socio-historical Study of Cervical Cancer from the Late Nineteenth Century to the Late Twentieth Century', PhD, Victoria University of Wellington, 1997.

Wilson, Alison Margaret, 'Primetime Television Coverage of the Cartwright Inquiry, 1987–1988', MA, University of Auckland, 1993.

Index